Recent Approaches to Economic Dynamics

DYNAMISCHE WIRTSCHAFTSTHEORIE

Herausgegeben von
Peter Flaschel und Michael Krüger

Band 2

Verlag Peter Lang
Frankfurt am Main · Bern · New York · Paris

Peter Flaschel/
Michael Krüger (eds.)

Recent Approaches to Economic Dynamics

Conference Volume
(Bielefeld: 7-9th October, 1987)

Verlag Peter Lang
Frankfurt am Main · Bern · New York · Paris

Library of Congress Cataloging-in-Publication Data

Recent approaches to economic dynamics : conference vol., (Bielefeld: 7 - 9th October, 1987) / Flaschel, P. ; Krüger, M. (eds.). - Frankfurt am Main ; Bern ; New York ; Paris : Lang, 1988
(Dynamische Wirtschaftstheorie ; Bd. 2)
ISBN 3-631-40474-3

NE: Flaschel, Peter [Hrsg.]; GT

ISSN 0930-2174
ISBN 3-631-40474-3

PREFACE TO THE SERIES

With the new series "Dynamische Wirtschaftstheorie" the editors intend to encourage and support contributions to dynamic macro- and micro-economic analysis. The main fields of publications should cover work on business and growth cycle theories, stability of market systems, new directions in dynamic economic theory, and applications of new methods to theoretical as well as empirical problems of economic dynamics.

Conditions for the publiction of dissertations can be obtaind on request from the editors. Manuscripts - no less than 100 and no more than 400 pages in length - should be sent to:

Prof. Dr. Peter Flaschel
Universität Bielefeld
Fakultät für Wirtschaftswissenschaften
Universitätsstraße
Postfach 8640
D-4800 Bielefeld 1
West Germany

CONTENTS

B. Problems of Micro- and Macrodynamics

C. Investment, Asset Markets, and Economic Policy

D. Empirical Models and Input-Output-Analysis

Preface

This volume contains the proceedings of the Conference on Dynamic Economic Theory, held October 7 - 9, 1987 at the University of Bielefeld. The conference grew out of a project of the University of Bielefeld on Macrodynamics, 1986-87. Within this time two workshops were held and gave rise to the initiative to organize a conference on Economic Dynamics, a subject which seems to gather way in recent times. A subsequent call for papers received great response. Participants came mainly from West Germany. Also participating were several collegues from Austria, Denmark, and the Netherlands.

Since the beginning 1980's we observe a revival of interest in economic dynamics. This re-orientation of the economics profession may partly be due to some dramatic changes in the structure of advanced economies and a remarkable new drive towards economic expansion. The availability of new methods for describing and analyzing dynamic processes may explain, to some extent, the recent attraction, too. Furthermore it is an old conviction that understanding an intrinsically dynamic society like the capitalist society requires a dynamic theory — suffice it to recall the "magnificent dynamics" (Baumol) especially of Marx and Schumpeter. Thus the need and the always latent interest for a dynamic economic theory!

Turning to the contents of this volume the reader will find in section A an interesting diversity of business cycle models: KRAFT and WEISE present a mathematical treatment of Spiethoff's theory. His ideas on the psychological climate of investment and the likelihood of cumulative processes are interpreted within the synergetic approach to dynamics. Next, starting from the inertia of investment, the implications of a second-order accelerator for a business cycle model are examined by HILLINGER and WESER, whereas FRANKE and WEGHORST investigate the chaos generating properties in a nonlinear multiplier-accelerator model. KRÜGER offers a model of cyclical accumulation of capital with some Kaldorian and some Marxian properties. The recent contribution by Grandmont is the point of departure in ECKWERT's paper; he offers an overlapping generations model in a two-country-world. FLASCHEL investigates a recent symmetry assertion on Classical and Keynesian fiscal policies in an extended Goodwin-model. Goodwin's model is also the basis of GLOMBOWSKI's analysis of a political business cycle.

The papers in section B deal with various problems of micro- and macrodynamics. The first paper by FEICHTINGER points to the usefulness of a generalized maximum principle for endogenous cycles. SCHITTKO and ECKWERT apply the Filippov-solution concept to the analysis of discontinuous vector fields and discuss an economic model which uses this method. The formation of expectations in the social environment is addressed in SCHNABL's contribution; the Verhulst equation and Feigenbaum-diagrams are used for the analysis of such a process. Schumpeterian entrepreneurs within Goodwin's model, that is, the introduction of superior technologies and the competition among different techniques are discussed in ENGLMANN's article. Finally, MENSCH relates Schumpeterian dynamics with the problem of micro-macro-modelling.

In section C we find a set of seven papers with special emphasis on investment, asset markets, and economic policy. SKOTT examines a Keynesian economy with endogenous money supply. The impact of tax reforms on investment decisions are discussed by FUNKE within a Tobinian framework. WENZEL, KRISTOF, and TSCHERNIG investigate governmental financing in a continuous-time version of the IS-LM-Model. A new expectations hypothesis is analyzed by GROTH, who also embeds this discussion into the IS-LM framework. If a small open economy is facing some nominal, or, alternatively, real wage rigidities, what will be the effects of different policy strategies? This problem is addressed by VAN DE KLUNDERT and VAN DER PLOEG. An extended Kouri model is presented in the next paper, in which TIVIG offers a critical evaluation of the asset-market-approach. Finally, NECK deals with the problem of time-inconsistency of stabilization policies in the framework of dynamic game theory.

The last section contains econometric and input-output-models. KUHBIER and WOLTERS give a critical assessment of three simple business cycle models and their relevance for describing the post-war development of the Federal Republic of Germany. FROHN's article is dealing with a small macroeconometric model for the FRG, too. Special emphasis is given to the problem of reducing possible prediction errors. Some aspects of the RWI-business cycle model are discussed in HEILEMANN's paper, in particular its wage-price dynamics are analyzed in greater detail. MEYER is focussing on lags within input-output-analysis, especially on a production lag relative to demand and a lag due to the discrepancy between the actual and the desired capital stock. Next, RITSCHL establishes a set of conditions which guarantees the consistency of a dynamic input-output model. Finally, the input-output framework is used by KATTERMANN and KURZ in order to determine the implications of technical change on the rate of employment and income distribution.

Last not least it is a pleasure to acknowledge gratefully the financial support of the STIFTUNG VOLKSWAGENWERK for a successful conference.

Peter Flaschel

Michael Krüger

List of Participants

Böhm, Volker	Universität Mannheim
Eckwert, Bernhard	Universität Augsburg
Englmann, Frank	Universität Tübingen
Feichtinger, Gustav	Technische Universität Wien
Flaschel, Peter	Universität Bielefeld
Franke, Reiner	Universität Bremen
Frohn, Joachim	Universität Bielefeld
Funke, Michael	Wissenschaftszentrum Berlin
Gärtner, Wulf	Universität Osnabrück
Glombowski, Jörg	Tilburg University
Groth, Christian	University of Copenhagen
Heilemann, Ullrich	RWI Essen
Hillinger, Claude	Universität München
Jungeiliges, Jochen	Universität Osnabrück
Kattermann, Dieter	Universität Bremen
van de Klundert, Theo	Tilburg University
Kirkpatrick, Grant	Institut für Weltwirtschaft, Kiel
Klemisch-Ahlert, Marlies	Universität Osnabrück
Kraft, Manfred	Universität-GHS-Paderborn
Kristof, Kora	Universität München
Krüger, Michael	University of Massachusetts at Amherst
Kuhbier, Peter	Freie Universität Berlin
Kurz, Heinz	Universität Bremen
Lorenz, Hans-Walter	Universität Göttingen
Malliaropulos, Dimitris	Freie Universität Berlin
Maussner, Alfred	Universität Erlangen-Nürnberg
Meiners, Jochen	Universität München

Mensch, Gerhard	Technische Universität München
Meyer, Ulrich	Universität Bamberg
Neck, Reinhard	Wirtschaftuniversität Wien
Paparoditis, Stathis	Freie Universität Berlin
Ritschl, Albrecht	Universität München
Schittko, Ulrich	Universität Regensburg
Schmid, Michael	Universität der Bundeswehr Hamburg
Schnabl, Hermann	Universität Stuttgart
Schwarz, Angelika	Freie Universität Berlin
Schödiauer, Gerhard	Universität Bielefeld
Skott, Peter	University of Aarhus
Sommer, Albrecht	Freie Universität Berlin
Stephan, Gunter	Universität Heidelberg
Tivig, Thusnelda	Universität Konstanz
Tschernig, Rolf	Universität München
van Wegberg, Marc	Rijksuniversiteit Limburg
Weghorst, Wolfgang	Universität Bremen
Weise, Peter	Universität-GHS-Kassel
Wenig, Alois	Fernuniversität Hagen
Wenzel, H.-Dieter	Universität München
Weser, Thilo	Universität München
Witt, Ulrich	Universität Mannheim
Wolters, Jürgen	Freie Universität Berlin

A. Business Cycle Models

Cumulative Processes, Disproportionalities, and Spiethoff's Theory of the Business Cycle

By Peter Weise and Manfred Kraft
Universität-GHS-Kassel and Universität-GHS-Paderborn

1. Introduction

Firstly, in the following the main point of Spiethoff's theory of business cycle shall be brought forth. This exists in the fact that during cumulative processes disproportionalities between the sector of investment goods and that of consumption goods arise, whereby business cycles are caused. The motor of these movements are the entrepreneurs who by their investment decisions create a psychological atmosphere, tempting them to over- and underinvest. Secondly, by means of a synergetic model the main point of Spiethoff's business cycle theory shall generally be shown.

2. Spiethoff's Causes for the Business Cycle

Spiethoff sees a remarkable difference between the ups and downs of the economic development in the times before and since the 19th century. While the causes for the ups and downs until the 19th century are mainly to be found in coincidential exogenous shocks such as wars, good or bad harvests, epidemics etc., arising with the beginning of the 19th century endogenous causes develop, forcing the economy in up to then unknown systematic oscillations. These causes, of which Spiethoff names three, exist mainly in a cessation of up to that point existing restrictions:[1]

(1) Individual restrictions: The human being underlying traditional and ethic behavioural restrictions, changes to an economizing person, who wants to make profits and who is tempted by promising expectations of optimism and by unpromising expectations of pessimism.

(2) Materialistic restrictions: The production of goods is taking leave of slow organic procedures of growth of resources, and is basing itself on multiplyable materials and free labour enabling a rapid widening of production.

(3) Institutional restrictions: Traditional economic property rights, exchange-relationships and self-financing are mainly replaced by free economic constitutions, market relationships and credit-financing, whereby it becomes possible to invest to a great extent mainly in the sector of investment goods.
These three causes have the effect that an economy can rapidly grow on account of investments. Yet as the future is uncertain, continously new products, new technologies, and new forms of organization are caused, as well as new markets being opened up; as future markets being non-existent for a complete a-priori-coordination of investment decisions, the expectations concerning the future play an important part. The investors orientate towards a climate of investment, which again they themselves create by their investments, and hereby drive the economy to cumulative processes. This leads to an over- or underinvestment in the sector of investment goods in relation to the production of consumption goods. The effect is a stoppage of the boom turning into a depression, in the case of which the produced disproportionalities between investment goods and consumption goods are abolished on a very low level, so that a new boom can begin. By the generation of a psychological climate of investment, leading via cumulative processes to disproportionalities, the economy has got involved in an endogenously caused business cycle. This explanation of the business cycle by Spiethoff shall in the following be analyzed generally by means of a synergetic model. Especially it shall be aimed at showing that the generation of a psychological climate of investment is compatible with individual rationality and that in the case of plausible assumptions from an economic point of view a business cycle, including disproportionalities between the investment and the consumption goods sector, in fact exists.

3. Psychological Climate of Investment, Cumulative Processes and Individual Rationality

Apart from Spiethoff also other authors point out the meaning

of the psychological climate of investment as well as cumulative processes as an explanation for the business cycle movement. So, for example Keynes formulated:
"By a cyclical movement we mean that as the system progresses in, e. g., the upward direction, the forces propelling it upwards at first gather force and have a cumulative effect on one another but gradually lose their strength until at a certain point they tend to be replaced by forces operating in the opposite direction; which in turn gather force for a time and accentuate one another, until they too, having reached their maximum development, wane and give place to their opposite."[2]
And: "...it is an essential characteristic of the boom that investments which will in fact yield, say 2 per cent. in conditions of full employment are made in the expectation of a yield of, say, 6 per cent., and are valued accordingly. When the disillusion comes, this expectation is replaced by a contrary "error of pessimism", with the result that the investments, which would in fact yield 2 per cent. in conditions of full employment, are expected to yield less than nothing; and the resulting collapse of new investment then leads to a state of unemployment in which the investments, which would have yielded 2 per cent. in conditions of full employment, in fact yield less than nothing."[3] Similarly Pigou expresses this:
"First, among businessmen, even when engaged in different occupations, there often exists a certain measure of psychological interdependence. A change of tone in one part of the business world diffuses itself, in a quite unreasoning manner, over other and wholly disconnected parts. An expansion of business confidence "propagates itself by that sympathetic and epidemic excitement which so largely sways communities of men". There comes into play a quasi-hypnotic system of mutual suggestion: One with another, soul with soul / They kindle fire from fire."[4]
By means of the so-called game of majority, Schwartz adopts a comparable standpoint. In this game each player has two possible moves, stay and go. If a player chooses "stay" he receives nothing, regardless of the action of the other players.

If a player chooses "go" and a majority of all players chooses "go", too, he receives a positive pay-off, but if a majority chooses "stay", his pay-off is negative. There are two equilibrium points stay-stay and go-go. Individuals suffer losses for not following the majority as the majority generates positive effects for all its members. Thus the majority dictates the behaviour of all its individual members, but is conversely set up by its members individually. Individuals themselves via majority building create the situations they are enclosed in. Individual and collective rationality diverge.[5]

In many other business cycle theories explanations, such as those by Spiethoff, Keynes, Pigou and Schwartz do not appear. In those theories it is mostly implied that a typical actor or an aggregation of actors but not a multitude of individuals makes decisions and acts; and that the information can be processed instantaneously on markets, and are at the decider's disposal, so that activities are performed in the case of a-priori coordination of decisions; and that a cumulative mutual contagion of the individuals cannot occur. Yet there are reasons to assume that for the explanation of the business cycle the fact that a great number of individuals are acting is important, as well as the manner in which individuals make decisions in an uncertain world.

There exist not futures markets for investments in new products and new technologies which could treat information correctly, and neither a closed decision space nor adequate numerical probabilities exist; the future is not risky but remains uncertain. Investment acts must take place without pre-reconcilable choices. In such a situation decisions are based on heuristics and proxy variables including psychological moments:[6]

(1) Each investor takes notice of the successes and failures of the other investors and lets his own decisions depend on the actions of the others.

(2) Investors infer the future from the present; in the case of the present being very similar and definite for all, then

the estimation of the future will be very similar for all, too.
(3) During good times investors underestimate bad news in contrast to good news and vice versa.
(4) The investors summarize the information to a point of reference, against which they value the profitability of investments; this point of reference is the psychological climate of investment.
The consequence to be drawn from these four points is that all investors together create a climate of investment upon which, in turn, they orientate and that this climate of investment inheres a bias for the intensification of optimism and pessimism thus causing cumulative processes. Because each investor lets his behaviour depend on one variable, i. e. the climate of investment, which is created sequentially by all together, self-intensifying processes occur, being a substantial element of the theories of the business cycle by Spiethoff, Keynes, Pigou and Schwartz as we have pointed out above. This behavioural assumption is also essential for our business cycle model, which explains the business cycle by cumulative processes and disproportionalities between the industries of investment goods and those of consumption goods.

4. A Stochastic Two-Sector Business Cycle Model

Consider an economy consisting of 2 N investors: 2 N_c investors of the consumption goods industry and 2 N_i investors of the investment goods industry. Thus 2 N = 2 N_c + 2 N_i. Each investor can choose between two alternative strategies:

- s^+: to invest one unit of real capital for producing consumption goods or investment goods, respectively,
- s^-: not to invest in real capital, i. e. to invest in liquid assets.

Assume $n_{\alpha j}(t)$ to be the number of investors in industry α, (α = c, i), choosing at point of time t strategy j, ($j = s^+, s^-$). Then

$$n_{cs^+}(t) + n_{cs^-}(t) = 2\ N_c, \text{ for all } t,$$

$$n_{is^+}(t) + n_{is^-}(t) = 2\ N_i, \text{ for all } t.$$

Define $n_\alpha(t) = n_{\alpha s^+}(t) - N_\alpha$, $(\alpha = c, i)$, then

$$- N_c \leqq n_c(t) \leqq N_c; \; - N_i \leqq n_i(t) \leqq N_i, \text{ for all } t.$$

The vector $n(t) = (n_c(t), n_i(t))$ then defining a discrete time path describes completely the state of the economy and its temporal evolution.

We assume that behaviour of the individual producers is determined by transition rates p^α_{kj} describing the probability per unit of time of changing the strategy j to k in the infinitesimally small time interval $[t, t + \Delta t]$. Let us assume further that Δt may be set such that only one change of strategies is possible within Δt and that the Markovian assumption is fulfilled. Let transition rates p^α_{kj} depend upon the psychological assessment of the state of economy, $n(t)$, at point of time t by producers in industry α. We assume that each investor on the one hand lets his behaviour depend on how the investors belonging to his own branch of industry and, on the other hand, how the investors of the respective other branch of industry are acting. Let for the sake of simplicity this dependence be linear. Then we may write the linear psychological climate functions

$$u(n(t)) = \hat{\varkappa}_c n_c(t) + \hat{\sigma}_c n_i(t); \; v(n(t)) = \hat{\varkappa}_i n_i(t) + \hat{\sigma}_i n_c(t)$$

for producers of consumption goods and investment goods, respectively. We may specify the transition rates as

$$p^c_+(n) = p^c_{s^+s^-}(n) = \gamma_c e^{u(n)}; \; p^c_-(n) = p^c_{s^-s^+}(n) = \gamma_c e^{-u(n)},$$
$$p^i_+(n) = p^i_{s^+s^-}(n) = \gamma_i e^{v(n)}; \; p^i_-(n) = p^i_{s^-s^+}(n) = \gamma_i e^{-v(n)}.$$

Let us assume that transitions can occur between adjacent states of the economy only, i. e. behaviour of the producers changes sequentially. Then the motion of the economy is described by the master equation

$$(dp(n;t)/dt) = (dp(n_c, n_i; t)/dt) =$$
$$= (n_{cs^+} + 1)p^c_-(n_c + 1, n_i)p(n_c + 1, n_i; t) - n_{cs^+}p^c_-(n)p(n;t)$$
$$+ (n_{cs^-} + 1)p^c_+(n_c - 1, n_i)p(n_c - 1, n_i; t) - n_{cs^-}p^c_+(n)p(n;t)$$
$$+ (n_{is^+} + 1)p^i_-(n_c, n_i + 1)p(n_c, n_i + 1; t) - n_{is^+}p^i_-(n)p(n;t)$$

$+ (n_{is^-} + 1)p_+^i(n_c, n_i - 1)p(n_c, n_i - 1;t) - n_{is^-}p_+^i(n)p(n;t).$

In the case of N being very large, in order to describe the state of the economy we can define quasi-continuous variables $(\bar{x}, \bar{y})$ as $\bar{x} = n_c/N_c$; $\bar{y} = n_i/N_i$.

If we can assume further that it is sufficient to consider the motion equations of the means $E(x(t)) = x(t)$ and $E(y(t)) = y(t)$[7], we obtain the approximate equations

$$\begin{aligned}(dx/d\tau) &= (\tanh (u(x,y) - x) \cosh (u(x,y))\\ (dy/d\tau) &= \lambda(\tanh (v(x,y) - y) \cosh v(x,y))\end{aligned} \tag{1}$$

where $\lambda = \gamma_i/\gamma_c$; $\tau = 2\gamma_c t$,

$u(x,y) = \varkappa_c x + \sigma_c y$; $v(x,y) = \varkappa_i y + \sigma_i x$,

$\varkappa_\alpha = \hat{\varkappa}_\alpha N_\alpha$; $\sigma_\alpha = \hat{\sigma}_\alpha N_\alpha$, where $\alpha = c,i$.

We see that the motion of the economy depends on the psychological climate assessment of the states of the economy by producers in industries c and i. At every point of time psychological assessments initiate individual actions resulting in interaction processes which change the states of the economy. In order to analyze the behaviour of the economy further it is necessary to consider the psychological climate functions more closely.

$\varkappa_i$, $\varkappa_c$ describe cumulative effects in the investment goods and consumption goods industries, respectively. It is assumed that increasing production generates optimism thus fostering production even more, and vice versa. It is plausible to assume further that the degree of optimism or pessimism is higher in the investment goods industry than in the consumption goods industry. Thus we postulate $\varkappa_i > \varkappa_c > 1$. Then $\varkappa_i y$, $\varkappa_c x$ denote the strength and the direction of these forces.

σ_i, σ_c describe the effects of the industries on one another. σ_i measures the influence of consumption goods industry on investment goods industry. Due to low cost of producing investment goods in the region of underproduction of consumption goods, i. e. $x < 0$, and vice versa, production of investment goods will be positively or negatively influenced, respective-

ly. Analogously σ_c measures the influence of the investment goods industry on consumption goods industry. Production of investment goods generates income thus fostering production of consumption goods. Thus we postulate $\sigma_i < 0 < \sigma_c$. Strength and direction of these influences are measured by $\sigma_i x$ and $\sigma_c y$.
We assume that the unit of goods is defined such that $x = y = 0$ is an equilibrium of the economy which means that in $(0,0)$ there is no disproportion of production of consumption and investment goods. Then γ_i, γ_c denote the fluctuations in this point of equilibrium and the time scale on which decision processes take place. We postulate $\gamma_i, \gamma_c > 0$.
Description of the behaviour of the economy requires discussing the system of differential equations (1) with respect to the admissible parameter values discussed above. Especially, we are interested in the existence of equilibrium points, stability and limit cycles. It is not difficult to see that equilibrium points are the solutions of the system of equations

$$\tanh(u(x,y)) - x = 0, \quad \tanh(v(x,y)) - y = 0$$

from which we obtain

$$\begin{aligned} y &= \frac{1}{\sigma_c}\left(\frac{1}{2}\ln\left(\frac{1+x}{1-x}\right) - \varkappa_c x\right) = f(x) \\ x &= \frac{1}{\sigma_i}\left(\frac{1}{2}\ln\left(\frac{1+y}{1-y}\right) - \varkappa_i y\right) = g(y). \end{aligned} \tag{2}$$

A typical diagram for admissible parameter values is the following:

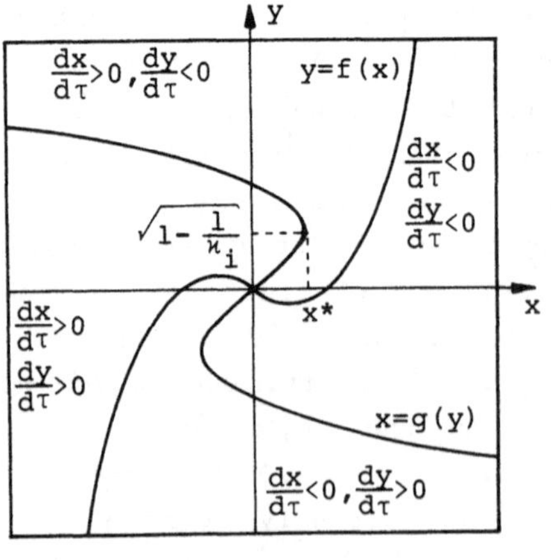

From an inspection of (1) it turns out that all the flux lines on the boundary points of the box $[-1 \leqq x, y \leqq 1]$ point inwards. The origin (0,0) is always an equilibrium point. A sufficient condition for (0,0) to be the unique equilibrium point is

$$\varkappa_c \geqq \frac{1}{2x^*} \ln \left(\frac{1+x^*}{1-x^*}\right) = \frac{\text{areatanh } x^*}{x^*},$$

$$\varkappa_i \geqq \frac{1}{2y^*} \ln \left(\frac{1+y^*}{1-y^*}\right) = \frac{\text{areatanh } y^*}{y^*},$$

where

$$x^* = \frac{1}{\sigma_i} \left(\frac{1}{2} \ln \left(\frac{1+\sqrt{1-\frac{1}{\varkappa_i}}}{1-\sqrt{1-\frac{1}{\varkappa_i}}}\right) - \varkappa_i\sqrt{1-\frac{1}{\varkappa_i}}\right),$$

$$y^* = \frac{1}{\sigma_c} \left(\frac{1}{2} \ln \left(\frac{1+\sqrt{1-\frac{1}{\varkappa_c}}}{1-\sqrt{1-\frac{1}{\varkappa_c}}}\right) - \varkappa_c\sqrt{1-\frac{1}{\varkappa_c}}\right).$$

Using a linear stability analysis to investigate the behaviour of the system (1) in the vicinity of (0,0) we have to examine the derivative of the r.h.s. of the system (2) at (0,0). We get

(0,0) is instable if

$$\gamma_c\varkappa_c + \gamma_i\varkappa_i \geqq \gamma_c + \gamma_i \text{ or } \varkappa_c + \varkappa_i - \varkappa_c\varkappa_i \geqq 1 - \sigma_c\sigma_i .$$

We see at once that this condition is fulfilled. Then due to a theorem by Poincaré-Bendixson there must exist a stable limit cycle, structurally stable for a comparatively large domain of admissible parameter values. Thus we have derived the existence of a business cycle:

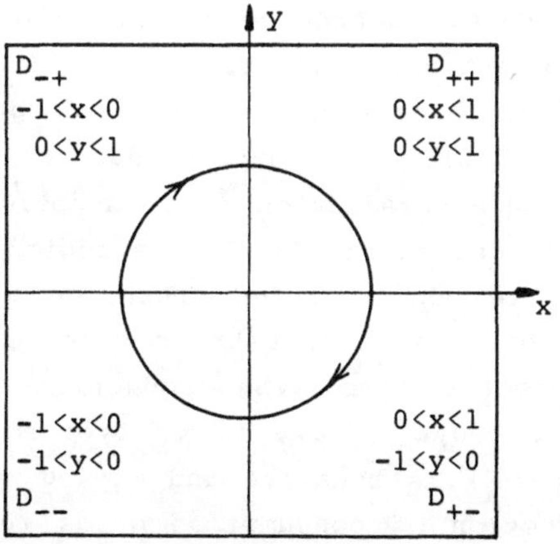

This business cycle denotes not a specific but rather the normal manner in which the economy considered above behaves. For the values of the parameters $\varkappa_i, \varkappa_c, \sigma_i, \sigma_c$: $(\varkappa_i, \varkappa_c > 1; \sigma_i < 0 < \sigma_c)$ exhibit rather meaningful economic hypotheses. In many theories of the business cycle cumulative effects or selfenhancing processes are stressed upon. These are described by $\varkappa_i$, $\varkappa_c$ which can be interpreted as "own multipliers" of the industries i and c. Positive effects are ceteris paribus prone to enhance themselves, and vice versa. Parameters σ_i, σ_c reflect the fact that both the industries i and c influence one another. Increasing production of investment goods always fosters production of consumption goods, and vice versa, since consumption goods are purchased by income. Investment goods, however, are produced in the expectation of a future yield and are financed by means of investment funds. But the cost of producing investment goods is low in the case of underproduction and high in the case of overproduction. Thus underproduction of consumption goods fosters production of investment goods and also conversely. Parameters σ_i, σ_c, therefore, may be interpreted in a similar manner as accelerator and multiplier, respectively, in some models of the business cycle.

The four parameters describe the degree and direction a business climate will evolve into, generated by the multitude of individual decision acts and, on the other hand, encouraging every producer to produce or not to produce. In the course of time this business climate gives rise to "errors of optimism and pessimism" causing over- and underproduction. In the region $(x < 0, y < 0)$, i. e. underproduction in both the industries, $\sigma_i x > 0$: Underproduction of consumption goods fosters production of investment goods. In the region $(x < 0, y > 0)$ $\sigma_i x, \varkappa_i y$, and $\sigma_c y$ are positive: Selfenhancing forces in industry i and positive effects on industry c cause overproduction of investment goods and increasing production of consumption goods. In region $(x > 0, y > 0)$, i. e. overproduction in both the industries, $\sigma_i x$ is negative and $\varkappa_i y$, $\sigma_c y$, $\varkappa_c x$ all are positive: Overproduction of consumption goods retards production of investment goods. Finally, in the region ($x > 0$, $y < 0$) $\sigma_i x$, $\kappa_i y$, and $\sigma_c y$ are negative: Selfenhancing forces in industry i in the opposite direction and negative effects on industry c bring about underproduction of investment goods and decreasing production of consumption goods. Going to region ($x < 0$, $y < 0$) again a further business cycle is apt to start up.

It turns out that the business cycle contains a sequence of disproportionalities of production of investment and consumption goods, a fact heavily stressed upon in some theories of the business cycle, too. The equilibrium proportion ($x = 0$, $y = 0$) will not be arrived at. To see this point more exactly let us consider two special cases. Firstly, let us assume $0 < \kappa_i, \kappa_c < 1$ and $\sigma_i < 0 < \sigma_c$.

A typical diagram is as follows:

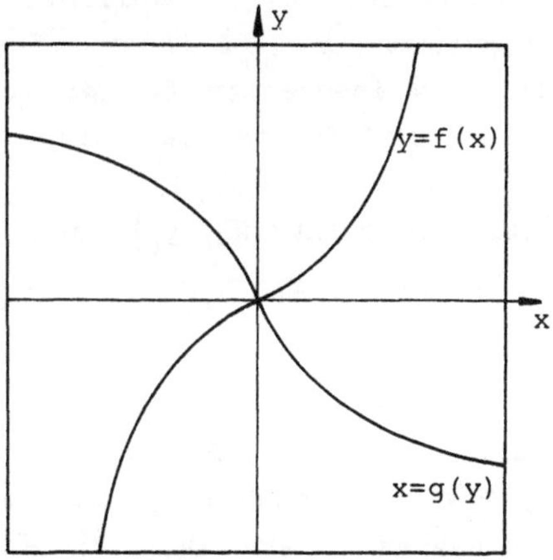

Because of $g'(y) < 0$ for all y and $f'(x) > 0$ for all x, the origin (0,0) is the unique equilibrium point. Furthermore, if κ_i, $\kappa_c < 1$ we conclude that $\gamma_c\kappa_c + \gamma_i\kappa_i < \gamma_c + \gamma_i$ and $\kappa_i + \kappa_c - \kappa_i\kappa_c < 1 - \sigma_i\sigma_c$.

Thus (0,0) is stable and no cycle exists. Productions of investment and consumption goods are in correct proportion to one another and each perturbation (caused by fluctuations) leads to a motion back to the equilibrium point. Secondly, let us assume $0 < \kappa_c < 1 < \kappa_i$ and $\sigma_i < 0 < \sigma_c$.

A typical diagram is as follows:

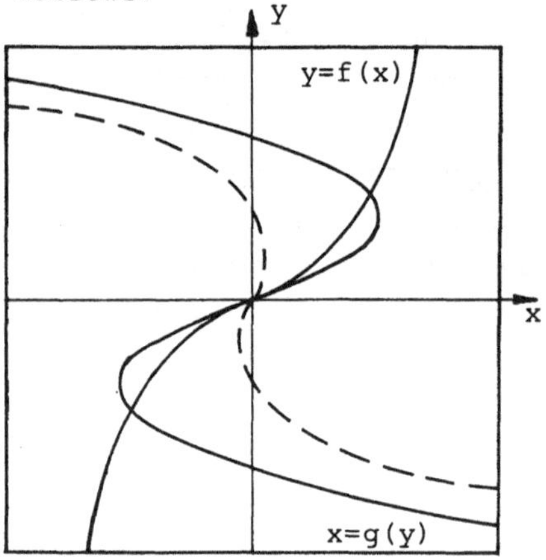

For (0,0) to be the unique equilibrium point in this case a necessary condition is $\varkappa_i \leqq (1-\sigma_c\sigma_i-\varkappa_c)/(1-\varkappa_c)$. If $\varkappa_i$ does not satisfy this inequality two or four further equilibrium points exist in addition to (0,0) which is always an equilibrium point.

For an equilibrium point (x_e, y_e) to be stable we obtain

$$-2(\gamma_c\sigma_c\sqrt{1-x_e^2}\, f'(x_e) + \gamma_i\sigma_i\sqrt{1-y_e^2}\, g'(y_e)) < 0$$

and $f'(x_e)\, g'(y_e) < 1$.

We conclude that for $\varkappa_i$ sufficiently large, σ_c, $|\sigma_i|$ sufficiently small, and $\varkappa_c$ sufficiently large (but smaller than 1) the economy may get locked in a "semi-slump" or a "quasi-boom" according to Keynes, respectively the first rise or the lack of capital according to Spiethoff.[9] The driving force $\sigma_i x$ then is not strong enough to push the economy into selfenhancing processes in the upward or downward direction. Thus the system persists in either one of the stable equilibria.

5. Conclusion

We have used a synergetic approach to depict the influence of a business climate on individual producers of investment and consumption goods. Generated by the total of producers this climate alters its strength and direction in the course of time, thus modifying the behaviour of each producer. Cumulative effects, errors of optimism and pessimism, and disproportions of investment goods and consumption goods production may occur. Assuming reasonable values for the parameters we obtain an asymptotically and structurally stable business cycle.

FOOTNOTES

[1] Cf. Spiethoff [**9**, pp. 23 - 85 and pp. 171 - 200]; see also Kraft and Weise [**6**].

[2] Cf. Keynes [**3**, pp. 313 - 314].

[3] Cf. Keynes [**3**, pp. 321 - 322].

[4] Cf. Pigou [**7**, p. 86 and pp. 87 - 98].

[5] Cf. Schwartz [**8**, pp. 62 - 65]; see also Kraft and Weise [**5**] and Weise and Kraft [**11**].

[6] Compare for a general view about conclusions drawn from the psychological literature Kaish [**2**].

[7] This is the case when the distribution of the stochastic variable $(\bar{x}(t), \bar{y}(t))$ is highly concentrated around the mean $(x(t), y(t))$ because then $(x(t), y(t))$ describes the most probable path. Cf. Haken [**1**], and Weidlich and Haag [**10**].

[8] For further sufficient conditions as well as details concerning the dynamic aspects of this model see Kraft, Landes and Weise [**4**].

[9] Cf. Keynes [**3**, p. 322]; cf. Spiethoff [**9**, pp. 80 - 85].

REFERENCES

[1] HAKEN, H.: Advanced Synergetics. Berlin: Springer, 1983.

[2] KAISH, S.: "Behavioral Economics in the Theory of the Business Cycle". in Handbook of Behavioral Economics. Behavioral Macroeconomics. Volume B, ed. by B. Gilad and S. Kaish. Greenwich: JAI Press, 1986.

[3] KEYNES, J. M.: The General Theory of Employment, Interest, and Money. London: Macmillan, 1936; reprinted 1967.

[4] KRAFT, M., T. LANDES AND P. WEISE: "Dynamic Aspects of a Stochastic Business Cycle Model", Methods of Operations Research, 53 (1986), 445 - 453.

[5] KRAFT, M. AND P. WEISE: "Das Spiel um Stabilität: Konjunktur und Prisoner's Dilemma", Jahrbücher für Nationalökonomie und Statistik, 194 (1979), 441 - 461.

[6] KRAFT, M. AND P. WEISE: "Eine Formalisierung von Spiethoffs Theorie der wirtschaftlichen Wechsellagen", Schweizerische Zeitschrift für Volkswirtschaft und Statistik, 123 (1987), 531 - 542.

[7] PIGOU, A. C.: Industrial Relations. Second Edition. London: Macmillan, 1929; reprinted 1967 by A.M.Kelley Publ.

[8] SCHWARTZ, J. T.: Lectures on the Mathematical Method in Analytical Economics. New York: Gordon and Breach, 1961.

[9] SPIETHOFF, A.: Die wirtschaftlichen Wechsellagen. Tübingen: Mohr, 1955.

[10] WEIDLICH, W. AND G. HAAG: Concepts and Models of a Quantitative Sociology. Berlin: Springer, 1983.

[11] WEISE, P. AND M. KRAFT: "Minsky's View of Fragility: A Game Theoretic Interpretation", Journal of Post Keynesian Economics, III (1981), 519 - 527.

BUSINESS CYCLE MODELS ON THE BASIS OF THE SECOND ORDER ACCELERATOR

C. Hillinger and T. Weser

Seminar for Mathematical Economics, University of Munich,
D-8000 Munich 22, Fed. Rep. of Germany

1. Introduction

This paper presents a type of business cycle model in which the inertia of investment and production plays a key role. Because of adjustment costs these variables are changed only slowly and approach their desired levels gradually. Formalization of this idea leads to the concept of a second order accelerator and implies cycles of constant or time varying amplitudes. These cycles are generated without interaction with a multiplier.

The analysis is focused on the dynamics of equipment investment and its interaction with demand. Production and inventory models are discussed by Hillinger (1986a).

The plan of the paper is as follows: Section 2 specifies our view of business cycles and relates it to empirical findings. In section 3 the concept of the second order accelerator is introduced and an elementary model of the equipment cycle is presented. In section 4 this model is expanded to take account of cyclical growth. Section 5 gives a summary and conclusions.

2. Business Cycles: Definition and Empiry

We define "business cycles" as *periodic* deviations of macroeconomic variables from their equilibrium trends. The relevance of this definition is supported by a long historical tradition and by contemporary empirical evidence.

For about a century, dating from the work of Juglar (1862), the dominant perception of business cycles has been that they are periodic. Gradually, it was recognized that there are several cycles primarily associated with the different components of the capital stock: a 3 - 4 year inventory cycle, a 7 - 10 year equipment cycle and a 20 year building cycle. The lengths of these cycles are plausibly explained by the speeds with which the different capital stocks adjust to their optimal levels.

Many theoretical models, most of them based on accelerator multiplier interaction, have been proposed to explain the periodicity of business cycles as an inherent property of economic dynamics. Seminal contributions were those of Samuelson (1939), Metzler (1947) and Hicks (1950).

Fisher (1925) and Slutsky (1937) denied that business cycles have a periodic aspect and regarded them as statistical fluctuations. This view rose to prominence with the Equilibrium Theory of Business Cycles of Lucas (1975). One reason for the change of predominant opinion was the application of spectral analysis of time series in a way that is improper to identify business cycles (Hillinger 1986a). Particularly, insufficient detrending and excessive smoothing had led to the structureless "typical spectral shape" (Granger, 1966).

After many years of intensive research on this problem, which has been performed by one of us (C.H.), we must emphasize that there is empirical evidence for the traditional view. Given adequate methodology, many economic time series show clear cycles. Results and methodology are described in Hillinger (1986a, 1986b).

Figure 1 is typical of the evidence relevant to the existence of business cycles at the macro level. The figure shows the spectra of German GNP and the two components most important for business cycle analysis: fixed and inventory investment. The data are subjected to polynomial detrending. The polynomial order and adjusted R^2 of the trend are shown at the bottom of the figure. The precise period of the principal peaks are similarly indicated. The horizontal axis gives the periods of the cycles in years. The vertical axis, which usually indicates the explained variance or "power" associated with each cycle, has been normalized to indicate the corresponding R^2.

The spectra fully confirm the traditional cycle view: Almost all of the spectral power is concentrated in the traditional business cycle range of 4 - 10 years. In fixed investment the longer periods dominate, in inventory investment the shorter ones. It is therefore reasonable to regard the longer periods as fixed investment or equipment cycles, the shorter periods as inventory cycles. It may be added that demand is stable and shows only weak cyclical contributions.

The theoretical model presented in the main part of this paper explains the long cycle in equipment investment and studies its interaction with demand in the context of a growing economy. It shows that equipment investment may be subject to strong cyclical fluctuations even in the presence of stable demand. The model may be extended to include production, inventories and the short cycle (Hillinger, 1987).

FIGURE 1 SPECTRA GERMANY, 1960-1984
GNP, FIXED INVESTMENT AND INVENTORY INVESTMENT
QUARTERLY DATA, SEASONALLY ADJUSTED
LOGARITHMS (EXCEPT INVENTORY), DETRENDED

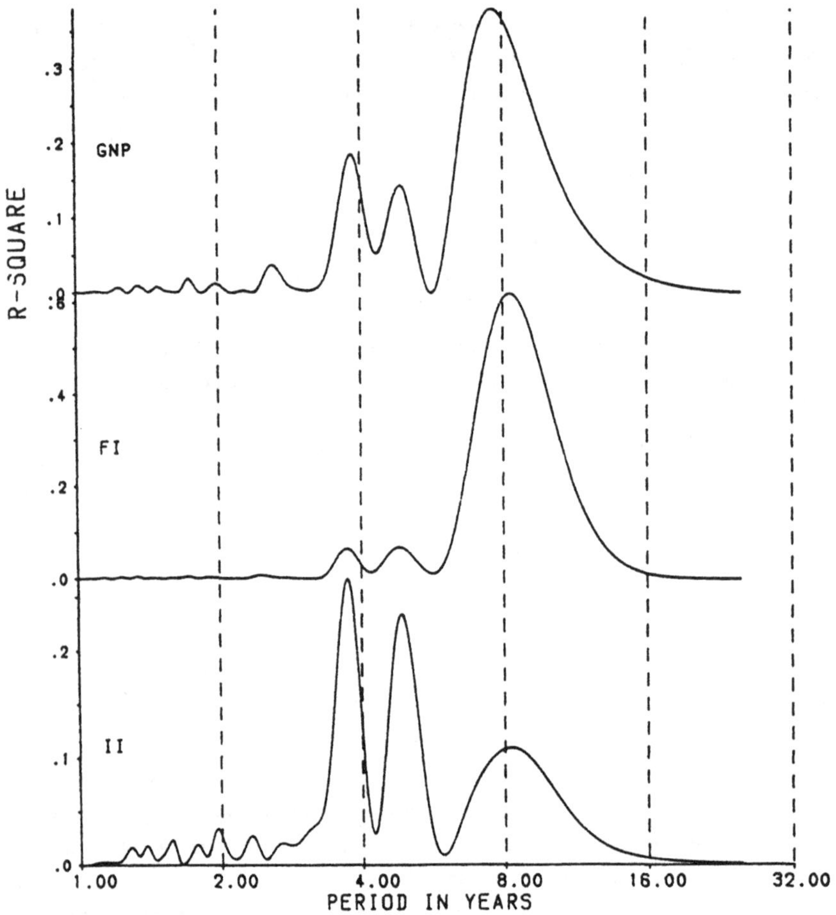

	MAXIMA			TREND:	ORDER	RSQADJ
GNP	3.84	4.87	7.62		3	0.992
FI	3.74	4.82	8.26		2	0.907
II	3.74	4.87	8.26		2	0.011

SOURCE: DIW

3. Second Order Accelerator

The adjustment process of equipment capital K_e towards its desired level K_e^* is commonly described in an accelerator equation of the form

$$I_e = DK_e = b_e(K_e^* - K_e) \tag{1}$$

where D is the differential operator with respect to time, I_e = DK_e is net equipment investment and b_e the speed of adjustment.

This formulation couples net investment I_e directly with desired equipment capital K_e^*. Since the magnitude of K_e^* depends on the expected future profitability of investment and such expectations are notoriously volatile, it follows that investment itself is a highly volatile and erratic magnitude. What is actually observed, however, is not a high degree of short run volatility, but rather a cyclical movement during which the actual capital stock is sometimes above and sometimes below its equilibrium value.

A very simple and compelling explanation is that investment can only be marginally altered at any point in time since most of the investment taking place currently is the result of past commitment. Machines are typically ordered in advance and are often constructed to specification. Contracts for new plants and associated equipment may have to be awarded some years in advance and cannot be cancelled without excessive costs.

A natural formalization of the preceding line of thought results if the left hand side of (1) is reinterpreted as desired investment I_e^* rather than actual investment I_e, and actual investment is assumed to be adjusted toward its desired level. With c_e being the (positive) speed parameter of this adjustment process the new model of the accelerator is

$$I_e^* = b_e(K_e^* - K_e) \tag{2}$$

$$DI_e = c_e(I_e^* - I_e) \tag{3}$$

This implies a second order differential equation for K_e:

$$D^2 K_e = c_e[b_e(K_e^* - K_e) - DK_e] \quad (4)$$

The solution to (4) may be written in the form

$$K_e(t) = K_e^h(t) + K_e^p(t) \quad (5)$$

where $K_e^p(t)$ depends on K_e^* (particular solution) and may be interpreted as the equilibrium solution. $K_e^h(t)$ denotes a solution to the homogenous part of the equation. The characteristic polynomial associated with (4)

$$P_e(x) = x^2 + c_e x + b_e c_e \quad (6)$$

has the roots

$$x_{1,2} = \frac{1}{2}\left[-c_e \pm \sqrt{c_e^2 - 4b_e c_e}\right], \quad (7)$$

The solution is always stable and periodic for $c_e<4b_e$. Under the assumption that changes of real investment are slow in comparison with desired changes in capital, that is

$$b_e \gg c_e \quad (8)$$

equation (4) may be approximated by

$$D^2 K_e = a_e(K_e^* - K_e) \quad , \quad a_e = b_e c_e \quad (9)$$

The homogenous part of equation (9) implies oscillations of constant amplitude around an equilibrium path depeending on K_e^*:

$$K_e^h(t) = A_e \cos(w_e t - S_e) \quad , \quad w_e = \sqrt{a_e} \quad (10)$$

where A_e and S_e are given by initial conditions. The period T_e of the oscillations is

$$T_e = \frac{2\pi}{\sqrt{a_e}}. \quad (11)$$

The simplicity of this formulation makes it very useful for further theoretical analysis. In contrast of other business cycle models the approximative constancy of amplitude is obtained under weak and in no way restrictive assumptions.

Since DK_e has the dimension of a rate of change or velocity, D^2K_e is an acceleration. It is therefore particularly appropriate to refer to (4) and (9) as accelerator equations. To be more specific we shall call it the **second order accelerator**.

The following section uses the second order accelerator as a building block in the construction of a closed macroeconomic model of cyclical growth.

4. Equipment Cycle and Equilibrium Growth

The economy is viewed as consisting of two sectors:

(a) an investment sector determining the target stock K_e^* and the actual stock K_e of equipment capital and net equipment investment I_e and

(b) an absorbing sector determining absorption Z_e.

Total output Q is devided into net equipment investment I_e and absorption Z_e :

$$Q = I_e + Z_e \tag{12}$$

To explain investment, equation (9) is modified to the form

$$DI_e = D^2K_e = a_e(K_e^* - K_e) + f_e Z_e, \qquad f_e > 0 \tag{13}$$

The rationale for the Term f_eZ_e is given by growth expectations: Firms know that investment must respond not only to an actual capital deficit K_e^* - K_e, but also to the secular growth of economy. The exact value of f_e is determined by an equilibrium condition; this will be discussed below.
The target capital stock is assumed to be proportional to absorption:

$$K_e^* = k_e Z_e, \qquad k_e > 0 \tag{14}$$

Regarding the absorption sector a target level Z_e^* proportional to output Q is assumed to be:

$$Z_e^* = z_e Q, \qquad 0 < z_e < 1 \tag{15}$$

Absorption itself is determined by an adjustment equation with a trend term:

$$DZ_e = d_e(Z_e^* - Z_e) + g_e Z_e, \qquad d_e, g_e > 0, \tag{16}$$

This implies that the absorbing sector also has expectations about secular growth. Actually, as absorption is known to be much more stable than investment, the growth aspect is regarded as the dominating contribution to the dynamics of absorption.

Setting

$$s_e := 1 - z_e \tag{17}$$

one obtains the complete model in the form of a matrix-vector differential equation:

$$Dy_e = A_e y_e \tag{18}$$

with

$$y_e = \begin{pmatrix} K_e \\ I_e \\ Z_e \end{pmatrix} \quad and \quad A_e = \begin{pmatrix} 0 & 1 & 0 \\ -a_e & 0 & a_e k_e + f_e \\ 0 & d_e z_e & g_e - d_e s_e \end{pmatrix}. \tag{19}$$

The associated characteristic polynomial

$$P_e(x) = det(x E_3 - A_e) \tag{20}$$

where E_3 denotes the 3x3 unit matrix can be written in the form

$$\begin{aligned} P_e(x) = {} & (x - g_e)[x^2 + d_e s_e x + a_e(1 - d_e k_e z_e) + d_e(g_e s_e - f_e z_e)] \\ & + d_e[g_e(g_e s_e - f_e z_e) + a_e(s_e - g_e k_e z_e)] \end{aligned} \tag{21}$$

The eigenvalues of the system are the roots of $P_e(x)$. Their calculation can be simplified by assuming the existence of an equilibrium solution where the actual and desired values of investment and demand are the same

$$K_e = K_e^*, \quad Z_e = Z_e^* \tag{22}$$

From

$$\overline{K}_e = k_e \overline{Z}_e \ , \ \overline{Z}_e = z_e \overline{Q} \ , \ \overline{Q} = \overline{I}_e + \overline{Z}_e \tag{23}$$

and

$$D\overline{K}_e = \overline{I}_e \ , \ D\overline{I}_e = f_e \overline{Z}_e \ , \ D\overline{Z}_e = g_e \overline{Z}_e \tag{24}$$

the following restrictions for the parameters g_e and f_e are obtained:

$$g_e = \frac{s_e}{k_e z_e} \ , \ f_e = k_e g_e^2 = \frac{s_e g_e}{z_e}. \tag{25}$$

The equation for g_e is the well known Harrod-Domar formula, since $s_e = 1-z_e$ is the equilibrium propensity to save, and $k_e z_e$ is the equilibrium capital/output ratio. Using (25) the characteristic polynomial is simplified to the form

$$P_e(x) = (x - g_e)[x^2 + d_e s_e x + a(1 - d_e k_e z_e)] \tag{26}$$

This factorization appears quite natural, as the equilibrium growth rate has to be an eigenvalue of the system by the assumptions above. This real eigenvalue is denoted by x_1.

The subsequent discussion will be concentrated on the cyclical component of the solutions, i.e. the roots x_2, x_3 of the quadratic part of (26):

$$x_{2,3} = \frac{1}{2}\left[-d_e s_e \pm i\sqrt{4a_e(1 - d_e k_e z_e) - d_e^2 s_e^2}\right] \tag{27}$$

The eigenvalues are complex for

$$4a_e(1 - d_e k_e z_e) > d_e^2 s_e^2 \tag{28}$$

The parameter d_e may be expected to be small because of the empirically observed stability of demand (This relation will be shown below). Therefore (28) may be regarded as being fulfilled, and the model has stable oscillatory solutions with the damping rate $d_e s_e/2$.

In the limit case $d_e=0$ the discussion is simplified considerably. One obtains for the system matrix:

$$A_e = \begin{pmatrix} 0 & 1 & 0 \\ -a_e & 0 & a_e k_e + f_e \\ 0 & 0 & g_e \end{pmatrix}, \tag{29}$$

for the characteristic polynomial:

$$P_e(x) = (x - g_e)(x^2 + a_e) \tag{30}$$

and for the eigenvalues:

$$x_1 = g_e \quad , \quad x_{2,3} = \pm i w_e = \pm i\sqrt{a_e}. \tag{31}$$

In this case demand does not oscillate and grows exponentially with the rate g_e. Equipment investment and capital are given as superpositions of exponential growth and oscillations, and the oscillations may be described by the elementary second order accelerator (9).

In the general case $d_e>0$ the oscillations are discussed studiing the eigensolutions $y_{2,3}(t)$ to the eigenvalues $x_{2,3}$ of the system. They may be written in the form

$$y_{2,3}(t) = y^0_{2,3} \exp(x_{2,3} t), \tag{32}$$

where $y^0_{2,3}$ denote eigenvectors of A_e, i.e. nontrivial solutions to the equation

$$(x_{2,3} E_3 - A_e) y^0_{2,3} = 0 \tag{33}$$

As d_e may be expected to be small, the discussion will be limited to first order accuracy in d_e, i.e. contributions of high powers of d_e (d_e^2, d_e^3 etc.) will be neglected. In this approximation the eigenvalues $x_{2,3}$ given by (27) can be rewritten as

$$x_{2,3} = -\frac{d_e s_e}{2} \pm i w_e \left(1 - \frac{d_e k_e z_e}{2}\right) \tag{34}$$

The eigenvector equation (33) leads to

$$x_{2,3} K^\circ_{e2,3} - I^\circ_{e2,3} = 0 \tag{35}$$

$$a_e K^\circ_{e2,3} + x_{2,3} I^\circ_{e2,3} - (a_e k_e + f_e) Z^\circ_{e2,3} = 0 \tag{36}$$

Inserting (35) into (36) and using the approximation

$$(x_{2,3})^2 = -a_e(1 - d_e k_e z_e) \mp i w_e d_e s_e. \tag{37}$$

yields

$$Z^\circ_{e2,3} = \frac{a_e k_e z_e \mp i w_e s_e}{a_e k_e + f_e} d_e K^\circ_{e2,3} \tag{38}$$

Equation (38) describes the relation between the oscillations in absorption and capital stock. Two conclusions may be drawn:

(a) The oscillation in Z_e is small (first order in d_e) in comparison with the amplitude of the oscillations in K_e.

(b) For realistic parameter values the real part of the numerator in (38) is by far greater than the imaginary part, and the ratio itself may be regarded as approximatively real (and positive). Therefore the oscillations of Z_e and K_e are roughly in phase. More precisely, from the sign of the imaginary part it can be seen that Z_e is slightly lagging.

The relation between investment and capital stock is given by (35) and (34):

$$I^\circ_{e2,3} = \left[-\frac{d_e s_e}{2} \pm i w_e \left(1 - \frac{d_e k_e z_e}{2} \right) \right] K^\circ_{e2,3} \tag{39}$$

The right hand side factor is mainly imaginary with a small (first order in d_e) real correction. This leads to a phase difference of about $\pi/2$ between the oscillations in investment and capital stock.

The eigenvectors $y^o_{2,3}$ can be determined using (38) and (39):

$$y^0_{2,3} = \begin{pmatrix} 1 \\ -\frac{d_e s_e}{2} \pm i w_e \left(1 - \frac{d_e k_e z_e}{2} \right) \\ d_e \frac{a_e k_e z_e \mp i w_e s_e}{a_e k_e + f_e} \end{pmatrix} \tag{40}$$

The complete phase information can be illustrated in a phase diagram, which shows the orientation of the eigenvector components given by (40) in the complex plane. Figure 2 shows such a phase diagram for x_2 (The corresponding diagram for x_3 is complex conjugate and contains the same information).

Figure 2: Phase diagram

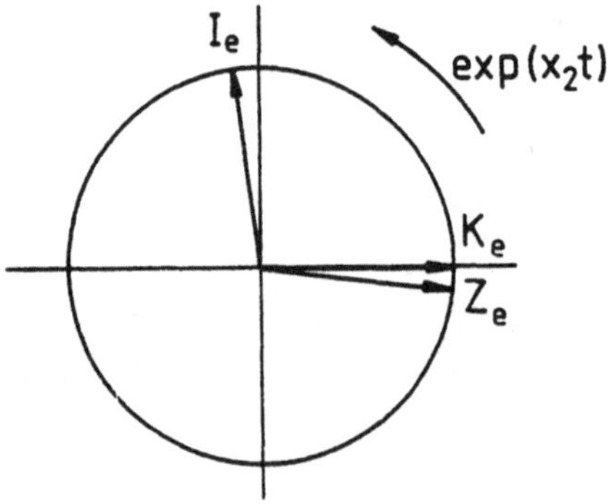

I_e equipment investment
K_e equipment capital stock
Z_e demand

Equipment investment leads by roughly one forth period before capital stock and demand.

As has been shown the parameter d_e determines on the one hand the damping rate of the oscillation (27), on the other hand the amplitude of the cycle in demand (38). This leads to a surprising conclusion of high political relevance: Fluctuations in investment and demand ($d_e>0$) imply damped cycles which are thereby stabilized whereas stabilizing demand ($d_e=0$) yields undamped investment cycles and acts destabilizing in this sense.

5. Summary and Conclusions

The basic ideas of the equipment cycle model presented in this paper can be summarized as follows:

The idea of inertia of investment is formalized in the concept of the second order accelerator. This leads to cycles of constant or decreasing amplitudes in equipment investment and capital stock. This concept may be integrated into a growth model including an absorption sector. Equilibrium growth is possible under Harrod Domar conditions, and the equipment cycle appears as a sinoidal deviation from the growth path. Autonomous growth of the absorption sector implies constant amplitudes of the investment cycle. If absorption may be (slightly) in disequilibrium and is subject to a first order adjustment process, cycles of decreasing amplitudes appear in all variables. The cyclical movement in equipment investment leads by roughly one forth period before the corresponding cycle in absorption.

The basic features of the model (demand: stable or weakly fluctuating, equipment investment: one clear cycle of approximately constant amplitude) correspond to the empirical observations.

REFERENCES

Fisher, I.: "Our unstable Dollar and the So-Called Business Cycle", Journal of the American Statistical Association (1925), 191-192

Granger, C.W.J.: "The Typical Spectral Shape of an Economic Variable", Econometrica 34 (1966), 150-161

Hicks, J.R.: "A Contribution to the Theory of Trade Cycles", Oxford: University Press, 1950

Hillinger, C.: "An Econometric Model of Mild Business Cycles", Manchester School of Economics and Social Studies (1966), 269-284

-------------: "Inventory Cycle and Equipment Cycle Interaction", in "Inventories in Theory and Practice" (Proc. Third International Symposium on Inventories, Budapest 1984), Amsterdam: Elsevier, 1986

-------------: "Theorie und Empirie der Konjunkturzyklen", Konjunkturpolitik 32 (1986), 101-129

-------------: "Business Cycle Stylized Facts and Explanatory Models", Journal of Economic Dynamics and Control 11 (1987), 257-263

Juglar, C.: "Des Crises Commerciales et leur Retour Periodique en France, en Angleterre et aux Etats-Unis", 1862

Lucas, R.E.: "An Equilibrium Model of the Business Cycle", Journal of Political Economy 83 (1975), 1113-1144

Metzler, L.A.: "The Nature and Stability of Inventory Cycles", Review of Economic Statistics (1947), 113-129

Samuelson, P.A.: "Interaction between the Multiplier Analysis and the Principle of Acceleration", Review of Economic Statistics (1939), 75-78

Slutsky, E.: "The Summation of Random Causes as the Source of Cyclic Processes", Econometrica (1937), 105-146

On a Stabilizing But Chaos Generating Mechanism in Behavioural Functions: Observations in a Multiplier-Accelerator Model

by

R.Franke and W.Weghorst

University of Bremen [1)]

I. Introduction

In many dynamic models where agents are not blessed with perfect foresight we encounter the problem of overshooting in individual reactions. In economic models, for example, this applies to firms which have to build up and expand their capital stock in order for production to meet future demand. If predictions of the latter turn out to have been too optimistic, they will be left with idle capacity. It might then have been better to accept (costly) over-utilization of capacity because of too low an estimation of demand[2)]. Which of these two potential dangers is taken for worse will, among other things, depend on an attitude which for short we may call risk-aversion. At a similar cross-roads is every agent who has to take a decision on an irreversible action now, but can base it only on a more or less sophisticated prediction of the future performance of some economic variables.

It is our first concern in this paper to present a simple and elementary way in which this element of risk-aversion can be incorporated into the corresponding behavioural function, as sort of a downward revising correction mechanism. In particular, modelling remains deterministic and no subjective probability or maximization of expected utility enters the scene. Although our formalization is with respect to the above situation of a firm urged to decide on investment in fixed capital, the basic idea may be transferred to model, in a succint way, the outcome of other decision processes in an explicitly dynamic setting.

Our second concern is to study the effects of these individual decisions at the macro level, where investment via the multiplier determines output and this in turn, via a change in predictions of demand, exerts an influence on investment in the next period. That is, we undertake our study in a one-sectoral model of the multiplier-accelerator type. Only that with our investment function the accelerator is more elaborated than in the textbook versions. It brings in a non-linearity and we end up with a fairly complicated dynamic system comprising four state variables, an analytical treatment of which appears almost impracticable. So we have to resort to computer simulation experiments. Employing not too implausible empirical data for the technological coefficients involved, it turns out that, from a certain degree of risk-aversion on, the negative feedback effects inherent in the system are sufficiently strong to generate persistent cyclical behaviour. Loosely speaking, we could state that we have obtained a floor and a ceiling in a far more endogenous way than usual (actually, we cannot derive a formal expression giving the upper- and lower-bounds).

Even more remarkable is our finding that the motions themselves display a considerable complexity. We are able to confirm this intuitive impression of "chaos" by means of spectral analysis. It may be interesting to investigate whether, or under what conditions, similar things will happen in different settings adopting the presented idea of endowing agents with a mechanism to guard against overshooting.

The paper is organized as follows. In the next section we set out the dynamic equations governing the motions of the economy, leaving yet unspecified the determination of investment. We take up a route of input-output modelling that was opened up in an article by Duchin and Szyld (1985). Its main feature is the dropping of the full capacity utilization hypothesis in order to overcome the serious deficiencies of the standard dynamic Leontief input-output model when it comes to disequilibrium. Our formulation differs in that we content ourselves with a one-sector version[3]. On the other hand, besides our extension of the investment function we explicitly take account of capital stocks. Incorporating them as a state variable permits replacement investment to fall short of depreciation, so that also a decrease of productive capacity can be considered (in Duchin and Szyld it can only grow). In section III we shortly mention how the economy behaves if we adopt Duchin and Szyld's simple prediction rule of future demand from which they derive the volume of investment. Since the outcome seems unacceptable, we then introduce our proposal for modifying the investment function. In section IV we present the main dynamic properties of the model thus defined, those directly observable and those obtained by spectral analysis.

II. The dynamic equations

The model is formulated in discrete time. Let us begin with productive capacity. Denote by b the quantity of the one good, inasfar as it is used as a capital good, that at the beginning of a period has to be installed in order for the sector to produce, during that period and at "normal" utilization, one unit of output. S(t) being the total capital stock, capacity K(t) for period t is defined by

$$K(t) \quad = \quad S(t)/b \tag{1}$$

A crucial role is played by the capacity $K^*(t+1)$ that the sector wishes to have under its control for production in the next period, t+1. Assuming that the gestation period coincides with the time period underlying, the corresponding decisions have to be made at the beginning of period t. They are based on information available to this date, i.e., on observations from period t-1, t-2, ... For the moment being we are content with expressing this idea by writing

$$K^*(t+1) \quad = \quad F[\ \dots\ (t-1)] \tag{2}$$

The desired capital stock, $S^*(t+1)$, is then given by

$$S^*(t+1) \quad = \quad b\ K^*(t+1) \tag{3}$$

Investment derives from the difference between $S^*(t+1)$ and the capital stock transferred from the preceding period. We suppose that it depreciates at a constant rate δ, independently of the degree of utilization. Thus gross investment demand z(t) for period t amounts to

$$z(t) \quad = \quad \max \{\ 0\ ,\ S^*(t+1) - (1-\delta)S(t)\ \} \tag{4}$$

With our assumption below it can actually be satisfied, so that next period's capital stock is

$$S(t+1) \quad = \quad (1-\delta)S(t) + z(t) \tag{5}$$

Turning from fixed to circulating capital, we state that in order to produce one unit of output, a $(0<a<1)$ units of the good are required as intermediate inputs. As concerns the other components of demand, since our main interest is with investment we employ the simple hypothesis that they, apart from autonomous demand $\gamma>0$, depend linearly and without any time lag on current production x(t). The latter is governed by a non-negative coefficient c. In this way, total demand adds up to $z(t) + ax(t) + \gamma + cx(t)$. We assume that production adjusts to this level. Accordingly, x(t) is determined by the equation

$$x(t) = (a+c)\,x(t) + z(t) + \gamma \qquad (6)$$
$$= [z(t) + \gamma] \,/\, (1-a-c)$$

As announced in the Introduction, this may result in under- or over-utilization of capacity $K(t)$. Formally, $u(t) := x(t)/K(t)$ will in general deviate from unity. It has to be noted that *a priori* investment demand might turn out so high that even with greatest effort $x(t)$ is no longer feasible. As a matter of fact, this is what we shall see happen in the next section and which will prompt us to reconsider the function representing investment demand.

It is immediately verified that eq.s (1)-(6) form a well-defined recursive system: having specified certain initial values, all variables are unambiguously determined period for period.

III. The investment function

Having presented the basic rules for the economy to move through time, we must now specify the form of function F. Since via eq.s (3) and (4) it readily determines the level of investment demand, we may call it directly the investment function. We first want to see what comes of if we follow the approach of Duchin and Szyld (1985): having to decide on $K^*(t+1)$ investors strive for normal capacity utilization, and for this purpose put forward a hypothesis on the rate of growth at which the last observed output $x(t-1)$ will expand in the next two periods. Duchin and Szyld propose to take some kind of average of the last two growth rates and specify

$$f(t-1) := [x(t-1) + x(t-2] \,/\, [x(t-2) + x(t-3)] \qquad (7)$$

as the expected growth factor. At high values of $f(t-1)$ the sector may nonetheless be afraid of building up a capital stock which, after a weakening of the boom, will lead to a long period of under-utilization. To guard against these excess capacities, "a maximum annual rate of expansion of capacity" (*loc. cit.*, p.6), $\bar{g}$, is imposed. To translate this into formal terms, define

$$K_a^*(t+1) := [f(t-1)]^2\, x(t-1) \qquad (8)$$
$$K_b^*(t+1) := (1+\bar{g})\, K(t)$$

and put

$$F[\ \ldots\ (t-1)] \quad = \quad \min \{ K_a^*(t+1)\ ,\ K_b^*(t+1) \} \qquad (9)$$

As this is the first proposal of an investment function we consider, let us system (1)-(9) shortly call process P1. We remark that in essence eq.(7) pictures an accelerator mechanism bringing in, in particular, a non-linearity. On the other hand, eq.(6) clearly shows the multiplier. So process P1 can succintly be characterized as a non-linear multiplier-accelerator process which takes account of the evolution of outputs and capacities without the hypothesis of full capacity utilization.

Despite the one-good assumption, owing to the de-coupling of output and capacity and to the lag structure in the planning of future capacity the process cannot be reduced to one variable. The best that can be achieved is a reduction to four state variables: $K(t)$, $x(t-1)$, $x(t-2)$, $x(t-3)$. That is, in order to compute $K(t+1)$ and $x(t)$ one must know the values of these four variables.

The positive autonomous demand term γ implies an equilibrium notion of stationarity. From $K^*(t+1) = K(t+1) = K(t)$ and $x(t) = x(t-1) = \ldots$, we infer via eq.s (2) and (7)-(9) that $K(t) = [f(t-1)]^2\ x(t-1) = x(t)$. Hence, in equilibrium utilization of capacity is indeed "normal". Furthermore, $S^*(t+1) = S(t+1) = S(t)$ and eq.s (3)-(6) give $x(t) = [\delta bK(t) + \gamma]/(1-a-c)$, from which the equilibrium level of output is calculated as

$$x^s \quad = \quad \gamma\ /\ [1-a-c-\delta b] \qquad (10)$$

("s" for stationary).

As regards the dynamic properties of process P1, upon closer scrutiny a negative as well as a positive feed-back loop could be detected, the first one indicating a stabilization effect, the second pointing to destabilizing tendencies. The skeleton of eq.s (1)-(9) gives no further information as to which of the two mechanisms will prove dominant. Just as with the ordinary multiplier-accelerator processes, knowledge of the different parameters involved is inevitable. For our numerical simulations we choose the following values.

Since in the FRG the capital-output ratio during the period 1950-1975 fluctuated around 2 (cf. Deschauer, 1977, pp.49ff, and Hirschhäuser, 1981, p.72, n28), we straightforwardly put b=2. The rate of depreciation is obtained by assuming that the average life-time of investment is 30 years (cf. the table recorded in Oest, 1979, p.61). Letting capital over this span of time shrink to a small fraction, namely 5%, of its initial volume, by solving the equation $(1-\delta)^{30} = 0.05$ we get $\delta=0.095$. As concerns a and c, only the sum a+c matters (cf. eq.(6)). We leave statistical estimates aside and determine it in such a way that claiming full utilization and setting $\gamma=0$ would lead to a uniform growth rate of $g^*=5\%$. This rate might be interpreted as the economy's potential for growth. From the corresponding equation $x = (a+c)x + (\delta+g^*)S = [a+c+b(\delta+g^*)]x$ we derive $a+c=0.71$. The maximal rate of expansion in eq.(8) is put equal to 12%, autonomous demand is set at $\gamma=10$. The other steps accepted, the latter is no more than choosing the unit in which the one good is measured. These values are maintained throughout the remainder of the paper. As an immediate consequence, eq.(10) determines output and capacity in the stationary equilibrium as $x^s = 100$.

With respect to initial conditions, in all simulation runs we standardize them by supposing that capacity at time t=1 is equal to its equilibrium value and that previous outputs are slightly below their normal value, the reduction uniformly being 1%. Thus, we have

$$S(1) = 200 \ , \ K(1) = 100 \ , \quad x(0) = x(-1) = x(-2) = 99$$

Letting now process P1 unfold we make the following observations. In spite of the modest size of perturbation, within only two periods the system falls into a severe recession where gross investment has completely ceased and has caused a fall of output to its minimum level, which according to eq.(6) is given by $x^o := [0 + \gamma]/(1-a-c) = 34.48$. This goes on for 10 periods until the capital stock has deteriorated to a level with capacity near x^o. Maintaining production at that level then at last requires positive (replacement) investment, and the resulting increase in production initiates the upswing. As a matter of fact, capacity expansion is stimulated so strongly that right from the start the upper-bound in eq.(9) takes effect, limiting its growth rate to 12%. This continues for all times and with some delay also output growth adjusts to that speed. In addition it has to be noted that utilization is extremely high. Temporarily it even exceeds 200% and then tends to u = 148%.

This is clearly an unsatisfactory result: not only the high values of u and the high rate of growth at which the system eventually expands, but rather the fact that the latter is imposed from the outside. If we had admitted a higher rate $\bar{g}$ in eq.(8) the system would have grown even faster, at this new growth rate. This phenomenon is in contradiction to the role the rate $\bar{g}$ was originally supposed to play, namely to guard against a long period of capacity under-utilization that is expected to follow a too impetuous phase of expansion.

There are certainly lots of arguments why economies do not grow so unwaveringly. For instance, even if over-utilization is always feasible, at any rate it means higher costs for producers. At the same time it gives some room for charging higher prices. These are important points a more thorough analysis would have to take up. Extensions in this direction are far beyond the scope of the present contribution. We concentrate on another mechanism which already in our limited framework is able to exert a stabilizing influence.

The basic idea is the fear of investors that capacities will turn out idle. However, the risk they perceive will not necessarily increase with the level of the expected growth rate as it is calculated in eq.(7). More important to us seems the over-all picture of past growth, i.e., whether it has been relatively uniform or whether it has been subject to strong fluctuations. So, in our concept producers take account of two components when deciding over future capacity: the average of output rates of growth observed in the past, and their dispersion. The average of past growth rates, calculated by using a formula like eq.(7), is taken as an upper limit for the capital stock to be built up. Or let us say it is their target capital stock. Whether, or to what extent, it will be maintained depends on the evaluation of the risk to be left with idle capacity in a more or less distant future. This risk is measured by the dispersion of past growth rates: the greater the dispersion, the greater the risk. We suppose that producers are risk averse and revise downwards the investment that would correspond to establishing the target capital stock. That is, the greater the risk, the greater the revision. To wit, producers (as a whole) are quite aware that in this way they might miss some chances, but nevertheless they are more afraid of the risk and prefer to wait for the growth trend, as computed by eq.(7), to be confirmed.

To translate this general idea into formal language, three points have to be made precise:

(a) How is the "dispersion" of past growth rates defined?
(b) What term is it that will be revised?
(c) What is the functional specification of the revision? In particular, it should allow for different degrees of risk-aversion.

To answer the first question we employ the simplest measure of dispersion possible, namely the absolute value of the difference of the last two growth rates. We write "d" to indicate "dispersion" and define

$$d(t-1) := |x(t-1)/x(t-2) - x(t-2)/x(t-3)| \qquad (12)$$

As for (b), we refer directly to net investment that would result from the average growth rate of eq.(7). Accordingly, define

$$q^*(t) := [f(t-1)]^2 \cdot x(t-1) - K(t)$$
$$= \left[\frac{x(t-1) + x(t-2)}{x(t-2) + x(t-3)} \right]^2 x(t-1) - K(t) \qquad (13)$$

With respect to point (c), we choose to multiply $q^*(t)$ by a "correction term" h, which is a function of dispersion and takes on values between 0 and 1 (for simplicity, it is to apply to positive and negative net investment alike). In compliance with the above discussion, h should satisfy three postulates: It attains the value 1, i.e. no correction, if dispersion is zero; it approaches zero as dispersion becomes large; and h decreases with dispersion increasing. On the other hand, we introduce a non-negative risk-aversion parameter ρ, rising values of which are to indicate a higher degree of risk-aversion. In the special case ρ=0 producers are not risk-averse. This means that independent of the dispersion, h is identically unity (we better avoid the expression "risk-neutral"). At large values of ρ, h is always close to zero, so that capacity can only grow at very small rates. This is the background of the following definitions,

$$h(\rho;d) := \frac{1}{1 + \rho d} \qquad (14)$$

$$K^*(t+1) = K(t) + h[\rho;d(t-1)]\, q^*(t) \qquad (15)$$

The dynamic process that has arisen from our discussion is thus defined by eq.s (1),(3)-(6),(12)-(15). For short, we call it process P2. We emphasize that in the investment decision no more exogenous upper rate of growth is involved.

IV. Dynamic effects of the correction term

Parameters and initial conditions are taken over from the preceding section. A first visual impression of the properties of process P2 can be gathered from Figures 1 and 2. Having underlying a value of ρ=500, Fig.2 shows how capacity K(t) evolves over the first 1000 periods, whereas Fig.s 1a and 1b display the short-run characteristics of the time path of K(t)

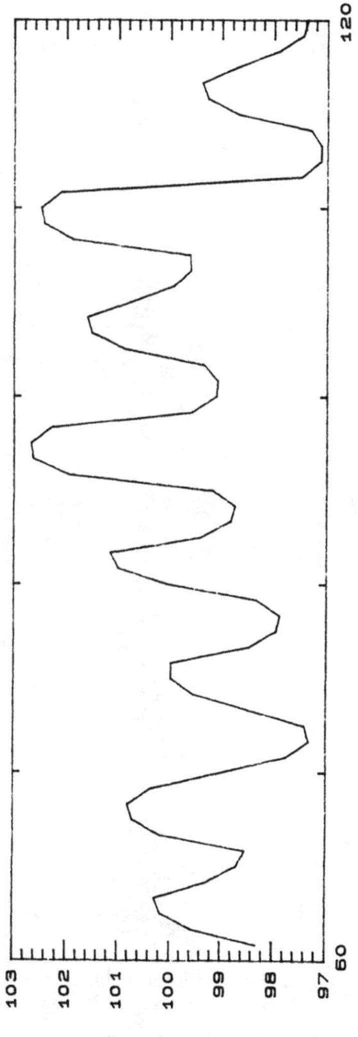

Fig. 1a (Capacity)

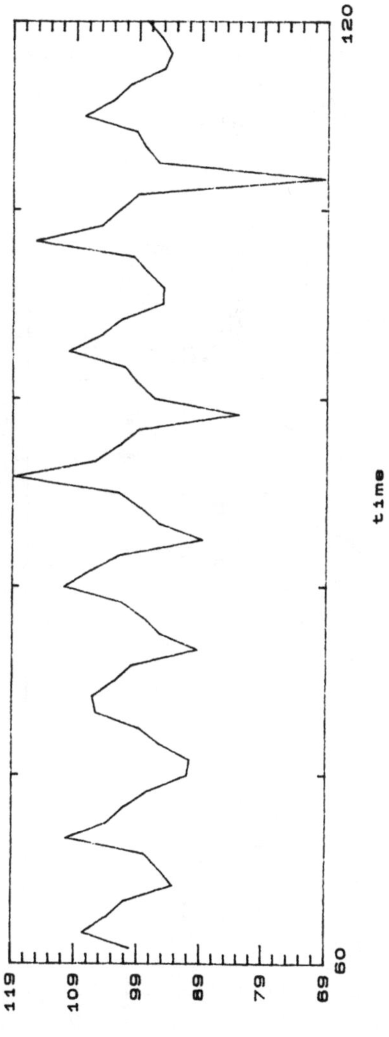

Fig. 1b (Output)

Fig. 2 (Capacity)

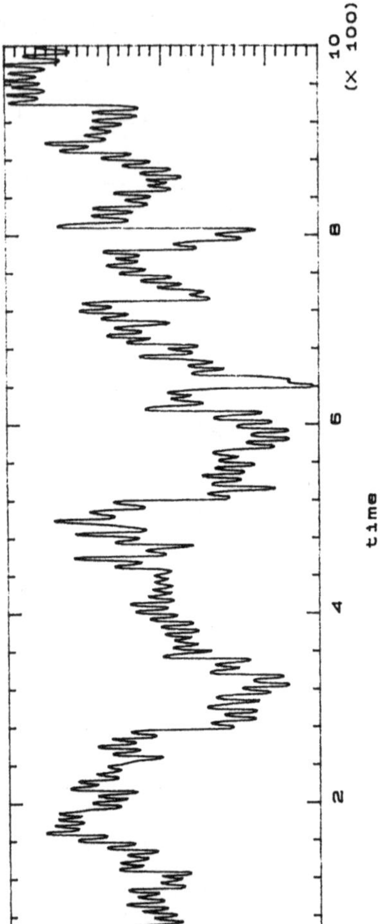

Fig. 3 (Power Spectrum of K)

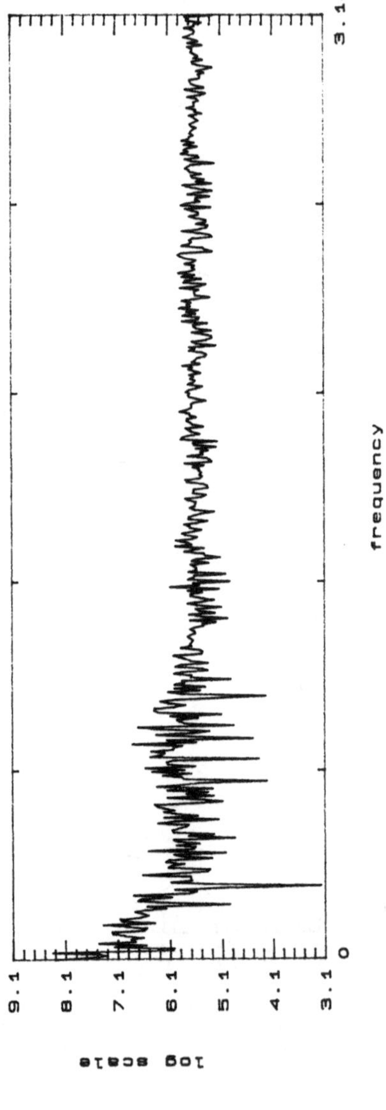

and x(t), respectively (we chose the span of time from t=60 to t=120).

Before we are dealing with the irregularities in Fig.2 that are leaping to the eye, let us take the diagrams as an illustration to the answer of the question whether the correction term introduced in eq.s (14),(15) is capable of checking the explosive tendencies which we have encountered in process P1. In fact, a sufficiently high level of the risk-aversion parameter ρ does not only pose, in an endogenous manner, an upper-bound on the system's rates of growth, but also K and x themselves are bounded from above. The stationary equilibrium, on the other hand, is still unstable. Consequently, cycles are generated and there is no tendency for them to dampen out[4]. Let us have a closer look how different levels of ρ take effect.

Already at low values of ρ some interesting phenomena of cyclical behaviour can be observed. We leave them aside since the system keeps on exploding, that is, we have an overflow of the computer's capacity to process real numbers. An overflow, within the first 1000 periods at least, is prevented from values of ρ=30 on, but over a certain range of ρ in some periods the output rates of growth are still completely out of range. Though fluctuations are wild, the fall of K is nevertheless cushioned. It no more gets in the vicinity of the minimum output value x^o = 34.48 (cf. preceding section). Even for ρ=30, for example, K(t) is always above K=67. Apart from that, the lower-bound that K(t) attains in the course of the process gradually increases with ρ.

At the same time, rising values of ρ steadily reduce the system's impulse to grow. It finally ceases at about ρ=400. The mean value of capacity is then near the stationary equilibrium value K^s=100 (we have mostly found it slightly above this mark, but at least we have as yet not sufficient information to conclude that this is more than just a matter of chance). Further results for some selected values of ρ are given in Table 1, which is based on simulation runs over 1000 periods. The results for ρ=100 are bracketed to point out that here still some long-run growth is present. Besides the mean value, the table reports the mean troughs and peaks (that is, we listed all peaks or troughs and calculated the respective averages). As will be expected, with ρ rising the peaks decline on an average and the troughs shift upwards. Correspondingly, the (average) amplitudes of the fluctuations are continuously decreasing. Similar observations, by the way, apply to the time paths of capacity utilization.

K \ ρ	100	500	1000	2500	5000
Mean Value	(161.6)	100.2	100.8	100.4	99.7
Mean Trough	(152.6)	98.3	99.9	100.0	99.6
Mean Peak	(178.4)	102.2	101.7	100.8	99.9
Mean Cycle	8.74	8.32	8.02	8.18	8.22

Table 1: Averages of capacity K over 1000 periods

We mention in addition that sufficiently high values of ρ can also ensure gross investment z(t) to remain positive. At $\rho=500$, z(t) falls down to zero only in exceptional cases (four times within 1000 periods), from $\rho=1000$ on this event has no longer occurred.

Whereas the parameter ρ brings about monotonous reactions in the the level of capacity, its average cycle length (the time between two successive peaks) exhibits no systematic tendency: for the stationary economies it is always between 8 and $8^1/_3$ periods (for $\rho<100$, lower values than 8.84 or even 8.00 are obtained).

A look at Figures 1a and 1b might suggest that capacity and output move more or less in step, though output can be seen as taking the lead and the amplitudes of its fluctuations are considerably larger. Calculating the mean cycle length of x, however, reveals that there is no simple kind of synchronization between x and K. At least on an average the cycles of x are always shorter than those of K.

Turning to the long-run dynamics of process P2 as presented in Fig.2, we first of all underline that it is by no means a particular example. Similar pictures are obtained for all other values of the parameter ρ that are able to limit the system from above (at least we have detected no counter-example). Likewise, our results do not appear to depend on the specific definition of the function $h=h(\rho,d)$ in eq.(14).

To the naked eye, the predominant impression of Fig.2 is disorder: After some more or less regular cycles in the short-run, a kind of jump is observed. It seems, however, hardly predictable when, in which direction, and with what strenght it will occur. Moreover, in the regular cycles we sometimes find a slight trend upwards or downwards, at other times successive peaks and troughs change less systematically. Apart from that, for fifty or even hundred periods the system may constantly move above or below the benchmark that is given by the stationary equilibrium value $K^*=100$, which as we know roughly coincides with the long-run time average. On this evidence we feel inclined to say that the trajectories of process P2 behave in an erratic manner, or that the system exhibits complex dynamics (trajectories of output or capacity utilization are not shown, but their fluctuations seem even wilder).

It may nevertheless be premature to speak of "chaos" [5] since the naked eye is not a very reliable jugde. Numerical examples show that even for cycles of relatively low period it may be impossible to distinguish regular time series from completely irregular series by simple inspection[6]. To confirm the impression of chaos we employ a more appropriate means of investigation which has proven very useful for studying (bounded) fluctuations, namely spectral analysis.

To give a shortened outline of the basic principles of spectral analysis the following notes may suffice[7]. Given a time series of a (scalar) variable observed at times $t=1,\ldots,N$, we consider the so-called Fourier frequencies

$$\omega_k := 2\pi k/N \quad , \qquad 0 \le k < N/2$$

and calculate the corresponding "intensities" $I_k = I(\omega_k)$ (they are always non-negative). The power spectrum diagram is the graph of these intensities plotted against the ω-axis.

Now, suppose that we shall find just one frequency ω_k with a large intensity $I(\omega_k)$, whereas all other intensities are relatively close to zero. We would have then identified a dominant sinusoidal motion with frequency ω_k. It corresponds to a period (wave lenght) of $T_k = 2\pi/\omega_k = N/k$, so that in the span of the data it would execute k cycles. Should we find several frequencies ω_k with comparatively large intensities or, in other words, should we find several peaks at frequencies ω_k in the power spectrum, then the time series can approximately be represented by additive sinusoids with corresponding frequencies (their contribution to the total sum is closely related to the height of their peaks). From the number of peaks we could thus readily tell the number of sinusoids involved (to be precise, we count the number of those peaks the corresponding periods $T_k = N/k$ of which are no multiples of each other, and we should be aware of possible "leakages"). Obviously the complexity of motions increases with the number of these peaks, though basically they are still periodic.

On the other hand, an irregular or erratic time series can be viewed as involving periods of all orders, so that the power spectrum displays "peaks" at each frequency. The appearance of such a broad band (it is also called a continuous spectrum), whether with some sharp peaks on top of it or not, is quite often taken as synonymous of chaos[8)].

After these preparatory remarks we can have a look at Fig.3 where we have plotted the power spectra of our simulation runs with $\rho=500$. The main result can be recognized at a glance: a broad band is indeed present. In this sense we are hence entitled to say that our time series exhibits chaotic behaviour. Moreover, this is not only true with respect to the particular value of $\rho=500$, the phenomenon remains if the parameter is increased. By employing a special measure (i.e., average entropy) that can distinguish sharp spikes from broad band features, it could in addition be argued that with ρ rising the fluctuations in K become even more irregular[9)].

Notes

1) Our work originated in the project "Technologischer Wandel und Beschäftigung" at the University of Bremen, which is supported by the Stiftung Volkswagenwerk. The main results here presented are extracted from a much longer article with a slightly different emphasis, Franke and Weghorst (1987), which we have submitted to *Metroecomica*. We wish to thank the editor S.Parrinello for his kind permission to have this shorter paper included in this volume. In fact, it may be seen as an announcement of, or an invitation to, the more detailed version.

2) If utilization comes up against limiting factors, it is sales opportunities that will have been missed.

3) A generalization to n sectors is nevertheless straightforward; see Franke and Weghorst (1987).

4) To avoid any misunderstanding, by "cycle" we simply mean a motion from one peak to the next. The more special meaning of a closed orbit (in the phase space) plays no role in this paper.

5) An informative introduction to "chaotic" phenomena is

given in Ruelle (1980).

6) This latter sentence is a quote from Gabisch and Lorenz (1987, p.173).

7) Most textbooks of spectral analysis right from the start refer to a stochastic context. One exception is Bloomfield (1976).

8) The idea of detecting chaos in dynamical systems by analyzing the power spectra of some of its variables plays an important role in physics and chemistry, especially in the study of turbulent systems; see, e.g., Ruelle (1980, pp. 132-136). A great number of further references can be found in Garrido and Simo (1983, p.7).

9) The details are provided in Franke and Weghorst (1987).

References

P.Bloomfield (1976), Fourier Analysis of Time Series: An Introduction, New York: J.Wiley

P.Deschauer (1977), Der Kapitalkoeffizient im Prozeß struktureller Anpassung, Darmstadt (Diss.)

F.Duchin, D.Szyld (1985), "A dynamic input-output model with assured positive outputs", Metroeconomica, 37, 269-282

R.Franke, W.Weghorst (1987); "Complex dynamics in a simple input-output model without the full capacity utilization hypothesis", Working Paper of the Research Group "Technologischer Wandel und Beschäftigung", University of Bremen

G.Gabisch, H.W.Lorenz (1987), Business Cycle Theory, Berlin: Springer

L.Garrido, C.Simo (1983), "Prologue: Some ideas about strange attractors", in L.Garrido (ed.), Dynamical Systems and Chaos, Berlin: Springer; pp. 1-28

K.F.Hirschhäuser (1981), Konjunkturanalyse mit Hilfe von System-Dynamics-Simulationsmodellen, Göttingen: Vandenhoeck und Ruprecht

W.Oest (1979), Ein dynamisches sektoral disaggregiertes Wirtschaftsmodell für die Bundesrepublik Deutschland zur Abschätzung alternativer Entwicklungsmöglichkeiten, Berlin: Duncker und Humblot

D.Ruelle (1980), "Strange attractors", The Mathematical Intelligencer, 2, 126-137

A GENERAL MODEL OF CAPITAL ACCUMULATION [1]

Michael Krüger
University of Massachusetts
Economics Department
Amherst, MA 01003

1. Introduction

In this paper we shall discuss a model of cyclical accumulation of capital which might, to some extent, be considered as a mathematical restatement of both Marxian and Kaldorian ideas. According to Marx the process of capital accumulation is governed by a persistent class conflict over the conditions of production and distribution. Moreover, the "law of motion" of the modern capitalist society, i.e., the cyclical accumulation of capital including the recurrence of crises, is derived from the irreconcilable conflict between capital and labour.

In particular, such cyclical behaviour arises in two ways from the employment rate (i.e., from the reserve army of labour): On the one hand, variations in the rate of employment affect the rate of exploitation and hence, through the profit rate, the accumulation of capital[2]. On the other, the rate of employment influences the accumulation of capital because it has an effect on work intensity[3]. The employment-profitability mechanism has its impact on distribution, whereas the employment-work intensity relation works more on production. We shall discuss these elements subsequently in the presentation of our model in the next section.

---

[1] Constructive suggestions from Donald W. Katzner are gratefully acknowledged.

[2] It should be noted that for the derivation of cyclical fluctuations of capital induced by distributional changes it is important to relate the <u>percentage rate of change</u> of the rate of exploitation rather than its level with some measure of the reserve army, for instance the rate of unemployment.

[3] See Marx (1974), chapter 25, pp. 589-597. We should emphasize this relation has, in more recent years, been investigated at the empirical level by American radical economists. Consult, for instance, Bowles/Gordon/Weisskopf (1983), chapter 6, or Weisskopf (1986).

Our model also incorporates certain aspects from the Kaldorian (-Keynesian) perspective. First, various forms of biased technical progress are permitted. The general concept of technical progress employed includes the Marxian form of technical progress[4] as a special case and implies Harrodian technical progress in equilibrium. Second, our model contains the standard (Keynesian) production reaction to mismatched plans of savers and investors. Third, we use the well-known Kaldorian saving function[5] in which the saving rate is determined by the distributional shares of income.

Integrating ideas and relations from different strands of thought may be justifiable as long as the underlying asumptions are not mutually excluding each other[6]. But instead of discussing the potential fruitfulness (?) of the matter or of other aspects of integrational procedures to a larger extent, we leave it to the reader to decide whether the presented model describes some interesting features of capital accumulation.

The model we develop below may also be regarded as a descendant of Goodwin's model of the growth cycle. In other words we shall consider the employment-distribution relation as the principal cycle-generating mechanism. But the following modifications are present as well:

---

[4] Suffice it to mention that his famous law of the rising organic composition of capital is due to his conviction of a prevailing biased (labour-saving) technical progress under capitalism.

[5] It should be clear that within the Marxian theory there exists no saving function in the modern sense, because Marx did not distinguish sharply between saving and investment.

[6] We are, of course, aware of the fact that Marxian and Keynesian/Kaldorian ideas are not in general related in a complementary way. The real wage, for instance, oscillates procyclically in the Marxian theory and definitely countercyclically in a Keynesian and/or Kaldorian framework.This rather crucial difference between the two approaches towards a theory of capital accumulation has been clearly pointed out by Kaldor in his brillant Peking-Lecture. Cf. Kaldor (1956). The essential Marxian ideas on the general law of capitalist accumulation have been restated in a rigorous way by Goodwin (1967) in his seminal contribution on growth cycles.

1) Workers may reduce their work effort (or work intensity) when the rate of employment rises, thereby increasing the resistance to work in the process of production.

2) The Marxian concept of "absolute overaccumulation of capital" is introduced . That is, during the upswing of the accumulation cycle, a point may be reached at which the level of profits ceases to rise. At this point, the rate of accumution should, according to Marx, fall sharply and may even become negative[7].

3) Technical progress is endogenously determined by means of a technical progress function. In equilibrium (as noted above) technical progress is neutral in the Harrodian sense. But out of equilibrium the capital coefficient may rise or fall over time. We could impose parameter restrictions such that technical progress implies labour productivity growth and a rise in the capital coefficient. This would be the case of Marxian technical progress. But then we could not establish equilibrium results compatible with equilibrium growth along Kaldorian lines because the equilibrium solution would not exist. For that reason the "Marxian technical progress function" will not be considered here.

4) Production is determined by a trend term and, additionally, by an adjustment mechanism with respect to excess demand.

5) Changes in the real wage are determined by assumptions concerning the formation of the money wage and the setting of the output price.

2. A Model of Capital Accumulation[8]

2.1 The Equations of the Model

We present the entire set of equations first and shall comment on the definitions and assumptions afterwords. The model is formulated in discrete time. We consider a one commoditiy, closed economy without governmental activity. There are only two markets: the product and the labour market.

---

[7] See Marx (1972), pp.250-254.

[8] The model presented here is a slightly generalized version of two previously published models. See Glombowski and Krüger (1986) and (1987).

The model consists of the following twenty equations:

$$y_t = Y_t / L_t \quad (1)$$

$$\psi_t = P_t / \Gamma_t \quad (2)$$

$$v_t = K_t / Y_t \quad (3)$$

$$\sigma_t = P_t / K_t \quad (4)$$

$$\theta_t = Y_t / P_t \quad (5)$$

$$\beta_t = L_t / A_t \quad (6)$$

$$\vartheta_t = w_t / y_t \quad (7)$$

$$w_t = l_t / p_t \quad (8)$$

$$\pi_t = (1 - \vartheta_t) Y_t \quad (9)$$

$$L_t = h_t \theta_t \Gamma_t \quad (10)$$

$$s_t = S_t / Y_t \quad (11)$$

$$s_t = s_\pi - (s_\pi - s_w) \vartheta_t \quad (12)$$

$$\Delta A_t / A_t = n \quad (13)$$

$$\Delta K_t / K_t = \begin{cases} c_1 + c_2 \pi_t / K_t, & \text{if } \pi_t / \pi_{t-1} \geq R \\ z, & \text{if } \pi_t / \pi_{t-1} < R \end{cases} \quad (14)$$

$$\Delta (P_t / \Gamma_t) / (P_t / \Gamma_t) = m_1 + m_2 \, \Delta (K_t / \Gamma_t) / (K_t / \Gamma_t) \quad (15)$$

$$\Delta (K_t / \Gamma_t) / (K_t / \Gamma_t) = n_1 + n_2 \vartheta_t \quad (16)$$

$$\Delta h_t / h_t = - d_1 + d_2 \beta_t \quad (17)$$

$$\Delta l_t / l_t = - a_1 + a_2 \beta_{t+1} + a_3 \, \Delta p_t / p_t \quad (18)$$

$$\Delta p_t / p_t = - b_1 + b_2 \theta_t + b_3 \, \Delta l_t / l_t \quad (19)$$

$$\Delta Y_t / Y_t = g + \delta (\Delta K_t / Y_t - S_t / Y_t) \quad (20)$$

Some of the definitions and equations need no detailed comments. Thus equation (1) is just a standard definition of actual labour productivity to be distinguished from "techni-

cal" labour productivity (η). This variable is defined in equation (2) as maximum production (P) over labour requirements (Γ), in other words it is a ratio between maximum production and conditions of maximum work intensity. We allow for changes in capacity utilization (θ) and work intensity (h) and therefore (y) and (η) will in general be different.
Equation(3) defines the actual capital coefficient, whereas (4) defines the technical capital coefficient (σ), relating the capital stock (K) and maximum net output. In equation (5) the rate of capacity utilization (θ) is defined. If capacity is partly unutilized, then $\theta\Gamma$ units of labour are required to produce actual output Y under conditions of maximum work intensity. Now let h be a labour inefficiency index which is equal to one if work intensity is at its maximum. h will rise in proportion to a declining work intensity. From equation (10) it then may be seen that the actual level of employment (L) is linked to labour input with full capacity and maximum work intensity by the rate of capacity utilization and the labour inefficiency index.

The relations between actual and 'technical' versions of the capital coefficient, labour productivity, and capital intensity may be summarized as follows:

$$v_t = K_t/Y_t = (K_t/P_t)(1/\theta_t) \tag{21}$$

$$y_t = Y_t/L_t = (P_t/\Gamma_t)(1/h_t) \tag{22}$$

$$K_t/L_t = (K_t/\Gamma_t)(1/h_t\theta_t) \tag{23}$$

Taking these modifications into account it may readily be checked that equation (15) is a modified version of Kaldor's technical progress function.
Equation (6) defines the rate of employment (β) in the usual way. (7) is the definition of the wage share in income (or the real wage (w) over labour productivity; w in turn is defined as the nominal wage (l) over the price level (p). Profits (π) are defined in equation (9), and (11) is the average saving ratio (s).
In (12) we apply the Keynesian/Kaldorian saving hypothesis. With regard to the growth of the labour supply (A) we stipulate the standard assumption of exponential growth at a constant rate (n).
Consider now equation (14). The accumulation rate is determined linearly by the rate of profit, until the level of profits ceases to rise. When this "absolute overaccumulation" occurs, the accumulation rate will decline "sharply". In other words the system will switch to the second regime.

A 'mechanization function' is introduced in equation (16). Changes in the technical capital intensity are induced by the wage share in the sense of inducing a higher degree of

mechanization whenever the wage share is "high".
Equation (17) tells us that a rising rate of employment will raise the labour inefficiency index, thereby reducing work intensity. Moreover, any reduction of the reserve army will have an impact on nominal wages to rise.
In addition the rate of inflation tends to increase nominal wages, because workers shall try to receive a compensation for rising prices (equation 18). Conversely, rising nominal wages may enforce a tendency towards higher prices. Note that equation (19) shows a positive influence of capacity utilization on the rate of inflation also. It is assumed that capitalists tend to raise prices with rising costs and rising demand[9]. Finally, the percentage rate of change of output is determined by a trend factor $1+g$ [10] ,and the pace of production will in disequilibrium accelerate or decelerate due to excess demand or supply. The 'production adjustment coefficient' δ measures the speed of adjustment of the growth rate of Y on deviations between planned savings and planned investment.

Now the reader may have checked that we have introduced not only a set of 20 equations in 20 variables, but some 20 parameters too. Therefore it might be convenient to discuss the restrictions imposed on these parameters separately. This is done in the next subsection.

2.2 Parameter Restrictions

The parameter restrictions introduced below are due to some theoretical reasons and, of course, to convenience. Because there are such a large number of parameters and so many restrictions imposed on these, only a few of the restrictions, namely those which play a prominent role for the equilibrium and the disequilibrium solutions of the model, are presented and discussed. The remaining restrictions belong to the standard arsenal of cyclical growth models, and there seems to be nothing special about them. The following (incomplete) set of restrictions is considered here[11]:

---

[9] This equation has been introduced by Solow and Stiglitz (1968).

[10] As will be seen below the trend rate of growth g will be determined in equilibrium.

[11] A more detailed discussion on the restrictions may be found in Glombowski and Krüger (1987).

$$0 \leq s_w < s_\pi \leq 1 \tag{24}$$

$$0 \leq n \tag{25}$$

$$0 < c_1 \ , \ 0 < c_2 \tag{26}$$

$$0 < R \leq 1 \ , \ z < c_1 \tag{27}$$

$$0 < m_1 \ , \ 0 < m_2 < 1 \tag{28}$$

$$0 \leq n_1 \ , \ 0 < n_2 \tag{29}$$

$$0 < d_1 \ , \ 0 < d_2 \tag{30}$$

$$0 \leq a_1 \ , \ 0 < a_2 \ , \ 0 < a_3 \leq 1 \tag{31}$$

$$0 \leq b_1 \ , \ 0 < b_2 \ , \ 0 < b_3 \leq 1 \tag{32}$$

$$0 < a_3 b_3 < 1 \tag{33}$$

$$0 < \delta \tag{34}$$

$$0 < m_1 - n_1(1-m_2) < n_2(1-m_2) \tag{35}$$

$$0 < c_2 - (s_\pi - s_w) \ , \ \underline{\text{or}} \tag{36a}$$

$$c_2 - (s_\pi - s_w) < 0 \tag{36b}$$

$$(n_1+n_2)(1-m_2) < m_1 \tag{37}$$

$$d_1 m_1(1+n)+(1+d_1 n)(1-m_2) \leq d_2(m_1(1+n)+n(1-m_2)) \tag{38}$$

$$\begin{aligned} 0 < \ & (1+g)(1-b_1-b_3(a_1+a_3))+a_2(1+n-b_3(1+g))(d_1 m_1 \\ & *(1+n)+(1+d_1 n)(1-m_2))-(1+n)(1-a_1-a_3(b_1+b_3)) \\ & \leq b_2 d_2(1+g-a_3(1+n))(m_1(1+n)+n(1-m_2)) \end{aligned} \tag{39}$$

The assumption (24) needs no justification: it is the classical Kaldorian assumption of different saving out of wages (s_w) and profits (s_π). The inequality (25) is but the standard assumption of a growing working population.
Turning to restriction (26) of the coeffifients of the accumulation function, it may be questioned whether positive values for the unexplained part of accumulation might be overly restrictive. In fact we could easily take non-positive values of c_1 into account without affecting the functioning of the model, but working with $c_1 > 0$ is a convenient simplification. The assumption for R in (27) may be weakened and we could analyse the case $R > 1$ within the framework of the model as well.
Note that to express the Marxian idea of absolute overaccumulation in a strict sense would require that $R = 1$. But here we prefer to analyze the case of mildly declining

profit levels before the switch point to the alternative accumulation regime. However, if the switching point is reached, capital reacts with a "drastic" reduction in the rate of accumulation. This is the economic background for the second inequality in (27).
The two inequalities of (28) state the Kaldorian idea of Harrodian neutrality of technical progress in equilibrium, and changing phases of labour saving and capital saving technical progress in disequilibrium[12], respectively. Therefore, in order to establish the desired properties of technical progress we need, at least when applying a linear version of the (modified) technical progress function[13], to impose the restrictions (28).
The range for the coefficients $n_{1,2}$ (cf. (29)) of the "mechanization function" (16) has been chosen for convenience. One might argue, however, that declining values for technical capital intensity should not be excluded and that n_1 should be permitted to be negative. However, since we are discussing both the short-term and long-term implications of our model, we consider the potential phases of a declining technical capital intensity as irrelevant.
The conditions (30) - (22) deserve no specific comment. Observe, however, that (33) precludes a wage - price spiral turning into hyperinflation.

By (36a) we assume that a rising profit share (π_t/Y_t) will induce a rise of the investment share relative to the saving rate or, in other words, an incremental rise of π_t/Y_t induces an incremental rise of ($\Delta K_t/Y_t$ - S_t/Y_t). We should note the fact that Kaldor and probably most Keynesians would postulate the opposite effect: Other things being equal, an increasing wage share would stimulate demand and an expansion of production. This is the basis for inequality (36b). Clearly, assumption (36a) is linked more closely to a "profit-squeeze"

---

[12] For a more complete analysis of different types of technical progress and its implications, based on Kaldor's concept of the technical progress function, see for instance van der Ploeg (1983).

[13] We are aware of the fact that a linear technical progress function may be transformed into a standard neoclassical production function. This is a rather old insight and may indeed be traced to a publication by Black some 25 years ago . But to be sure, in this article we certainly are not concentrating on differences with neoclassical growth or business cycle theory.

version of the approach to cyclical accumulation of capital along Marxian lines.
As far as the remaining restrictions are concerned, conditions (35) - (39) serve to guarantee economically meaningful equilibrium values for the wage share, the capital coefficient, the employment rate, and the rate of capacity utilization[14]. With (2.35) we ensure a positive wage share which is less than 1 in equilibrium, conditions (36a) and (37) are needed for a strictly positive capital coefficient, and (38) provides the parameter restrictions necessary for a strictly positive equilibrium employment rate. Such employment rates may be compatible with full employment or with unemployment: From a Kaldorian perspective of long-run processes of economic growth it would be appropriate to work with the full employment assumption, whereas the Marxian concept of an industrial reserve army of labour would instead imply unemployment equilibrium. In this paper, particularly when studying the disequilibrium paths of the model by means of simulations, we apply the Marxian idea of a positive reserve army of labour in equilibrium. Finally, the inequalities in (2.39) describe the the conditions required for a strictly positive rate of capacity utilization. Moreover, capacity utilization is bounded from above, i.e., there is a full capacity limit on output.

3. Some Properties of the Model

The model can be reduced to a system of four nonlinear difference equations in β, λ, θ, and v :[15]

$$\text{(I)} \quad \begin{aligned} \beta_{t+1} &= \beta_t f_1(\beta_t, \lambda_t, v_t) \\ \lambda_{t+1} &= \lambda_t f_2((\beta_t f_1(\beta_t, \lambda_t, v_t)), \beta_t, \lambda_t, v_t, \theta_t) \\ \theta_{t+1} &= \theta_t f_3(\lambda_t, v_t) \\ v_{t+1} &= v_t f_4(\lambda_t, v_t) \end{aligned}$$

The system (I) refers to the regime $\pi_t/\pi_{t-1} \geq R$; under the condition $\pi_t/\pi_{t-1} < R$ the relevant variables are governed by

---

[14] Note that in order to guarantee a strictly positive equilibrium value of the capital coefficient, when analyzing some implications of the Kaldorian case (36b), we should replace, for example, the condition $0 < c_1$ in (26) by the inequality $c_1 < 0$.

[15] The necessary steps are found in the appendix.

the somewhat simpler system

$$
\text{(II)}\qquad
\begin{aligned}
\beta_{t+1} &= \beta_t F_1(\beta_t, \lambda_t, v_t; z) \\
\lambda_{t+1} &= \lambda_t F_2((\beta_t F_1(\beta_t, \lambda_t, v_t; z)), \beta_t, \lambda_t, v_t, \theta_t; z) \\
\theta_{t+1} &= \theta_t F_3(\lambda_t, v_t; z) \\
v_{t+1} &= v_t F_4(\lambda_t, v_t; z)
\end{aligned}
$$

3.1 Equilibrium

Under the assumptions and the restrictions introduced in the previous section a unique and economically meaningful equilibrium solution can be shown to exist:[16]

$$\lambda_e = (m_1 - n_1(1-m_2))/n_2(1-m_2)$$

$$v_e = ((c_2 - s_\pi)(m_1 - (n_1+n_2)(1-m_2)) + s_w(m_1 - n_1(1-m_2)))/c_1 n_2(1-m_2)$$

$$\beta_e = (d_1 m_1(1+n) + (1+d_1 n)(1-m_2))/d_2(m_1(1+n) + n(1-m_2))$$

$$
\begin{aligned}
\theta_e = {} & ((1+n)(1-a_1-a_3(b_1+b_3)) - (1+g)(1-b_1-b_3(a_1+a_3)) \\
& + a_2(1+n-b_3(1+g))(d_1 m_1(1+n) + (1+d_1 n)(1-m_2))) \\
& / b_2 d_2(1+g-a_3(1+n))(m_1(1+n) + n(1-m_2))
\end{aligned}
$$

Firstly we observe that in equilibrium the rates of growth of the "technical" capital coefficient and "technical" capital intensity must be equal: the respective rates of growth are determined by $m_1/(1-m_2)$.
Secondly we assume that in equilibrium the levels of planned investment and planned savings are equal: $\Delta K_t = S_t$[17]. Then the percentage rate of change of output coincides with the trend g which in turn is determined by $(m_1(1+n)+n(1-m_2))/(1-m_2)$. In other words we rediscover the familiar standard solution for $\Delta Y_t/Y_t$, i.e., the equilibrium growth path of output is determined by the growth rates of technical progress and labour suply.

We assume in addition that the labour inefficiency index h is constant in equilibrium. But then it follows immediately

---

[16] Here we show the equilibrium values for system (I) only.
[17] This seems to be a reasonable equilibrium assumption in order to avoid unbounded accumulation or decumulation of unplanned inventory changes.

that the pace of growth of 'actual' and 'technical' labour productivity will coincide in eqilibrium:
$m_1/(1-m_2) = \Delta y_t/y_t = \Delta \psi_t/\psi_t$.

To sum up: In equilibrium the distinction between "technical" and "actual" variables becomes essentially irrelevant, because all three "actual" and "technical" variables, the capital coefficient, labour productivity, and capital intensity are proportional.

We complete our brief description of the equilibrium solution by stating that if the wage share is constant in equilibrium, so is the profit share. But if the wage share is constant, then, of course, the real wage must rise with the growth rate of labour productivity, i.e., $\Delta w_t/w_t = m_1/(1-m_2)$. Note that this ratio determines the difference between rising nominal wages and the rate of inflation without determining neither the former nor the latter rate. In fact our path of economic growth in equilibrium may be compatible with different levels of the inflation rate, depending ceteris paribus on the equilibrium values of both the employment rate and the rate of capacity utilization[18].
In equilibrium the rate of profit is constant; so is the rate of capital accumulation. That is to say, we have reestablished all standard properties of equilibrium growth models. From a Kaldorian perspective this path of economic growth should be stable, whereas the equilibrium solution should be neutrally stable or unstable from a Marxian perspective[19]. Now a careful stability analysis would show that the stability properties of the equilibrium solution depends to a large extent on the conditions (36a) and (36b). If the Marxian "profit-squeeze"-mechanism is at work, then the equilibrium solution tends to be unstable. Conversely, the Kaldorian case (36b) tends to stabilize the equilibrium solution[20].

In the next section we shall study the working of our model in

---

[18] In equilibrium the percentage change of the real wage is equal to $\Delta l_t/l_t - \Delta p_t/p_t = m_1/(1-m_2)$. Observe now that we may write : $\Delta l_t/l_t - \Delta p_t/p_t =$
$= (a_2(1-b_3)\beta_e - b_2(1-a_3)\theta_e - a_1(1-b_3) + b_1(1-a_3))/(1-a_3 b_3)$. This follows from equations (18) and (19).

[19] Recall that the equilibrium solution in Goodwin's original model is stable, but not asymptotically stable. In technical terms this form of a stable equilibrium is called neutrally stable.

[20] The analysis is available from the author upon request.

disequilibrium. But instead of examining the dynamics around a stable equilibrium growth path and describing the convergent path towards equilibrium we shall analyze processes around the Marxian equilibrium path. We shall see that an interesting disequilibrium solution may arise, i.e., a cyclical oscillation of capital accumulation which may be interpreted as a Marxian process of growth and crisis.

3.2 Recurrent Cycles

Unfortunately the systems of equations (I) and (II) cannot be studied by means of standard analytical methods: There is no mathematical theory available for nonlinear difference equations of the 4th order. We could, of course, restrict our analysis to the equilibrium solution and its local stability properties, but then we would be unable to study some interesting disequilibrium dynamics. Instead we shall apply simulation methods based on a structure of parameter values such that

a) the equilibrium solution is economically meaningful, and

b) the disequilibrium dynamics are in line with at least some "stylized facts" of growth cycles[21].

In particular we focus on one class of solutions that emerge from a specific set of parameter values and a range of initial conditions, and that exhibit recurrent growth cycles in output, and cyclical oscillations in the employment rate, the wage share, the rate of capacity utilization, and the capital coefficient[22].
Projecting the process from the four-dimensional state space into the two-dimensional state subspace spanned by the employment rate and the wage share we find a 'stable band' around the equilibrium point, as pictured in figure 1. If the

---

[21] In our view the following restrictions should be met:
1) The length of a cycle should be not less than three years (periods) and not more than ten years;
2) the intervals for oscillating variables like the employment rate, the wage share in income, and the rate of capacity utilization should be "small";
3) the cyclical upswings should on average be longer than the downswings, at least in the absence of stagnation.

[22] The specific set of parameter values and the Fortran 77 program for simulating systems (I) and (II) is available from the author upon request.

system starts close to the equilibrium point, it is driven into the band. Thus it is stable from the inside.

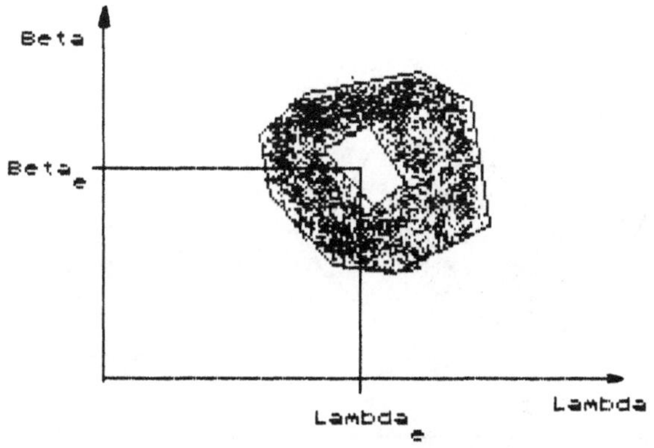

Figure 1

Once the system enters the band, it will never leave it. Within the band the oscillations around the equilibrium point are recurrent but not periodic; rather the time length of subsequent cycles changes in an unpredictable way. For instance, if the first cycle is complete after five periods then the second cycle may be shorter or longer. What most frequently occurs is a sequence of two or three longer cycles, say, six and eight periods, to be followed by one or two short cycles.

The shaded region is stable from outside too. If the point of departure of the system is "far away" from the equilibrium, the system will nevertheless approach the band and enter the stable region. What happens thereafter is essentially the same as in the case of entering the region from inside, that is to say, we observe a similar sequence of unpredictable changes of the length of the periods. Figure 2 shows an example of recurrent cycles with the initial point (*).

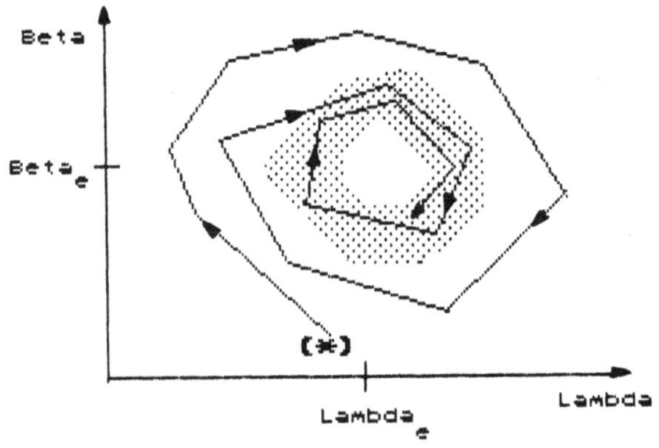

Figure 2

It remains to give a compact economic interpretation of the behaviour of the system within the band. The economic process is very similar to the workings of Goodwin's model, although in our model more variables are involved and therefore the interaction between the variables is more complex.
We start at a position of crisis, that is to say, the value of the employment rate is low and the magnitude of the wage share is below its equilibrium value. This implies a resonably high profit share and therefore the profit rate will determine a level of the employment rate such that labour demand exceeds labour supply. Thus the size of the reserve army will decline. Note that the improvement of the bargaining position of the labouring class will not automatically lead to rising real wages, but rather that the difference between the sum of productivity growth and the rate of inflation, i.e., the growth rate of nominal wages may be reduced. Eventually the continuation of this process will lead to a halt of the declining wage share and---if nominal wages tend to rise faster than productivity and prices---the profit share may then start falling. It should be observed that the profit share will be squeezed in two ways simultaneously: First, the lower the level of the reseve army, the faster is the rise in nominal

wages; and second, productivity growth tends to be reduced due to declining work intensity. If nominal wages rise faster than prices and productivity then the profit rate will fall. Moreover, the capital coefficient may also decline since, under conditions of the boom, production and demand will rise along with the rate of capacity utilization.

This Marxian profit squeeze will induce a decline in the rate of capital accumulation. Thus the labour demand will come to an end. As soon as labour supply exceeds labour demand the reserve army will tend to rise, thereby reversing the profit-squeeze process. Then the falling employment rate will enforce labour discipline in production and will weaken the bargaining power of labour. The sum of productivity growth and inflation will be greater than nominal wage growth and, at a later stage, even real wages may fall. In any case, the weaker bargaining position of labour will induce a higher profit rate due to a shift in the profit share; at the same instant the profitability of production will be raised via higher work discipline. Finally the "reinstallation" of appropriate profit rate levels may induce another cycle of capital accumulation.

4. Final Remarks

It is clear that our analysis is still rather preliminary and that our short sketch of the economic cycle cannot be considered to be the full story. This mainly stems from the fact that the dynamics of systems (I) and (II) are not fully understood. The specific solution also needs to be analyzed further. One reason is the following: the set of parameters has not yet been tested by means of sensitivity analysis. All we have done so far was to find the range of initial positions ($\beta(0)$, $\eta(0)$, v_e, θ_e) such that the system will be driven into the stable region. Having detected such a range is an encouraging preliminary result for the ongoing research on the dynamic properties of the model.

5. References

Bowles, S./Gordon, D. M./Weisskopf, Th. E., 1983, Beyond the Waste Land. A Democratic Alternative to Economic Decline, Garden City, New York,

Glombowski, J./Krüger, M., 1986, Some Extensions of a Classical Growth Cycle Model, Semmler, W., (Ed.), Competition, Instability, and Nonlinear Cycles, Berlin et al., pp. 212-251

Glombowski, J./Krüger, M., 1987, Generalizations of Goodwin's Growth Cycle, Batten, D./Casti, J./Johansson, B., (Eds.), Economic Evolution and Structural Adjustment, Berlin et al. pp. 260-290,

Kaldor, N., 1956, Capitalist Evolution in the Light of Keynesian Economics, Kaldor, N. Essays on Economic Stability and Growth, London 1960, pp. 243-258

Marx, Karl, 1974, Capital. A Critical Analysis of Capitalist Production, Vol. I, London

Marx, Karl, 1972, Capital. A Critique of Political Economy, Vol. III, London

Solow, R. M./Stiglitz, J. E., 1968, Output, Employment, and Wages in the Short Run, Quarterly Journal of Economics 82, pp. 537-560

Van der Ploeg, F., 1983, Predator-Prey and Neo-Classical Models of Cyclical Growth, Zeitschrift für Nationalökonomie, 1983, pp. 235 ff.

Weisskopf, Th. E., 1986, The Effect of Unemployment on Labour Productivity: An International Comparative Analysis, unpublished paper, University of Michigan, 33 pp.

Appendix

In this appendix it will be indicated how to reduce the model to a system of four difference-equations in β_t, λ_t, θ_t, and v_t; we shall develop the set of equations (*), that is, for the profitability regime $\pi_t/\pi_{t-1} \geq R$.
To start with the employment rate (β): From (6) and (13) we get

$$\beta_{t+1}/\beta_t = (L_{t+1}/L_t)/(A_{t+1}/A_t) = (L_{t+1}/L_t)/(1+n) .$$

Substituting from (17), (5), and (20) yields

$$L_{t+1}/L_t = (1-d_1+d_2\beta_t)(1+g+\delta(\Delta K_t/Y_t-s_t))(\Gamma_{t+1}/P_{t+1})/(\Gamma_t/P_t) ;$$

additional substitutions from (14), (13), (3), and (15) lead finally to

$$\beta_{t+1}/\beta_t = ((1-d_1+d_2\beta_t)(1-\delta s_w+g+\delta c_1 v_t+\delta(c_2-(s_\pi-s_w))(1-\lambda_t))) \\ /(1+n)(1+m_1+m_2(n_1+n_2\lambda_t)) . \qquad (I.1)$$

The equation for the capital coefficient (v) can now easily derived from (3), (14), and (20):

$$v_{t+1}/v_t = (K_{t+1}/K_t)/(Y_{t+1}/Y_t) = \\ = (1+c_1+c_2(1-\lambda_t)/v_t) \\ /(1-\delta s_w+g+\delta c_1 v_t+\delta(c_2-(s_\pi-s_w))(1-\lambda_t)) . \qquad (I.2)$$

As to the rate of capacity utilization (θ) we observe from (2.5) that

$$\theta_{t+1}/\theta_t = (Y_{t+1}/Y_t)/(P_{t+1}/P_t) \ .$$

Substituting from (20), (15), and (16) we get

$P_{t+1}/P_t = (1+m_1+m_2(n_1+n_2\lambda_t))(1+c_1+c_2(1-\lambda_t)/v_t)/(n_1+n_2\lambda_t)$, and then

$$\theta_{t+1}/\theta_t = (1-\delta s_w+g+\delta v_t+\delta(c_2-(s_\pi-s_w))(1-\lambda_t))(n_1+n_2\lambda_t) /(1+m_1+m_2(n_1+n_2\lambda_t))(1+c_1+c_2(1-\lambda_t)/v_t) \qquad (I.3)$$

The last equation for the wage share (λ) may be obtained by means of (8), (18), and (19):

$$\lambda_{t+1}/\lambda_t = ((w_{t+1}/w_t)(1-d_1+d_2\beta_t))/(1+m_1+m_2(n_1+n_2\lambda_t)) \ ;$$

the growth factor of the real wage (w) is by definition equal to $w_{t+1}/w_t = (l_{t+1}/l_t)/(p_{t+1}/p_t)$; the growth factors of the nominal wages and prices in turn are determined by

$l_{t+1}/l_t = (1-a_1-a_3(b_1+b_3)+a_2\beta_{t+1}+a_3b_2\theta_t)/(1-a_3b_3)$, and

$p_{t+1}/p_t = (1-b_1-b_3(a_1+a_3)+a_2b_3\beta_{t+1}+b_2\theta_t)/(1-a_3b_3)$. Therefore

the missing equation is

$$\lambda_{t+1}/\lambda_t = ((1-d_1+d_2\beta_t)(1-a_1-a_3(b_1+b_3)+a_2\beta_{t+1}+a_3b_2\theta_t)) /(1+m_1+m_2(n_1+n_2\lambda_t))(1-b_1-b_3(a_1+a_3)+a_2b_3\beta_{t+1}+b_2\theta_t) \qquad (I.4)$$

System (II) may be derived along the same lines; but the functions F_i are---under the regime $\pi_t/\pi_{t-1} < R$---much simpler than the functions f_i; this is due to the accumulation regime described by $\Delta K_t/K_t = z$. The equations are described by the formulas

$$\beta_{t+1}/\beta_t = (1-d_1+d_2\beta_t)(1-\delta s_w+g+\delta z v_t-(s_\pi-s_w)(1-\lambda_t)) /(1+n)(1+m_1+m_2(n_1+n_2\lambda_t)) \qquad (I.5)$$

$$v_{t+1}/v_t = (1+g+\delta z v_t)/(1-\delta s_w+g+\delta z v_t-(s_\pi-s_w)(1-\lambda_t)) \qquad (I.6)$$

$$\theta_{t+1}/\theta_t = (1-\delta s_w+g+\delta z v_t-(s_\pi-s_w)(1-\lambda_t))(1+n_1+n_2\lambda_t) /(1+m_1+m_2(n_1+n_2\lambda_t))(1+z v_t) \qquad (I.7)$$

The equation in the wage share (λ) remains invariant with respect to the two different accumulation regimes.

ENDOGENOUS BUSINESS CYCLES IN A MONETARY TWO-COUNTRY MODEL

by

Bernhard Eckwert

Universität Augsburg

Contents

* This research was supported by the Deutsche Forschungsgemeinschaft which is gratefully acknowledged.

Introduction

Until recently nontransitory endogenous economic fluctuations in the framework of a competitive equilibrium model seemed inconceivable to most economists. Short run fluctuations were attributed to market imperfections, exogenous shocks, or an adverse policy while in the long run competitive economies were believed to become typically stationary. Papers in this line of thought are Böhm/Keiding (1985), Buiter (1981), Mitsui/Schittko/Watanabe (1985), Okuno/Zilcha (1983) and others.
Not long ago Grandmont (1985) showed within the framework of an overlapping generations (OLG) model that cyclical behaviour of a competitive economy might well occur in the long run and that these cycles arise in a comletely endogenous manner. Grandmont carries out his analysis in the simplest possible formulation of the OLG-model. He considers a closed economy in which trade takes place at each period on a commodity and a money market. In models of this type real demands depend on the intertemporal price ratio $p_t/p_{t+1}=:\theta_t$ only and an equilibrium price sequence may be defined inductively by choosing θ_{t+1} in such manner that the current (time t) goods market clears for a predetermined value of θ_t.
It is the purpose of this paper to develop methods allowing for the investigation of higher dimensional OLG-models, i.e. of models comprising more than one commodity market. We conduct the analysis within the framework of a two country model with country specific goods markets. However, the employed techniques apply to any two-dimensional (and indeed any finitedimensional) OLG-structure.
As was already shown by several authors one good OLG-models trivially possess a stationary equilibrium in which the intertemporal price ratio is constant and equal to 1. Though in our two-dimensional framework a stationary equilibrium fails to exist, periodic equilibria can be shown to exist under mild assumptions. In this respect the model considered here behaves very much like the OLG-standard model analyzed by Grandmont. However, the mechanisms generating endogenous cycles are quite different.

"The origin of these endogenous deterministic business cycles will be seen to be the potential conflict between the wealth effect and the intertemporal substitution effect that are associated with real interest rate movements" (Grandmont (1985), p. 997).

We shall not make any assumption on the relative strength of the wealth and intertemporal substitution effects. Nonetheless periodic equilibria typically exist provided there is a competitive equilibrium at all. Hence the phenomenon of endogenous business cycles carries over to higher dimensional structures and it even seems to be reinforced.
Since people live for two periods and hence plan one period

ahead into the future their decisions will depend on price expectations. In this work individuals are assumed to have rational expectations meaning that they make no forecasting errors. Rationality of expectations, specific though this concept is, might be justifiable in the context considered here since we are concerned mainly with periodic equilibrium sequences along which the consumers' environment is repetitive. Individuals should be able to recognize this repetitiveness and to avoid forecasting errors accordingly.
A final remark on the OLG-structure that underlies the analysis of this paper is in order. As is well known finite economies do not assign positive values to financial assets. The OLG-model seems to be a suitable framework to address monetary issues since it carries a natural infinite structure. In addition the OLG-model is based on individual optimizing behaviour admitting the investigation of efficiency respectively optimality issues. Hence the most interesting questions in economic theory can be addressed to within this framework.
The paper is organized as follows. We outline in section 1 the micro structure of the model and state the equilibrium conditions and basic identities. In section 2 we define and analyze a function f that describes the dynamics of the economy. The existence of (periodic) equilibria will be discussed in chapter 3.

1. The Model

We shall consider a two-country-world in the framework of the simple overlapping generation model with two country specific goods. All agents live for two periods and for the sake of simplicity we shall assume that at each (discrete) point of time t there exists exactly one consumer in each country who is born at the beginning of period t-1. There will be accordingly two agents in every period and every country, one 'young' and one 'old'. The economy began operation at time $-\infty$ (i.e. there is no starting point) and continues over periods extending indefinitely into the future ...t, t+1, ...
We distinguish two country specific perishable consumption goods x and X and two imperishable fiat monies m and M, where x and m denote the consumption good and money of the home country (country 1), while capital letters refer to the corresponding foreign (country 2) quantities.
The money markets are not internationally integrated; agents hold national currency only. At each date there are competitive spot markets for the consumption goods and for money.
The consumer born at time t in country 1 is characterized by her or his consumption set $\mathbb{R}^4_+$, his utility function $u^t : \mathbb{R}^4_+ \to \mathbb{R}_+$ and his initial endowment $(e_1,e_2)_t \in \mathbb{R}^2_{++}$.

e_i, i=1,2, are the quantities of the home good at time t and t+1 the consumer can dispose of, while his equipment of foreign goods is zero. The contemporary's tastes in country 2 are represented by the utility function $U^t: \mathbb{R}_+^4 \rightarrow \mathbb{R}_+$ and he is provided with the endowment vector $(E_1,E_2)_t \in \mathbb{R}_{++}^2$, where E_i, i=1,2, are quantities of the foreign good at time t and t+1.
To keep the analysis as simple and transparent as possible we assume that all consumers born in country i (i=1,2) at different points of time are identical. Hence we write $u=u^t$, $U=U^t$, $(e_1,e_2)=(e_1,e_2)_t$ and $(E_1,E_2)=(E_1,E_2)_t$.
We shall make the further

Assumption 1.1: The utility functions u and U are twice continuously differentiable, strongly quasi-concave and strictly increasing in all arguments. In addition $\lim_{c_j \to 0} \partial v/\partial c_j\ (c_1,c_2,c_3,c_4) = \infty$ for v=u,U and j=1, ..., 4.

Let p_t and P_t be the money prices of the home and foreign consumption good in national currencies at time t and consider the decision problem that a young consumer in country 1 solves. His problem is to choose current consumption $x_1 \geq 0$ and $i_1 \geq 0$ of the home and the imported foreign good, to formulate a money demand $m^d \geq 0$ and to plan his future consumption $x_2 \geq 0$ and $i_2 \geq 0$ so as to maximize his utility function subject to the budget constraints

$$(1.1)\quad z_1 + \tau_t\, i_1 + \frac{m^d}{p_t} = 0 \quad \text{and} \quad z_2 + \tau_{t+1}\, i_2 = \frac{m^d}{p_{t+1}} \quad ,$$

$z_i = x_i - e_i$, i=1,2, denoting excess demand for the home good in the i-th half of the agent's life. τ_t stands for the terms of trade and is defined as the relative price of imports (expressed in home currency by means of the exchange rate π_t)[1] to exports, $\tau_t = \pi_t P_t/p_t$. In (1.1) all variables that are indexed by (t+1) should be regarded as expected magnitudes. In this paper individuals are supposed to have rational expectations, so that expected prices are in fact realized along any equilibrium path.
Under Assumption 1.1 the decision problem has a unique solution, and in all points $(p_{t+1}, P_{t+1}, p_t, P_t)$ with $m^d(\cdot) > 0$ the behavioural functions z_j, i_j and m^d (j=1,2) are continuously differentiable with respect to current and future prices. In addition the excess demands for goods z_j and i_j are homogeneous of degree zero and m^d is homogeneous of degree one in all arguments.
Foreign country's consumers are modelled in a completely analogous manner. The current and expected budget

constraints of an individual born at time t are

$$(1.2)\quad Z_1 + \frac{1}{\tau_t} I_1 + \frac{M^d}{P_t} = 0 \text{ and } Z_2 + \frac{1}{\tau_{t+1}} I_2 = \frac{M^d}{P_{t+1}} .$$

Elimination of the money variables in (1.1) and (1.2) yields

$$(1.3)\quad z_1 + \tau_t i_1 = - \frac{p_{t+1}}{p_t} (z_2 + \tau_{t+1} i_2)$$

$$(1.4)\quad Z_1 + \frac{1}{\tau_t} I_1 = - \frac{P_{t+1}}{P_t} (Z_2 + \frac{1}{\tau_{t+1}} I_2) .$$

The government's role in country i=1,2 is to finance a constant path of public consumption $g_i > 0$ by money creation. To keep matters as simple as possible we assume that both countries' governments consume domestically produced goods only. With $u_t := (p_{t+1}, P_{t+1}, p_t, P_t)$ the trade balance (in home currency) is defined by

$$b(u_t, u_{t-1}) := p_t\Big(I_1(u_t) + I_2(u_{t-1})\Big) - \pi_t P_t \Big(i_1(u_t) + i_2(u_{t-1})\Big) .$$

The model is completed by aggregate consistency of the individual choices. A competitive equilibrium is a sequence of prices $(p_t, P_t)_t$, $t \in \mathbb{Z}$, such that equations (1.5) - (1.8) are satisfied for all t. This should be regarded as a preliminary definition that will be slightly specialized below.

$$(1.5)\quad z_1(u_t) + I_1(u_t) + z_2(u_{t-1}) + I_2(u_{t-1}) + g_1 = 0$$

$$(1.6)\quad Z_1(u_t) + i_1(u_t) + Z_2(u_{t-1}) + i_2(u_{t-1}) + g_2 = 0$$

$$(1.7)\quad p_{t+1} z_2(u_t) + \pi_{t+1} P_{t+1} i_2(u_t) = m + \sum_{-\infty}^{t} p_i g_1 + b(u_i, u_{i-1})$$

$$(1.8)\quad P_{t+1} Z_2(u_t) + \frac{p_{t+1}}{\pi_{t+1}} I_2(u_t) = M + \sum_{-\infty}^{t} P_i g_2 - \frac{1}{\pi_i} b(u_i, u_{i-1})$$

Equations (1.5) and (1.6) describe equilibrium on the goods markets at time t. In view of (1.1) and (1.2) the left hand sides of (1.7) and (1.8) represent the national money demands. The right hand sides equal the national money supplies at time t, which depend on the capital streams between both countries that are induced by the trade balance realized in former periods. m and M are positive exogenous money supply components in country one and two, respectively.

Lemma 1.1 Let $(p_t, P_t)_t$ be a sequence such that the goods markets are cleared for all t. If the money markets are equilibrated at some time $\tilde{t}$, then money market equilibrium prevails for all t.
Proof: The result follows directly by combining the individual budget constraints and the equilibrium conditions (1.5) - (1.8). (Compare Eckwert (1987)).

q.e.d.

According to Lemma 1.1 money market equilibrium follows from cleared goods markets by Walras law (if necessary the national exogenous money supply components are to be chosen appropriately), so that the system (1.5 - 1.8) is not overdetermined.
For the model considered here neither a stationary eqilibrium (where prices do not depend on time) nor a weakly stationary equilibrium (where $p_{t+1}/p_t = P_{t+1}/P_t = \lambda$ for all t) exists. Since in a stationary equilibrium the price sequence $(p_t, P_t)_t = (p, P)_t$ is independent of time the money market equations (1.7), (1.8) cannot be satisfied if government consumption is positive. Along a weakly stationary equilibrium $(p_{t+1}/p_t = P_{t+1}/P_t = \lambda \neq 1)$ the world money supply $\tilde{M}_t := m + \pi M + \Sigma^t_{-\infty} p_i g_1 + \pi P_i g_2$ must satisfy the difference equation

$$(1.9) \quad \tilde{M}_{t+1} = \lambda \tilde{M}_t .$$

From the definition of the world money stock we derive however

$$\tilde{M}_{t+1} = m + \pi M + \sum_{-\infty}^{t+1} \lambda^i (p_0 g_1 + \pi P_0 g_2) = m + \pi M + \lambda \sum_{-\infty}^{t} \lambda^i (p_0 g_1 + \pi P_0 g_2) = \lambda \tilde{M}_t + (m + \pi M)(1 - \lambda)$$

contradicting 1.9.

2. Dynamics of the Model

To formalize and investigate the dynamics of our model we start out from the goods market equations (1.5) and (1.6) which give rise to an important observation.
Given current prices p_t, P_t the demands $g_1(u_t) := z_1(u_t) + I_1(u_t)$ as well as $h_1(u_t) := Z_1(u_t) + i_1(u_t)$ are locally constant with respect to p_{t+1} and P_{t+1} for sufficiently high values of future prices p_{t+1}, P_{t+1}. This follows from the fact that national money demands $m^d(u_t)$ and $M^d(u_t)$ vanish if future prices are sufficiently high as compared to current prices (note that initial endowments are strictly positive in

each period). A solution (p_{t+1}, P_{t+1}) to (1.5), (1.6) with vanishing national money demands will never be unique then. Let $\tilde{D}_1 \subset \mathbb{R}^4_+$ be the set of all points u_{t-1} for which a solution $z_{t+1} = (p_{t+1}, P_{t+1})$ to (1.5), (1.6) exists such that either $m^d(u_t)$ or $M^d(u_t)$ vanishes. We shall make an assumption on the multiplicity of solutions to (1.5), (1.6) if u_{t-1} does not belong to $\tilde{D}_1$.

<u>Assumption 2.1 (regularity):</u> For any $u_{t-1} \in \mathbb{R}^4_+ \setminus \tilde{D}_1$ there is at most one solution to (1.5), (1.6) in the variables p_{t+1}, P_{t+1}.

This assumption is not essential and there is actually no compelling reason to make it. Dealing without it would complicate the analysis considerably, however.
Let $\tilde{D}_2 \subset \mathbb{R}^4_+$ be the set of all points u_{t-1} for which a solution $z_{t+1} \in \mathbb{R}^2_+$ to the system (1.5), (1.6) exists and define $\tilde{f}: \tilde{D}_3 := \tilde{D}_2 \cap (\mathbb{R}^4_+ \setminus \tilde{D}_1) \to \mathbb{R}^2_+$, $u_{t-1} \to z_{t+1}$. By Assumption 2.1 $\tilde{f}$ is well defined. We assume $\tilde{f}$ to possess limits at all points of $D := \tilde{D}_2 \cap \text{cl}\, \tilde{D}_3$ and denote its unique continuous extension by f, $f: D \to \mathbb{R}^2_+$. Some remarks on the properties of our function f are in order.

<u>Remark 2.1:</u> The picture under f of any vector $u_{t-1} \in D$ of prices in periods t and t-1 is a vector of goods market equilibrium prices z_{t+1} for period t+1. D can be a proper subset of $\tilde{D}_2$, i.e. there might be some point $u_{t-1} \notin D$ for which a solution to (1.5), (1.6) in the variables p_{t+1} and P_{t+1} exists. However, if $(z_t)_t = (p_t, P_t)_t$ is a sequence satisfying (1.5) and (1.6) for all $t \in \mathbb{Z}$ such that $u_{t-1} = (z_t, z_{t-1}) \notin D$ for some t then at least one of the national money demands (and also the corresponding national money supply) vanishes at time t. This being contrafactual it does not seem to be a great loss if we exclude equilibria of that type from our analysis. Hence we can slightly specialize the definition of a competitive equilibrium given above. A <u>competitive equilibrium</u> is a sequence of prices $(z_t)_t, t \in \mathbb{Z}$, such that $(z_t, z_{t-1}) \in D$ and equations (1.5) - (1.8) are satified for all $t \in \mathbb{Z}$.

<u>Remark 2.2:</u> We note for later reference that if $(x_n)_n = (x^1_n, x^2_n, x^3_n, x^4_n)_n$ is a sequence in D tending to $(\bar{x}^1, \bar{x}^2, \bar{x}^3, \bar{x}^4)$ then there is no subsequence $(y^1_{n'}, y^2_{n'})_{n'}$ of $(y^1_n, y^2_n)_n := (f(x_n))_n$ such that both $y^1_{n'}$ and $y^2_{n'}$ diverge to $+\infty$.
Otherwise $m^d(y^1_{n'}, y^2_{n'}, x^1_{n'}, x^2_{n'})$ and $M^d(y^1_{n'}, y^2_{n'}, x^1_{n'}, x^2_{n'})$ are both zero for sufficiently large n' contradicting $x_{n'} \in \text{cl}\, \tilde{D}_3$ for all n'.

We shall use the mapping f for the description of our economy's dynamic evolution. To serve this purpose we proceed defining what I'm going to call the generalized iterates of f.

Definition 2.1: Let $(a,b) \in D$ and $D_2 := \{(a,b) \in D \mid (f(a,b),a) \in D\}$. If (a,b) is an element of D_2 the second generalized iterate of f at (a,b) is defined by ${}^2f(a,b) = f(f(a,b),a)$.
Let $D_{n+1} := \{(a,b) \in D_n \mid ({}^nf(a,b), {}^{n-1}f(a,b)) \in D\}$. If (a,b) is an element of D_{n+1} the (n+1)th iterate of f at (a,b) is defined by ${}^{n+1}f(a,b) = f({}^nf(a,b), {}^{n-1}f(a,b))$. To ease the notation we define ${}^\circ f(a,b) = a$, ${}^1f(a,b) = f(a,b)$ and $D_1 = D$.

The next lemma (for a proof compare Eckwert (1987)) shows that the generalized iterates of f describe the dynamics of our economy along an equilibrium path.

Lemma 2.1: Let $(z_t)_t = (p_t, P_t)_t$ be an equilibrium sequence. Then

(2.1) $z_{t+k} = {}^kf(z_t, z_{t-1})$

holds for all t and all $k \geq 0$.

A periodic sequence $(z_t)_t = (p_t, P_t)_t$, say with period k, is said to be a periodic equilibrium if it satisfies (1.5) - (1.8) for all t. It can be described by a periodic orbit, i.e. by its k consecutive values $(z_1^*, \ldots, z_k^*)$. Since the periodic orbit $(z_1^*, \ldots, z_k^*)$ contains all information on the equilibrium $(z_t)_t$ there is no need for a discrimination and we shall in fact occasionally identify them both.
We proceed now to define a function $g: D \to \mathbb{R}_+^4$ having the property that the fixed points of its iterates $g^n = \underbrace{g \circ \ldots \circ g}_{n}$ are closely related to the set of periodic equilibria of our economy.
Let $g: D \to \mathbb{R}_+^4$ be defined by $g(x,y) = (f(x,y), x)$, $x \in \mathbb{R}_+^2$, $y \in \mathbb{R}_+^2$. It is not hard to see that the iterates of g are given by

$$(2.2) \quad g^n(x,y) = \left({}^nf(x,y),\ {}^{n-1}f(x,y)\right), \quad (x,y) \in D_n \quad .$$

If (x,y) is a fixed point of g^n and if $g^i(x,y) \neq (x,y)$ for $i=1, \ldots, n-1$ holds we call $(v_1^*, \ldots, v_n^*)$, $v_i^* := {}^{i-2}f(x,y)$, $i=3,4,\ldots,n$, $v_1^* := y$, $v_2^* := x$, a cycle of g with period n.

The function g exhibits some important properties which can be exploited to obtain an existence result for periodic equilibria. First, however, we state some lemmata showing the close relationship between the cycles of g and the periodic price sequences along which goods markets clear.

Lemma 2.2: Let (a,b) be an element of the domain of ${}^{n+m}f$.
For all $m \in \mathbb{N}$, $n \in \mathbb{N}$, the equalities

$$(2.3)\quad {}^{n+m}f(a,b) = {}^{n}f \circ g^{m}(a,b) = {}^{n}f\left({}^{m}f(a,b),\ {}^{m-1}f(a,b)\right)$$

hold.

Proof: See Eckwert (1987).

A cycle of g was defined as an orbit $(v_1^*, \ldots, v_n^*)$ generated by the iterates of g at the point (v_2^*, v_1^*) with (v_2^*, v_1^*) being a fixed point of g^n. The next lemma shows that any vector (v_i^*, v_{i-1}^*) consisting of two consecutive elements of the orbit is a fixed point of g^n and hence generates the same cycle as (v_2^*, v_1^*). In particular the cycles $(v_1^*, \ldots, v_n^*)$ and $(v_i^*, \ldots, v_n^*, v_1^*, \ldots, v_{i-1}^*)$ are identical.

Lemma 2.3: Let $(v_1^*, \ldots, v_n^*)$ be a cycle of g with period n, $v_i^* = {}^{i-2}f(x,y)$, i=3, ..., n, $v_1^* = y$, $v_2^* = x$, and define $v_0^* := v_n^*$. Then (v_i^*, v_{i-1}^*) is a fixed point of g^n for all $i \in \mathbb{N}_n$.

Proof: See Eckwert (1987).

The cycles of g characterize the set of periodic goods market equilibrium price sequences. This is an immediate consequence of our next corollary and will be discussed below.
Before proceding we shall simplify our notation in order to ease the presentation in the following. To this end we define

Definition 2.2: Let $(v_1^*, \ldots, v_n^*)$ be a cycle of g. For $l \in \mathbb{N}$ we define v_l^* by $v_l^* := v_{l-k_l n}^*$, with $k_l \in \mathbb{N}_0$ being uniquely determined by the condition $1 \leq l - k_l n \leq n$.

Corollary: Under the assumptions of Lemma 2.3 the equation

$$(2.4)\quad {}^{kn+j}f(v_i^*, v_{i-1}^*) = v_{i+j}^*$$

holds for all $k \in \mathbb{N}$, $0 \leq j \leq n-1$, $1 \leq i \leq n$.

Proof: See Eckwert (1987).

Let $(z_t)_t$ be a periodic equilibrium with periodic orbit $(z_1^*, \ldots, z_k^*)$. Any term of the sequence satisfies $z_t = z_{t+nk}$, $n \in \mathbb{Z}$. Hence by Lemma 2.1 (z_2^*, z_1^*) is a fixed point of g^k and $(z_1^*, \ldots, z_k^*)$ is a cycle of g with period k. On the other hand if $(v_1^*, \ldots, v_k^*)$, $v_i^* = {}^{i-2}f(x,y)$ for i=3,4,...,k, $v_1^* = y$, $v_2^* = x$, is a cycle of g with period k then by the above corollary $(v_1^*, \ldots, v_k^*)$ is an orbit of the periodic goods

market equilibrium price sequence $(p_t,P_t)_t$ with $(p_t,P_t) = v_t^*$ for $t\geq 1$, $(p_o,P_o) = v_k^*$, $(p_t,P_t) = v_{t-tk}^*$ for $t\leq -1$.
We thus see that the set of periodic goods market equilibrium price sequences is characterized by the set of all cycles of our mapping g. For this reason we shall pay special attention to the cycles of g in the sequel.
All behavioural functions in (1.5) and (1.6) are homogeneous of degree zero in their arguments. This implies that f is homogeneous of degree one and it follows directly from the definition of the generalized iterates (Definition 1.2) that also ${}^nf$ and by that g^n is homogeneous of degree one for all $n\in\mathbb{N}$. Suppose $g^n(x,y) = (\lambda x,\lambda y)$, $\lambda>0$, holds at a point $(x,y) \in \mathbb{R}_+^4$ where g^n is defined.
We then have

$$(2.5)\quad g^{kn}(x,y) = \lambda^k(x,y),\ k\in\mathbb{N},$$

which means that in the course of a stretch of n periods prices are multiplied by the factor λ. Since all consumption demand functions are homogeneous of degree zero the equilibrium allocation at any time t is reproduced all n periods, i.e. at all dates $t + kn$, $k\in\mathbb{N}$. We shall call a price sequence satisfying equation (2.5) a weak cycle with period n.

Definition 2.3: Let $h:\mathbb{R}_+^4 \setminus\{0\} \rightarrow S^3$, $v \rightarrow \frac{1}{\|v\|}v$ and $p:D \rightarrow S^3$, $p = h \circ g$.[2] Since g is linear homogeneous the iterates $p^n:D_n \rightarrow S^3$ of p are given by $p^n = h\circ g^n$. If $(x,y)\in D_n$ is a fixed point of p^n we call $(v_1^*, \ldots, v_n^*)$, $v_i^*:={}^{i-2}f(x,y)$, $v_1^*:=y$, $v_2^*:=x$, $i=3, \ldots, n$, a weak cycle of g with period n.

If we define a sequence $(z_t)_t = (p_t,P_t)_t$ to be weakly periodic, say with period k, if $z_{t+k} = \lambda z_t$ holds for some $\lambda > 0$ and all $t\in\mathbb{Z}$, then there exists a relation between the weak cycles of g and the weakly periodic goods market equilibrium price sequences of our economy which is completely analogous to that one already discussed above between the cycles of g and the set of periodic goods market equilibrium price sequences. This follows from the fact that Lemma 2.3 and the corollary belonging to it also hold in an appropriately reformulated version for weak cycles of g. We state the analogues of Lemma 2.3 and the corollary below omitting the proofs.

Lemma 2.4: Let $(v_1^*, \ldots, v_n^*)$ be a weak cycle of g with period n, $v_i^*:={}^{i-2}f(x,y)$, $i=3,\ldots,n$, $v_1^*:=y$, $v_2^*:=x$. With $v_o^*:=(\|g^n(x,y)\|)^{-1}v_n^*$ the vector $h(v_i^*,v_{i-1}^*)$ is a fixed point of p^n for all $i\in\mathbb{N}$.

Corollary: Under the assumptions of Lemma 2.4 the equation

$$(2.6)\quad {}^{kn+j}f(v_i^*, v_{i-1}^*) = \begin{cases} \mu^k(v_{i+j}^*, v_{i+j-1}^*); & i+j \leq n \\ \mu^{k+1}(v_{i+j}^*, v_{i+j-1}^*); & i+j > n \end{cases}$$

holds for all $k \in \mathbb{N}$, $0 \leq j \leq n-1$, $1 \leq i \leq n$, where $\mu := \| g^n(x,y) \|$.

3. Existence of periodic Equilibria

In this section we address the issue how special weakly periodic equilibria are as opposed to ordinary competitive equilibria. To serve this purpose we answer the question: Under which circumstances does a weakly periodic equilibrium exist provided there is a competitive equilibrium? We shall find that under (almost) all economically sensible conditions there is a weakly periodic equilibrium if a competitive equilibrium exists.

To ease the analysis a bit we define the functions $g_j(\cdot) := = z_j(\cdot) + I_j(\cdot)$, $h_j(\cdot) := Z_j(\cdot) + i_j(\cdot)$, $j=1,2$, and adopt a further regularity assumption.

Assumption 3.1: Given $p_t \neq 0$ and $P_t \neq 0$ the mappings $g_2(p_t, P_t, p_{t-1}, P_{t-1})$ and $h_2(p_t, P_t, p_{t-1}, P_{t-1})$ possess limits for $(p_{t-1}, P_{t-1}) \rightarrow (0,0)$.

Referring to these properties in the sequel we proceed on the assumption that the functions $g_2(\cdot)$ and $h_2(\cdot)$ are defined at points $(p_t, P_t, 0, 0)$ with $p_t \neq 0 \neq P_t$.

The existence of a competitive equilibrium implies that the set $E \subset D$ consisting of all points, whose image under g^t lies in D for all $t \geq 1$ is not empty. However if g^t is defined at $x \in D$ it is also defined at $y = \frac{1}{\|x\|} x$ with $g(y) = = \frac{1}{\|x\|} g(x)$ from which we conclude $E \cap S^3 \neq \emptyset$.

Lemma 3.1: The set $E \cap S^3$ is compact, where $E := \{x \in D \mid g^t(x) \in D, t \in \mathbb{N}\}$.

Proof: See Appendix.

We are able now to show that the existence of a competitive equilibrium implies the existence of a weakly periodic equilibrium under a (mild) additional assumption.

Theorem 3.1: If a competitive equilibrium $z_t=(p_t,P_t)_t$ exists and if the cluster points of the sequence z_t form a finite set C, then a weakly periodic equilibrium whose period does not exceed the cardinality of C exists.

Proof: We show first that a fixed point $x=(x_1,x_2), x_i \in \mathbb{R}^2_+, i=1,2,$ of p^t exists with $t \leq$ Card C. All terms of the sequence $(p^n(z_1,z_0))_n$ are elements of $E \cap S^3$. Let x_1 be a cluster point of this sequence. Since $E \cap S^3$ is compact by Lemma 3.1 we have $x_1 \in E \cap S^3$ and by continuity $x_2=p(x_1)$ is another cluster point of $(p^n(z_1,z_0))_n$. By induction we obtain a sequence $x_1,x_2,\ldots$ of cluster points satisfying $x_{i+1}=p(x_i)$. Since the set of cluster points C is finite there exist positive integers k and l, $k \leq l$, such that $p(x_l)=x_k$. Thus $x:=x_k$ is a fixed point of p^{l-k+1}. The generalized iterates $^tf(x)$ define a weakly periodic goods market equilibrium price sequence $\tilde{z}_t$. Since z_t is a competitive equilibrium (with cleared money markets) it follows from Lemma 1.1 and a simple continuity argument that money market equilibrium prevails along the weakly periodic price sequence $(\lambda \tilde{z}_t)_t$ for suitably chosen $\lambda \in \mathbb{R}_+$.

q.e.d.

Corollary: A weakly periodic equilibrium exists provided the set $S^3 \cap E$ is finite and nonempty.

Proof: Suppose a weakly periodic equilibrium does not exist. By Theorem 3.1 there is a point $x \in S^3 \cap E$ such that the sequence $(p^n(x))_n$ possesses infinitely many cluster points. Being compact (compare Lemma 3.1) $S^3 \cap E$ contains all these cluster points and therefore has an infinite cardinality.

q.e.d.

Theorem 3.1 guarantees the existence of a weakly periodic equilibrium if $(p^n(x))_n$ possesses a finite set C of cluster points. Though it is not difficult to construct a sequence with infinitely many cluster points the case where C has infinite cardinality might be considered as rather pathological from an economic point of view. In addition by the above corollary the existence of a weakly periodic equilibrium is questionable only if the set $E \cap S^3$ is too large, i.e. if it is infinite. In that case however there should be a good chance that among the infinite many elements of $E \cap S^3$ there exists one point x such that the cluster points of the sequence $(p^n(x))_n$ form a finite set. If the exogenous national money supply components m and M are suitably chosen there would be a weakly periodic equilibrium then as well by Lemma 1.1.

Suppose $(z_t)_t$, $t \in \mathbb{Z}$, is a competitive equilibrium such that the conditions of Theorem 3.1 are satisfied, i.e. such that the cluster points of the sequence $(p^n(z_t,z_{t-1}))_n$ form a

finite set. By Theorem 3.1 there is a weakly periodic equilibrium $(\tilde{z}_t)_t$ and the proof of the theorem shows that the elements of the finite set $G:=\{h(\tilde{z}_t,\tilde{z}_{t-1})\mid t\in\mathbb{Z}\}$ are cluster points of the sequence $(h(z_t,z_{t-1}))_t$. Hence the terms of the sequence $(h(z_t,z_{t-1}))_t$ eventually come very close to the elements of G. If however $h(z_t,z_{t-1})$ is close to $h(\tilde{z}_{t'},\tilde{z}_{t'-1})$ then $h(z_{t+1},z_t)=h\circ g\circ h(z_t,z_{t-1})$ is close to $h(\tilde{z}_{t'+1},\tilde{z}_{t'})=$ $=h\circ g\circ h(\tilde{z}_{t'},\tilde{z}_{t'-1})$ by the continuity of g.
Loosely speaking we can therefore say that any competitive equilibrium $(z_t)_t$ of the above type (i.e. such that the requirement of Theorem 3.1 is satisfied) eventually becomes weakly periodic. Thus if we set a particular value on the asymptotic behaviour of an economy weak periodicity of equilibria seems to be rather a typical than a specialfeature.

4. Conclusion

Our analysis shows that the phenomenon of long-term cyclical economic behaviour carries over from the standard OLG-model to more general structures. As could be expected periodicity of competitive equilibria is not a special feature of the one-dimensional (one good) OLG-model. Moreover any competitive equilibrium that satisfies a certain (rather mild) qualification eventually behaves in a weakly periodic manner.
One point we did not address to is the question under what circumstances there may coexist different (weakly) periodic equilibria in an economy. Grandmont (1985) showed in the one-dimensional OLG-framework that many weakly periodic equilibria 'typically' coexist. Also the issue of efficiency of competitive, and in particular of weakly periodic, equilibria remains to be investigated. If the economy starts operation at time $-\infty$ efficiency of competitive equilibria seems to be less likely as compared to the case where it starts out from some finite point of time. Recently some progresson this issue could be achieved (compare Cass/Okuno/Zilcha (1979), Eckwert/Schittko (1987)).
Weakly periodic equilibria which exactly reproduce the picture of an economy all k periods will surely never be observed in reality. This fact indicates that the concept of a deterministic model generating deterministic business cycles is not fully appropriate. Business cycles whose length and intensity vary according to exogenous stochastic disturbances would be a more realistic and highly interesting outcome as they would yield insights into the interaction between stochastic and purely endogenous mechanisms.

Appendix

In this Appendix we prove Lemma 3.1.

Proof of Lemma 3.1: Let $(x_n)=(x_n^1,x_n^2,x_n^3,x_n^4)$ be a convergent sequence in $E\cap S^3$ tending to $\bar{x}=(\bar{x}^1,\bar{x}^2,\bar{x}^3,\bar{x}^4)$. Clearly $(p_{t+1}^n,P_{t+1}^n,p_t^n,P_t^n,p_{t-1}^n,P_{t-1}^n):=(f(x_n),x_n)$ satisfies equations (3.1) and (3.2) for all $n\in\mathbb{N}$. In the first part of the proof we show $\bar{x}\in D\cap S^3$. Obviously $\bar{x}^1$ is different from zero: $\bar{x}^1=0$ would imply that neither $\bar{x}^2$ nor $\bar{x}^3$ nor $\bar{x}^4$ can differ from zero since otherwise either $g_2(x_n)$ or $h_2(x_n)$ would diverge to $+\infty$ contradicting $x_n\in D$ for sufficiently large n. However, $x_n\in S^3$ for all n, so that at least one component of $\bar{x}$ must differ from zero. This shows $\bar{x}^1\neq 0$ and a completely analogous argument yields $\bar{x}^2\neq 0$.
Hence the functions $g_2(\cdot)$ and $h_2(\cdot)$ are defined at $\bar{x}$. Let $y_n=(y_n^1,y_n^2):=f(x_n)$ so that $g(x_n)=(y_n^1,y_n^2,x_n^1,x_n^2)$. The sequence $(y_n)_n$ is bounded:
First we note that it is impossible that one component of $(y_n)_n$ is bounded while the other is not. In that case either $(g_2(y_n^1,y_n^2,x_n^1,x_n^2))_n$ or $(h_2(y_n^1,y_n^2,x_n^1,x_n^2))_n$ would be unbounded contradicting $g(x_n)\in D$, $n\in\mathbb{N}$. Suppose both components of $(y_n)_n$ are unbounded and choose a subsequence $(y_{n'})_{n'}$ such that $(y_{n'}^1)_{n'}$ diverges to $+\infty$. By Remark 2.2 $(y_{n'}^2)_{n'}$ does not diverge to $+\infty$ so we can find a subsequence $(y_{n''})_{n''}$ of $(y_n)_n$ such that $(y_{n''}^1)_{n''}$ diverges to $+\infty$ while $(y_{n''}^2)_{n''}$ is bounded. This however is not possible as was already shown.
Let $\bar{y}=(\bar{y}^1,\bar{y}^2)$ be a cluster point of $(y_n)_n$. The components of $\bar{y}$ are both different from zero. Otherwise the terms of the sequence $(g(x_n))_n = (f(x_n),x_n^1,x_n^2)_n$ would leave the set D for sufficiently large n (recall $x_n^i \to \bar{x}^i \neq 0$, $i=1,2$) contradicting $x_n\in E$, $n\in\mathbb{N}$. Hence the functions $g_1(\cdot)$ and $h_1(\cdot)$ are defined at $(\bar{y}^1,\bar{y}^2,\bar{x}^1,\bar{x}^2)$ and since the equations (3.1) and (3.2) hold for all (y_n,x_n), $n\in\mathbb{N}$, they are satisfied as well at the point $(\bar{y},\bar{x})$ as a simple continuity argument shows.
We conclude $\bar{y}=f(\bar{x})$ and thus $\bar{x}\in D\cap S^3$.
Using the above result that the limit of any convergent sequence in $E\cap S^3$ lies in $D\cap S^3$ we show in the second part of the proof 'g^t is defined at $\bar{x}$ and $h\circ g^t(\bar{x})\in D\cap S^3$ for all $t\in\mathbb{N}$' from which $g^t(\bar{x})\in D$, $t\in\mathbb{N}$, and hence the assertion of the lemma immediately follows.
$t=1$: g is defined at $\bar{x}$ (see above) and $h\circ g(x_n)\in E\cap S^3$, $n\in\mathbb{N}$, by assumption. Continuity yields $h\circ g(x_n)\underset{n\to\infty}{\to} h\circ g(\bar{x})$ so that $h\circ g(\bar{x})\in D\cap S^3$ follows by the first part of this proof.
$t\to t+1$: The induction hypothesis implies $h\circ g^t(\bar{x})\in D\cap S^3$, hence $g^t(\bar{x})\in D$.
Furthermore $x_n\in E$ for all n so that $h\circ g^{t+1}(x_n)\in E\cap S^3$, $n\in\mathbb{N}$.
By continuity of $h\circ g^{t+1}$ and the first part of the proof we

conclude $h \circ g^{t+1}(x_n) \underset{n\to\infty}{\to} h \circ g^{t+1}(\bar{x}) \in D \cap S^3$.

q.e.d.

Footnotes

[1] In this paper the exchange rate must be independent of time, $\pi_t = \pi$, $t \in \mathbb{Z}$. In the face of a non constant exchange rate and rational expectations individuals can profit from arbitrage opportunities by holding foreign currency. Thus our assumption that people hold national currency only forces the exchange rate to be constant over time.

[2] Throughout the paper S^3 stands for the unit simplex in $\mathbb{R}^4$ and the norm $\|\cdot\|$ is defined by $\|x_1,x_2,x_3,x_4\| = \sum_{i=1}^{4} |x_i|$.

References

1. Böhm, V. and Keiding, H., 1985, Inflation and Welfare in International Trade, Discussion Paper 50 pp., Universität Mannheim, University of Copenhagen, October 1985.

2. Buiter, W.H., 1981 Time Preference and International Lending and Borrowing in an Overlapping - Generations Model, Journal of Political Economy 89, 769 - 797.

3. Cass, D., M. Okuno and I. Zilcha, 1979, The Role of Money in Supporting the Pareto Optimality of Competitive Equilibrium in Consumption - Loan Type Models, Journal of Economic Theory 20, 41 - 80.

4. Eckwert, B., 1987, Endogenous Business Cycles in a Monetary Two-Country Model, Discussion Paper, University of Augsburg.

5. Eckwert, B., Schittko, U.K., 1987, Existence and Optimality of Endogenous Business Cycles within a Productive Two-Country Growth Model, Discussion Paper, University of Augsburg.

6. Grandmont, J.-M., 1985,
On Endogenous Competitive Business Cycles,
Econometrica 53, 995 - 1045.

7. Mitsui, T., U.K. Schittko and S. Watanabe, 1985,
The Tobin Effect and Optimal Taxation in the Overlapping Generations Model, Discussion Paper, University of Kansas and University of Augsburg.

8. Okuno, M. and I. Zilcha, 1983,
Optimal Steady - State in Stationary Consumption - Loan Type Models, Journal of Economic Theory 31, 213 - 228.

FISCAL POLICY IN AN ACCELERATOR - AUGMENTED CLASSICAL GROWTH CYCLE

by
Peter Flaschel
Universität Bielefeld

0. Introduction

In this paper I shall reexamine Wolfstetter's (1982) assertion that a classical as well as a Keynesian fiscal policy is capable of stabilizing the classical growth cycle provided that there are problems of effective demand present in the modelling of this cycle and provided that both policies are exercised with sufficient strength. This assertion is a very surprising one, since Goodwin's classical growth cycle model attempts to model the formation and subsequent removal of a profit squeeze situation by means of the reserve army mechanism. A Keynesian policy (which does not consider this mechanism - as it has occurred in the past) should be expected to hinder, yet not remove this mechanism, thereby making the Marxian "Bereinigungskrise", longer and probably more severe - - in striking contrast to Wolfstetter's stability analysis and the above symmetry assertion.

In order to allow a thorough reexamination of Wolfstetter's provocative claim we shall here introduce several new aspects into his extension of the classical growth cycle by a government sector:

1. A representation of the cycle model must be found which is locally unstable and which takes account of the fact that the employment rate cannot exceed and the share of wages cannot reach the value "1". The well-known nominal versions of Goodwin's cycle model are still insatisfactory in this regard.
2. The inclusion of problems of effective demand into the classical growth cycle may destabilize this cycle further, yet it should not lead to its complete degeneration in the sense that fiscal policy is absolutely necessary to prevent its immediate collapse.
3. The above "symmetry assertion" must be tested for those magnitudes of fiscal policy parameters which at least roughly correspond to factual magnitudes.
4. A purely qualitative analysis - as it is undertaken in Wolfstetters's article - will in general be incapable of analyzing or solving tasks 2. and 3. in a satisfactory way. It thus has to be supplemented by simulation studies.

The above four points represent the problems we see in Wolfstetter's introduction of fiscal policy into the Goodwin model. In the following we will therefore investigate these four points step by step. Our central finding will be that there in fact is no symmetry between the above two types of fiscal policies in the context of the classical growth cycle. However, also the models used in this paper have their problematic features, as can be seen, e.g., from some of the simulations which follow. The question raised by Wolfstetter therefore needs and deserves further elaboration(far beyond the contribution of this paper).

1. A simple completion of the classical growth cycle

A standard extension of the well-known Goodwin model of cyclical (or "classical") growth is given by the following two dynamic equations [see Wolfstetter(1977, 3.4.1) for their detailed introduction]:

$$\hat{u} = -(\gamma + \alpha) + \rho v - (1 - \eta) \lambda[(1 + a)u - 1] \quad (1)$$

$$\hat{u} = s_p\sigma - (\alpha + \beta) - s_p\sigma u \quad (2)$$

These equations model the dynamic interaction of money-wage fixing, price settings, profits and accumulation in a technological world with fixed proportions and Harrod-neutral technical progress.

In these equations we denote by:

u the share of wages (steady state u^*)
v the employment rate (steady state v^*)
$\hat{u}, \hat{v}$ their rates of growth ($\dot{u}/u$, $\dot{v}/v$)
α the growth rate $\hat{y}$ of labour productivity ($y = Y/L$), a constant
β the growth rate $\hat{N}$ of (labour) population N , a constant
g^* $\alpha + \beta$
σ the output-capital ratio Y/K , a constant
s_p the rate of savings out of profits ($0 \leq s_p \leq 1$; $s_w = 0$), a constant
$\hat{m}$ the growth rate of money wages m
$\hat{p}$ the rate of inflation $\dot{p}/p$
a mark-up factor on wage-unit-costs, a constant .

The above dynamics is based in addition on the following structural equations and assumptions:

$\hat{m} = \rho v - \gamma + \eta \hat{p}$ [a money-wage phillips-curve] ,
$\hat{p} = \lambda[(1 + a)u - 1]$ [a mark-up theory of inflation] ,
where ρ, γ , η , γ denote positive parameters
[compare again Wolfstetter (1977) for details] . In sum, equations (1) and (2) thus state, that the percentage change of the share of wages u depends positively on the rate of employment v and (for $\eta < 1$) negatively on the level of u (because of the assumed mark-up theory of inflation) and that the percentage change of the rate of employment v is governed by savings out of profit per unit of capital $s_p\sigma$ (1-u) diminished by the growth rates α, β of the labor force and labor productivity.

It is easily shown by means of Olech's theorem that the above model's behaviour can be characterized as follows [see, e.g., Flaschel (1984)] :

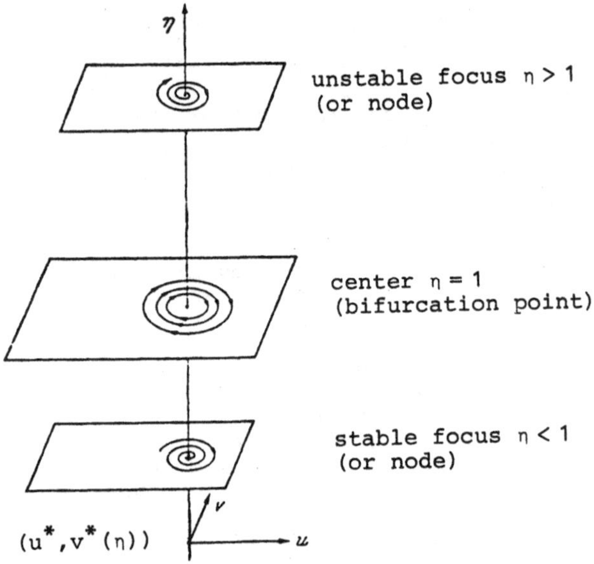

Fig. 1

The depicted behaviour is valid locally as well as globally and it is incomplete in the same way the unrestricted (linear) multiplier-accelerator model is not yet complete, i.e., it does not remain restricted to economically meaningful values of u, v in the case $\eta > 1$ (of explosive cycles) and it is not yet a complete theory of the "business cycle" in the case of $\eta < 1$ (implosive cycles).

It is my opinion that local instability (here $\eta > 1$) should be a key characteristic of a model of the Marxian theory of accumulation, yet that in addition to this feature reproductiveness, (i.e., outward stability) should also prevail, at least as long as the Marxian type of technological change is not yet included into such a model [as in the case in the above dynamics (1),(2)]. In order to introduce such forces into the above situation which attempt to model the idea of the reproducibility of capitalism along the lines described in Marx (1954, p. 582):

> "The rise of wages therefore is confined within limits that not only *leave intact* the foundations of the capitalistic system, but also secure its reproduction on a progressive scale" ,

it seems adequate in the first instance to abstract from goods-market problems (and also the bond market). The inclusion of these aspects would overload the model such that it would no longer provide a clear-cut description of the basic forces of this "classical" growth cycle. We therefore start with the following simple extension of the above model (1), (2):

$$\hat{u} = f(v) - (1 - \eta(u)) \cdot \lambda(u)[(1+a)\,u - 1] - \alpha \qquad (3)$$

$$\hat{v} = s_p i(v) \cdot (1 - u)\,\sigma - (\alpha + \beta) \qquad (4)$$

where

1. $f(v)$ fulfills $f'(v) > 0$, $\exists\, v_o \in (0,1): f(v_o) = 0$,
2. $\lambda(u) > 0$ fulfills $\lambda'(u) \geq 0$, $\lambda(u) \to \infty$ for $u \to 1$ and $\lambda(u) = 0$ if $u \leq \frac{1}{1+a}$,
3. $\eta(u)$ fulfills $\eta(u^*) > 1$, $\eta'(u) < 0$ (and $\eta(u) < 1$ for $u > u^*$ near '1', and where
4. $i(v) = \begin{cases} 1 \text{ for } 0 \leq v \leq 1 - \varepsilon, \ \varepsilon > 0 \text{ small} \\ 0 \text{ for } v = 1 \end{cases}$

(all functions are supposed to be sufficiently smooth).

Assumption 1. simply allows for a more general type of Phillips-curve than is involved in (1). The mark-up pricing rule behind the term $\hat{p}(u)$ in (3), i.e.

$$\dot{p} = \lambda(u)\left[\frac{(1+a)m}{y} - p\right]$$

is modified by assumption 2 insofar that it is now assumed that the adjustment speed λ of price changes (which are driven by the discrepancy between marked up wage-costs per unit of output and actual prices) increases (to infinity) as the share of wages increases (approaches 1). Furthermore, the occurence of deflation has been excluded from the present form of the model to avoid the existence of problematic types of steady states. Assumption 3 states that wage earners demand more than just the rate of inflation $\hat{p}$ in their money-wage claim if the wage share is at its steady value u^* ($\alpha > 0$!). This and the negative relationship between the aspiration factor η and the share of wages u can be motivated as follows:

Since the output-capital ratio σ is constant, we have

$$g = \hat{Y} = \hat{K} = s_p i(v)(1-u)\,\frac{Y}{K} = s_p i(v)(1-u)\,\sigma$$

Therefore $g \gtreqless g^* = \alpha + \beta$ if $u \gtreqless u^*$. It is thus implicitly assumed in 3. that η relates positively to the actual rate of growth g with $\eta = 1$ already occuring at a point below normal growth g^*. Assumption 4, finally, says that it is of no use to further accumulate capital if there is no labor force available for the new capital goods.

Model (3), (4) represents an elementary completion of (1), (2) by projecting the reaction patterns of the distributional conflict and the accumulation process into the price/wage-sector and a simple capital formation equation. Inflation is the means by which the rate of profit is defended against rising money wage claims and a simple fall in the rate of growth of capital is here assumed as reaction to the full employment barrier [1] [see, however, § 4 for a more elaborate approach].

<u>Proposition 1:</u> Assume $s_p = 1$ for simplicity and also that

$f(v^*) = (1-\eta(u^*)) \cdot \lambda(u^*)(Au^* - 1) + \alpha$,[2] $u^* = 1 - g^*/\sigma$ has a meaningful solution $(u^*, v^*) \in (0,1)^2$ within the range where $i(v)$ is still constant ($\equiv 1.$). Assume further: $\eta'(u^*) = 0$ (or small). The model (3), (4) then has at least one limit cycle around its unique and unstable steady state solution u^*, v^*.

This proposition represents a fairly obvious application of the Poincaré-Bendixon Theorem. Calculating the Jacobean of (3), (4) at the steady state gives[3]

$$J = \begin{pmatrix} \underbrace{-(1-\eta(u^*))}_{-} \; \underbrace{\hat{p}'(u^*)\cdot u^*}_{+} & \underbrace{f'(v^*)u^*}_{+} \\ \underbrace{-\sigma v^*}_{-} & 0 \end{pmatrix} \qquad (J_1)$$

i.e. we have det $J > 0$ and trace $J > 0$. The steady state (u^*, v^*) is thus an unstable node or focus.[4] And with regard to an analysis of global stability and the existence of limit cycles we will put up with the following phase portrait (and skip the details of the proof):

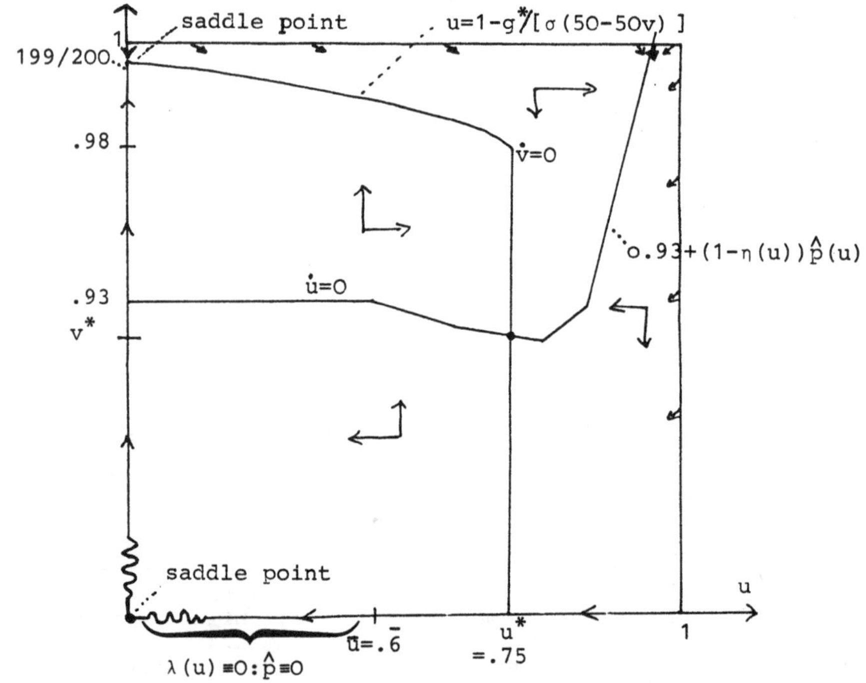

Fig. 2: This picture corresponds approximately to the numerical values of the simulation example of this section.

Remarks:

1.) Note, that the $\dot{u}=0$-isocline should not touch the horizontal axis [cf. assumption 3. in this regard].
2.) In contrast to Wolfstetter (1977, 1982) we do not assume here $A = 1/u^*$, i.e. a non-inflationary steady-state: $\hat{p}(u^*)=0$.
3.) $\hat{p} < 0$, i.e.: deflation, is however excluded in order to avoid that $\hat{u}$ may depend negatively on u for small u and v near full employment (this would imply a falling share of wages despite a situation near full employment). Furthermore, additional interior and stable points of rest are avoided thereby.

The above model shows explicitly (some of) the forces which keep the dynamics of Goodwin's growth cycle within economically meaningful bounds. Paired with the instability of the steady state this implies that each trajectory will be attracted by some closed curve in the relevant (u, v)-domain. The model provides a simple example for Marx's view on capitalistic accumulation as it is expressed in the above quotation from "Capital". Note that this explanation of positive profits is quite different in nature from the type of "explanation" that is offered by the so-called Fundamental Marxian Theorem [which in fact gives no explanation at all, but only establishes a relationship between two different basic magnitudes of Marx's "Capital"].

We conclude this preliminary analysis of a more appropriate representation of the "classical growth cycle" as it is given by the model (1), (2) with a simulation study of the model (3), (4):[5)]

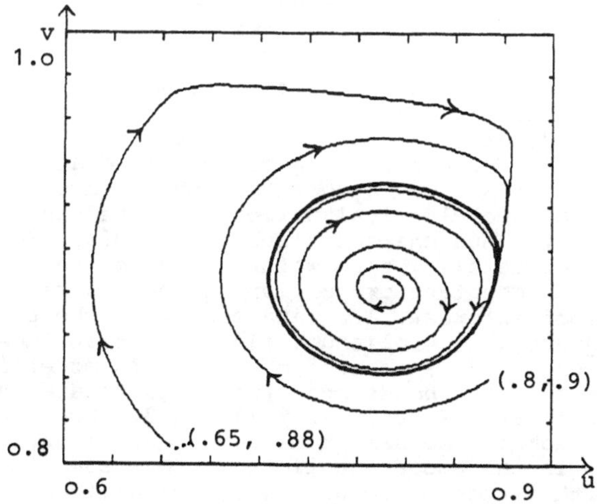

Fig. 3 : The limit cycle of a simple, but complete model of Marx's analysis of cyclical accumulation

Parameter values and functions:

$\alpha = .03$, $\beta = .02$, $\sigma = .2$, $a = .5$, $\bar{u} = .66$,

$f(v) = \rho v - \gamma$, $\rho = 1$, $\gamma = .9$,

$i(v) = -50v + 50$ on $[.98,1]$ and $\equiv 0$ or 1 outside this interval

$1 - \eta(u)$ the piecewise linear function with $h(0)=-.5, h(.72)=-.05,$
$h(.78)=-.05, h(1)=.5,$

$\lambda(u)$ the continuous function with $\lambda' \equiv 0$ on $[0,\bar{u}]$, $\lambda \equiv 4$ on $(.65, .75]$,
$\lambda' \equiv 400$ on $(.75,1]$.

2. Government stabilization policies

To discuss the effects of fiscal policies Wolfstetter (1982) introduces a public sector into the model (1),(2) in the following way:

$$G/Y = t + \mu(v^* - v) \qquad (5)$$

$$T = t(Y + i B) \qquad (6)$$

$$G + i B \equiv T + \pi \dot{B} \qquad (7)$$

The new symbols in these equations and the identity denote:

t	tax rate (a fixed parameter)
T	taxes
G	(unproductive) government expenditures
B	stock of government bonds
i	rate of interest
π	bond price
μ	reaction parameter .

Equation (5) describes an expenditure policy rule which implies a constant proportion t of government spending in national income for the steady state. And outside the steady state it implies $\widehat{G/Y} = -\mu\dot{v}$. A Keynesian policy would attempt to counteract the cycle and would therefore be characterized by $\mu > 0$, whereas a (neo)classical policy would reduce expenditure in the slump and is thus characterized by $\mu < 0$[see Wolfstetter (1982) for further details, also with regard to the following].

Taxes T are a constant proportion of the tax base $Y + iB$, with i the rate of interest. The parameter π is the price of new bonds $\dot{B}$. Since a portfolio decision is not included by Wolfstetter in the above extension of Goodwin's model it is necessary to assume that bonds and capital goods are characterized by the same rate of return $r = (1-u)Y/K$. The simplest way of doing this which at the same time avoids any misunderstanding of the terms π, i in the above equations is to assume fix-price bonds ($\pi = 1$) with a rate of return always equal to r . This is an admissable assumption, since all bonds are issued by the government, which thus simply has to guarantee that the prevailing rate of profit will always be assumed as point of reference for current interest payments $iB = rB$. Introducing the equations (5) - (7) into our model (3), (4) modifies it in the following way: Equation (3) remains unchanged since wage- and price-formation rules are by assumption not modified by government policy and since u is still defined by $mL/pY =$ $= \frac{m/p}{Y/L} = \frac{w}{y}$ [w the real wage, y output per head].

The growth of the capital stock is now, however, governed by[6)]

$$i(v)[(1-t)(Y + rB - wL) - \dot{B}] \qquad (8)$$

i.e. by invested profits (including interest) after tax minus that investment that has gone into government bonds instead of real capital formation. Note that - because of Say's law - there is no goods-market problem, i.e. the aggregation of all components of goods demanded will necessarily be identical with the supply of goods $Y \equiv (1-t)wL + [(1-t)(Y + rB - wL) - \dot{B}] + G$ $\equiv (1-t)(Y + rB) - \dot{B} + (1-t)(Y + rB) + \dot{B} - rB$ [the term in square brackets represents the capitalists' demand for consumption and investment goods].

Equation (8) implies by means of (5) - (7):

$$\begin{aligned} \dot{K} &= i(v)[(1-t)(Y + rB - wL) - (G + rB - t(Y + rB))] \\ &= i(v)[Y-(1-t)wL - tY - \mu(v^* - v)Y] \\ &= i(v)[(1-\delta)(Y-wL) + \mu(v - v^*)Y] \end{aligned}$$

On the basis of this modification the equation (4) now reads

$$\hat{v} = i(v)\sigma[(1-t)(1-u) + \mu(v - v^*)] - (\alpha + \beta) \qquad (9)$$

which together with

$$\hat{u} = f(v) - \alpha - (1-\eta(u))\lambda(u)[(1+a)u - 1] \qquad (10)$$

constitutes the new system of differential equations to be investigated.

Note first, that this new model is qualitatively identical to (3), (4) if $\mu = 0$[$u^* = 1 - g^*/\sigma$ is then to be replaced by $u^* = 1 - g^*/(\sigma(1-t))$]. And for an arbitrary parameter μ we get for the $\dot{v} = 0$ - isocline

$$u = (1-t-g^*/(\sigma \cdot i(v)) + \mu(v - v^*))/(1-t),$$

which gives nearly the same sort of isocline as in section 1 if $|\mu|$ is sufficiently small [the vertical part is now downward sloping if $\mu < 0$ and upward sloping if $\mu > 0$ and will pass through the same steady state value as before, i.e. the one belonging to $\mu = 0$]. The $\dot{u} = 0$-isocline is, of course, strictly the same as before, i.e. the outward bounds which it and $\dot{v}=0$ determine are of the same type as in figure 2. —

For the Jacobean of (3), (9) we get (under the assumptions of section 1):

$$J = \begin{pmatrix} \underbrace{-(1-\eta(u^*))p'(u^*)u^*}_{+} & \underbrace{f'(v^*)u^*}_{+} \\ \underbrace{-\sigma(1-t)v^*}_{-} & \underbrace{\sigma\mu v^*}_{\pm} \end{pmatrix} \qquad (J_2)$$

and therefore a further destabilizing force with regard to the steady state for the Keynesian policy rule ($\mu > 0$). The classical rule ($\mu < 0$), on the other hand, adds a stabilizing term to the characteristics of steady equilibrium.[7)]

In the case of the original Goodwin model (where $\eta \equiv 1$) these two policy alternatives consequently give rise to two sharply distinguished cases (of the type depicted in fig.1, with $\mu = 0$ now instead of $\eta = 1$), because we then have:

$$J = \begin{pmatrix} 0 & + \\ - & \sigma \mu v^* \end{pmatrix}$$

This sharp distinction also extends to larger regions in $(0,1)^2$. An alternative to the proof of this fact in Wolfstetter (1982) is given by the following:

Proposition 2: The continuous function $L: (0,1) \times (0,1-\varepsilon) \to \mathbb{R}$ defined by

$$L(u,v) = \int_{v^*}^{v} (f(\tilde{v})-\alpha)/\tilde{v}\, d\tilde{v} - \int_{u^*}^{u} [\sigma(1-t)(1-\tilde{u})-(\alpha+\beta)]/\tilde{u}\, d\tilde{u}$$

fulfills the following conditions

(1) $L(u^*,v^*) = 0$ and $L(u,v) > 0$ for $(u,v) \neq (u^*,v^*)$

(2) $\dot{L} < 0 (\dot{L} > 0, \dot{L} = 0)$ for $\mu < 0$ ($\mu > 0$, $\mu = 0$) in $(0,1) \times (0,1-\varepsilon) - \{(u^*,v^*)\}$

The well-known closed orbit structure of the case $\mu=0$ around the equilibrium u^*,v^* consequently determines domains which are positively (negatively) invariant with regard to the dynamics $\mu < 0$ ($\mu > 0$). By Theorem 2 in Hirsch/Smale (1974, p.196) this consequently implies the situation of global (in)stability of the type depicted in Wolfstetter (1982, p.381).

Proof: 1) Obvious, since $f(\tilde{v})/\tilde{v}$ and $-[\sigma(1-t)(1-\tilde{u})-(\alpha+\beta)]/\tilde{u}$ are both strictly increasing functions.

2.)
$$\begin{aligned}
\dot{L} &= [(f(v)-\alpha)/v] \cdot \dot{v} - [(\sigma(1-t)(1-u) - (\alpha+\beta))/u]\, \dot{u} \\
&= (f(v)-\alpha) \cdot \hat{v} - (\sigma(1-t)(1-u)-(\alpha+\beta)) \cdot \hat{u} \\
&= (f(v)-\alpha)[\sigma((1-t)(1-u)+\mu(v-v^*))-g^*]-[\sigma(1-t)(1-u)-g^*](f(v)-\alpha) \\
&= (f(v)-\alpha)\mu(v-v^*) = \mu[(f(v)-\alpha)(v-v^*)]
\end{aligned}$$

$$\begin{matrix} > \\ = \\ < \end{matrix}\; 0 \quad \begin{matrix} \mu > 0 \\ \mu = 0 \\ \mu < 0 \end{matrix} \quad \text{and } v \neq v^*.$$

Remarks: The instability of (u^*,v^*) in the case $\mu > 0$ is always supplemented by the fact that region $(0,1)^2$ is invariant if equation (3) [and not $\eta \equiv 1$] holds true. Furthermore, proposition 1 of section 1 need not hold true (if the interior equilibrium becomes non-unique). There may, for example, exist large values of $\mu > 0$ which allow for a stable equilibrium $(u^{**},v^{**}) > (u^*,v^*)$ which is highly inflationary[note that $1-\eta(u^*)$ will become positive for $u > u^*$ sufficiently large!].

The central result of this section:that Keynesian fiscal policy will be destabilizing while a classical policy will do the opposite (and may even eliminate the limit cycle of section 1, see the following simulations) is not too surprising. This is already noted in Wolfstetter (1982). A Keynesian policy (ceteris paribus) raises

$$\dot{B} = \mu(v^* - v)Y + (1 - t)rB$$

[in comparison to the steady state characterization $\hat{B} = \dot{B}/B = (1-t)r$],if $v^* > v$ and lowers it in the opposite case. It consequently supplies extra funds for physical capital accumulation in the boom and absorbes such funds in the depression - which clearly makes the process of overaccumulation (and underaccumulation) of capital on which this model's cycle rests more pronounced. By contrast, a classical rule would absorb funds in the boom and supply extra funds during the depression and thereby establishes a force which counteracts this cycle-maker. The following simulations (Fig. 4a -d) illustrate the various situations we have described above (note that in addition to the parameters of section 1 we have assumed for the new parameter t the value .2):

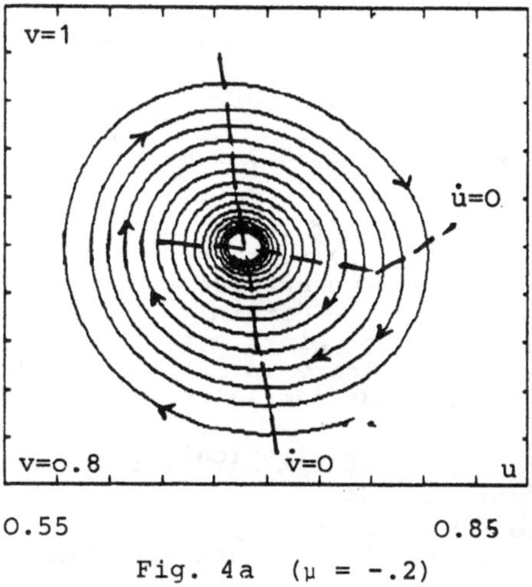

Fig. 4a (μ = -.2)

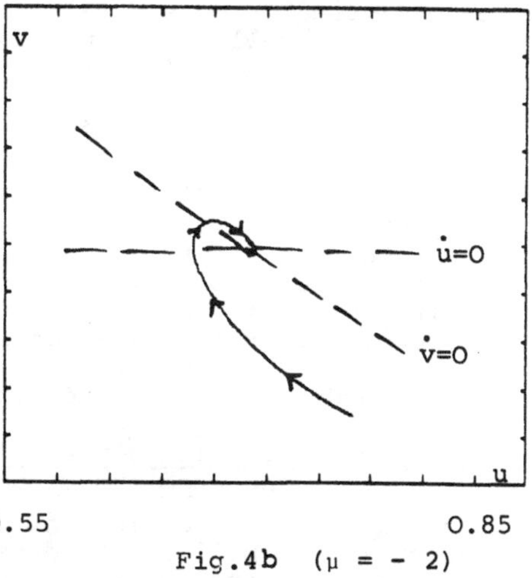

Fig.4b (μ = - 2)

Classical Policy

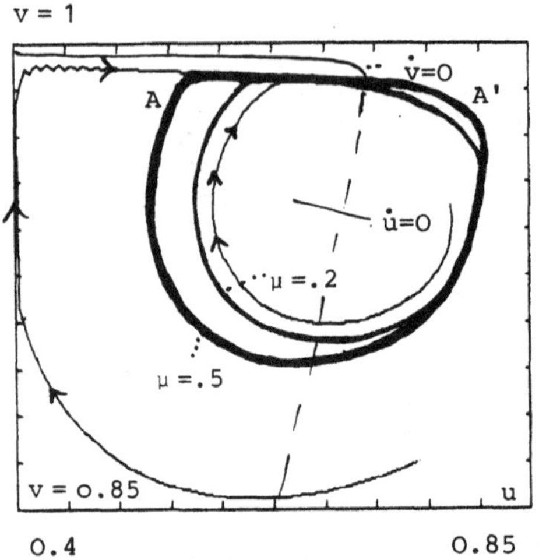

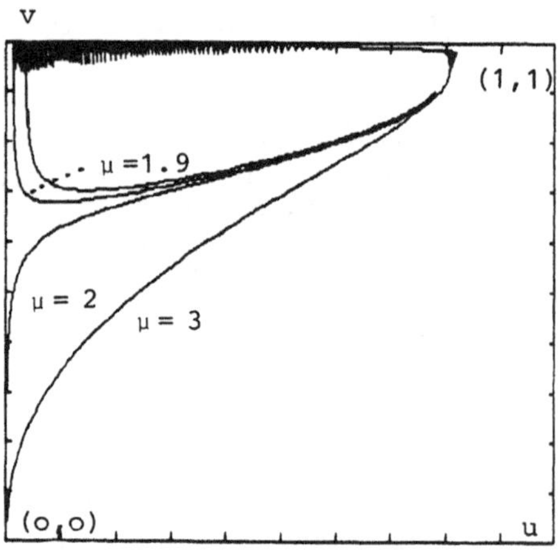

Fig. 4c, (μ = 2,.5) — Keynesian Policy — Fig. 4d (μ large)

In the above remark we noted that large $|\mu|$ may give rise to new and also problematic situations which should, on the one hand, be considered in more detail than we have done here. On the other hand, economically relevant magnitudes of $|\mu|$ can be estimated (very crudely) as follows:

$t = 30\,\%$ and $\max |v - v^*| \le o.1$,
$\max |\mu(v - v^*)\, Y| \le o.1 \cdot tY$, i.e.
$\max |\mu(v - v^*)| = \max |\mu|\; |v - v^*| \le o.1t$, i.e.
$o.1\; |\mu| \le o.1t$, i.e. $|\mu| \le o.3$

The problematic situations depicted in fig. 4d can therefore safely be excluded from consideration. However, these extreme situations do nevertheless indicate that the enforcement of the turning points in the above cycle within the domain $(o,1)^2$ (see the behaviour of the dynamics near the two points A,A' in fig. 4c) must still be considered as rather preliminary in its formulation. This should be kept in mind for later reconsideration.

3. Independent investment behaviour, the classical growth cycle, and problems of stabilization.

To allow for a proper test of the Keynesian policy rule Wolfstetter (1982, p.387) introduces the following investment function into the Goodwin growth cycle model:

$$\hat{K} = I/K = \frac{\varepsilon}{D + \varepsilon}\left[g^* - \left(1 - \frac{x}{x^*}\right)\right] \qquad (11)$$

where ε is an adjustment parameter and where x and x^* denote the actual and the desired degree of capacity utilization. Since x^* is exogenously given, we can simplify to $x^* = 1$ [$x = Y/Y_p$ where Y_p is given by σK].

The introduction of a new equation I = S (goods market equilibrium) and a new variable which allows for the realization of this equilibrium implies instead of model (3), (4) the following set of dynamic equations (see Wolfstetter (1982) for details of this approach):

$$\hat{u} = f(v) - \alpha - (1-\eta(u))\,\hat{p}(u) \quad (12)$$

$$\hat{v} = \sigma x(1-u) - (\alpha+\beta) + \hat{x} \quad (13)$$

$$\dot{x} = [\varepsilon(g^* - 1 + x - \kappa) + \sigma x\dot{u}]/(\kappa/x) \quad (14)$$

where $\kappa = \sigma x(1-u)$ and where $\hat{p}(u) = \lambda(u)\cdot[Au-1]$, $A = 1+a$ gives the rate of inflation.

The first equation is the same as before if - as it is assumed - actual output Y experiences the same rate of growth of labor productivity $Y/L = \alpha$ as normal output Y_p. It again describes the conflict over the distribution of actual income based on a money-wage Phillips-curve and a mark-up pricing rule. The second equation has been modified slightly in comparison to (4) since actual profits are now given by $(1-u)Y = x(1-u)Y_p$ and not by $(1-u)Y_p$ and since we now have

$$\hat{v} = \widehat{L/N} = \widehat{(Y/Y_p)\cdot(L/Y)\cdot(Y_p/K)\cdot K/N} = \hat{K} - \alpha - \beta + \hat{x}$$

by definition. The third equation, finally, is new and is derived in the following way: Goods Market equilibrium I = S implies $I/K = S/K$ or $\frac{\varepsilon}{D+\varepsilon}(g^* - 1 + x) = \sigma x(1-u)$. This in turn gives $\varepsilon(g^* - 1 + x - \sigma x\cdot(1-u)) = \sigma\dot{x}(1-u) - \sigma x\dot{u}$ and thus

$$\dot{x} = \frac{\varepsilon(g^* - 1 + x - \sigma x(1-u)) + \sigma x\dot{u}}{\sigma(1-u)}$$

Introducing the relative magnitude $\theta = v/x$ allows to rewrite (12) - (14) as follows

$$\hat{u} = f(\theta x) - \alpha - (1-\eta(u))\cdot\lambda(u)[Au-1] \quad (12')$$

$$\hat{\theta} = \sigma x(1-u) - (\alpha+\beta) = \sigma x(1-u) - g^* \quad (13')$$

$$\dot{x} = \frac{\varepsilon(g - 1 + x - \sigma x(1-u)) + \sigma x\cdot\dot{u}}{\sigma(1-u)} \quad (14')$$

Following Wolfstetter we assume the existence of a meaningful steady state solution (which for $A=1/u^*$ is the same as that of model (1),(2)). The Jacobean at these equilibrium values is given by ($\eta'(u^*) = 0$ as in section 1):

$$J = \begin{pmatrix} -\hat{p}(u^*)(1-\eta(u^*))u^* & f'(\cdot)u^* & f'(\cdot)\theta^* u^* \\ -\sigma\theta^* & 0 & \sigma(1-u^*)\theta^* \\ \frac{(\varepsilon-\hat{p}'(u^*))(1-\eta(u^*))u^*}{1-u^*} & \frac{f'(\cdot)u^*}{1-u^*} & \frac{\varepsilon(1-\sigma(1-u^*))+\sigma f'(\cdot)u^*\theta^*}{\sigma(1-u^*)} \end{pmatrix} \quad (J_3)$$

Comparing the Jacobean (J_3) with (J_1) of section 1 we see that destabilizing factors have entered the system even if we follow Wolfstetter and assume that the original system $\begin{pmatrix} J_{11} & J_{12} \\ J_{21} & J_{22} \end{pmatrix}$ is asymptotically stable ($\eta < 1$).

In the first instance this is due to the positive dependence of $\dot{x}$ on $x (J_{33} > 0!)$, a Harrodian knife-edge situation a la Phillips (1961). We have ($\eta < 1$):

$$\text{trace } J = - \hat{p}(u^*)(1-\eta)u^* + \frac{\varepsilon(1-g^*) + \sigma f'(\cdot)\theta^* u^*}{\sigma(1 - u^*)} > 0$$

if, e.g., the parameter ε is sufficiently large, which implies instability. And for det J we get

$$\begin{vmatrix} -\varepsilon & 0 & -\varepsilon(1-g^*)/\sigma \\ -\sigma\theta^* & 0 & g^* \\ J_{31} & \frac{f'(\cdot)u^*}{1-u^*} & \frac{\varepsilon(1-g^*)}{g^*} \cdot \theta^* \end{vmatrix}$$

$$= \varepsilon(1-g^*)\theta^* \cdot f'(\cdot)u^*/(1-u^*) + \varepsilon g^* \frac{f'(\cdot)u^*}{(1-u^*)} \cdot \theta^* = \varepsilon\theta^* f'(\cdot)u^*/(1-u^*) > 0 ,$$

which implies that the above model will be unstable --quite independent from any stabilizing influence from money illusion ($\eta < 1$). The question arises how this instability of the "private sector" of the economy can be removed. In analyzing this question Wolfstetter (1982) arrives at a surprising conclusion: Suppose that fiscal policy is introduced in the above model (12)-(14) in the way of section 2. Then - in striking contrast to section 2 - we get in this new situation that <u>both</u> policies will stabilize the economy provided that are exercised with sufficient strength ($|\mu| >> 0$).

An alternative proof of this result will be given below. In particular from a Marxian perspective, however, this result looks rather unconvincing: <u>If</u> the conflict between capital and labour is the central conflict which determines the evolution of a capitalistic system and <u>if</u> Marx's general law of accumulation is at least correct with regard to its analysis of the cyclical nature of capital accumulation, <u>then</u> the result should be that a policy which hinders the "Bereinigungskrise" by making the slump less deep (and which speeds up accumulation in the boom - the Keynesian policy rule $\mu > 0$!) should make accumulation less stable and not more stable!

Starting point of Wolfstetter's (1982,3.) analysis is the model (1),(2) of section 1 with $s_p \equiv 1$, η = const $\in (0,1)$, λ = const., $f(v) = \rho v - \gamma$ and $\hat{p}(u) = \lambda[u/u^* - 1]^p$. We shall follow him in this regard to not overload the presentation of his model when government activities are included. Transferred into the form of an explicit system of differential equations his model (3.2),(3.15), (3.18) then reads:

$$\hat{u} = \rho\theta x - (1-\eta)\lambda A^* u + (1-\eta)\lambda - (\gamma+\alpha) = a_2 v + a_1 u + a_o, \quad \theta x = v \tag{15}$$

$$\hat{\theta} = \sigma\kappa - (\alpha+\beta) = \sigma x[(1-t)(1-u) + \mu(x\theta - v^*)] - g^* \tag{16}$$

$$x = \frac{\varepsilon(g^* - 1 + x - \sigma\kappa)/\sigma + x((1-t)\dot{u} - \mu x^2 \dot{\theta})}{\kappa + \mu \cdot \theta \cdot x^2}$$

$$= \frac{\varepsilon(g^* - 1 + x - \sigma\kappa)/\sigma + x(1-t)(a_o + a_1 u + a_2 v)u - \mu x^2 \theta(\sigma\kappa - g^*)}{\kappa + \mu \cdot \theta \cdot x^2} \tag{17}$$

The last equation of this system follows from

$$\dot{\kappa} = \begin{cases} \dot{x}\kappa/x - x(1-t)\dot{u} + x\mu\theta\dot{x} + x^2 \cdot \mu\dot{\theta} \\ \varepsilon(g^* - 1 + x - \sigma\kappa)/\sigma \end{cases}$$

which gives

$$\dot{x}(\kappa/x + x\mu\theta) = \varepsilon(g^* - 1 + x - \sigma\kappa)/\sigma + x(1-t)\dot{u} - x^2\mu\dot{\theta}$$

or

$$\hat{x}(\kappa + \mu x^2\theta) = \varepsilon(g^* - 1 + x - \sigma\kappa)/\sigma + x(1-t)u[a_o + a_1 u + a_2 v] - \mu x^2\theta[\sigma\kappa - g^*].$$

The equilibrium values for this system are (for $A^* = 1/u^*$):

$$u^* = 1 - g^*/(\sigma(1-t)), \theta^* = v^* = \frac{\alpha+\gamma}{\rho} \text{ and } x^* = 1 .$$

To investigate the behaviour of system (15)-(17) near the steady-state and for large parameters μ the following simplification (approximation) of eq.(17) is very helpful: Dividing both the numerator and the denominator of this eq. by μ gives for μ sufficiently large the approximate equation:

$$\hat{x} = \frac{-\varepsilon x(x\theta - v^*) - x^2\theta(\sigma\kappa - g^*)}{\theta x^2 + x(x\theta - v^*)} =: \frac{Z}{N} \tag{18}$$

This equation is of help as a first approximation to the Jacobean of the full system. It gives in connection with (15), (16) :

$$\tilde{J} = \begin{pmatrix} -u^*(1-\eta)\lambda A^* & u^*\rho & u^*\rho\theta^* \\ -\theta^*(1-t)\sigma & \theta^*\sigma x^*\mu & \theta^*(g^* + \sigma \cdot \mu\theta^*) \\ (1-t)\sigma & -\tilde{J}_{22}/\theta^* - \varepsilon/\theta^{*2} & -\tilde{J}_{23}/\theta^* - \varepsilon/\theta^* \end{pmatrix}$$

Note here that the numerator N of (18) is zero at the equilibrium so that the partial derivatives of (18) are equal to Z_i/N with $N = \theta^* x^{*2}$, $i = u, \theta, x$. The characteristic polynomial of $\tilde{J}^1$ is thus given by $x^3 - (\text{trace } J)x^2 + qx - \det J$, where q is given by

$$q = \begin{vmatrix} \tilde{J}_{11} & \tilde{J}_{12} \\ \tilde{J}_{21} & \tilde{J}_{22} \end{vmatrix} + \begin{vmatrix} \tilde{J}_{11} & \tilde{J}_{13} \\ \tilde{J}_{31} & \tilde{J}_{33} \end{vmatrix} + \begin{vmatrix} \tilde{J}_{22} & \tilde{J}_{23} \\ \tilde{J}_{32} & \tilde{J}_{33} \end{vmatrix}$$

We get for the coefficients of this polynomial:

$$\text{trace } J = -u^*(1 - \eta)\lambda A^* - g - \varepsilon/\theta^* < 0 \quad (\text{if } \eta < 1), \text{ and}$$

$$\det \tilde{J} = \begin{vmatrix} -u^*(1-\eta)\lambda A^* & u^*\rho & u^*\rho\theta^* \\ -\theta^*(1-t)\sigma & \theta^*\sigma\mu & \theta^*(g^* + \sigma\mu\theta^*) \\ 0 & \varepsilon/\theta^{*2} & \varepsilon/\theta^* \end{vmatrix} = \begin{vmatrix} -u^*(1-\eta)\lambda A^* & u^*\rho & 0 \\ -\theta^*(1-t)\sigma & \theta^*\sigma\mu & \theta^* g^* \\ 0 & \varepsilon/\theta^{*2} & 0 \end{vmatrix}$$

$$= -u^*(1-\eta)\lambda A^*(\varepsilon/\theta^{*2})(\theta^* g^*) = -(1-\eta)\lambda A^* \cdot \varepsilon u^* g^*/\theta^* < 0 \text{ , and}$$

$$q = -u^*(1-\eta)\lambda A^*\theta^*\sigma\mu + u^*\rho\theta^*(1-t)\sigma + u^*(1-\eta)\lambda A^*(J_{23}/\theta^* + \varepsilon/\theta^*) + u^*\rho\theta^* J_{21}/\theta^* + \varepsilon g^*/\theta^*.$$

For $\eta < 1$ we consequently get the Routh-Hurwitz conditions:

$$\text{trace } J < 0, \ \det J < 0 \quad [q > 0].$$

And with regard to the final Routh-Hurwitz condition there holds

$$(-\text{trace } J)q + \det J > 0$$

since q still depends on μ and det J does not. For large values of μ this therefore confirms that the classical policy as well as the Keynesian rule make the model of section 3 asymptotically stable, which again establishes the assertion made in Wolfstetter (1982).

Remarks: 1) The assumption $\eta < 1$ is essential for the obtained result. For the original Goodwin model ($\eta=1$) computer simulations show that the classical rule introduces asymptotic stability into the model while the Keynesian rule does not (for any $\mu > 0$).

2.) If the parameter η is > 1 (so that the private sector (1), (2) is already instable in the context of Goodwin's model, see figure 1 , one can show for η sufficiently near to 1: trace J < 0, det J > 0 (and q > 0). Both policies consequently fail in this case, however large they are exercised.

Returning to Wolfstetter's analysis ($\eta < 1$) we want to argue now that his results are problematic in two respects:

a) As shown in section 2 the parameter μ should by economic reasons be less than .3 . This may prevent an applicability of Wolfstetter's result.

b) Left to itself ($\mu = 0$) the private sector may, of course, be not asymptotically stable. However, its type of instability should not be of a completely erratic nature (at least in the short-run).

Both aspects were out of reach for the type of qualitative analysis given in Wolfstetter (1982). However, the following figures (see the last page) based on computer simulations of the above model for $\alpha = .03$, $\beta = .02$, $\gamma = .9$, $t = .2$, $\varepsilon = .2$, $\sigma = 1/3$, $\rho = 1$, $\eta = .8$, $\lambda = 1$ clearly show that the above policy rules are far from being equivalent over economically relevant ranges of policy parameters μ and that the private sector (when left to itself, $\mu = 0$) will collapse within the time-span of one year. We consequently have to conclude that the above symmetry conclusion is doubtful and that the Goodwin model with independent investment behaviour needs some reformulation in order to approach this question on a firmer basis.

4. On the non-equivalence of classical and Keynesian stabilization policies

Inspecting eq.s (12)-(14) makes it obvious that there is no simple way of stabilizing this model with independent investment behaviour in the way we have solved this task for the model (1), (2) in section 1. Following a proposal of E. Wolfstetter we therefore start to improve the model of section 3 (with regard to its analyzed weaknesses) by making use of a simplifying limit case of this model ($\varepsilon = \infty$) . This reduces the model again to only two dynamical laws, whereby additional stabilizers of the private sector's reaction pattern can be designed much easier than for (12) - (14). Due to space limitations we shall, however, only provide a brief survey of the modifications to be made and the results obtained. As already stated, our first modification of model (12) - (14) concerns the investment function (11). It will now read:

$$I/K = g^* + c(x/x^* - 1) \qquad (19)$$

where $x^* = 1$ again represents normal capacity use. The new parameter c takes account of the fact that changes in capacity utilization are not necessarily transferred 1:1 into percentage investment plans. Furthermore, we have now assumed '$\varepsilon = \infty$', i.e., there is no adjustment lag in investment behaviour.

Using (19) the new savings-investment equilibrium reads (cf. (15) - (17); $s_p = 1$ again):

$$\frac{I}{K} = g^* + c(x-1) = \sigma x[(1-t)(1-u) + \mu(v-v^*)] = \frac{S}{K} \qquad (20)$$

i.e. we now have the following static relationship between x and u, v:

$$x = \frac{c - g^*}{c - \sigma[(1-t)(1-u) + \mu(v-v^*)]} \qquad (21)$$

This equation makes it possible to attempt a global analysis of the type introduced in section 1 for the case of Say's law. As in that situation, something concerning the full-employment ceiling must be said, however. Note here, that we did not use the function i(v) of section 1 to modify S/K near full employment as in section 1. This is no longer adequate in a model which differentiates between the act of saving and that of investment. It is, however, also not adequate to apply i(v) in the spirit of section 1 to simply limit investment behaviour I/K thereby. Goods-market equilibrium concerns value magnitudes (of I and S) whereas i(v) has been introduced to describe the development of the capital stock K as an index for employment near full employment. Such an index is of the nature of a physical magnitude, i.e., its growth is not necessarily properly described by I/K even in our one-good model with disembodied technical progress. When approaching full employment the same amount of investment I may (even in such a model) be assumed to result in less and less growth of K as an index of employment, speeding up instead the growth rate of labour productivity y = Y/L . In addition to (20) we consequently should simply assume:

$$\hat{K} = \dot{K}/K = i(v)[g^* + c(x-1)] \neq I/K \qquad (22)$$

in combination with $\alpha = \alpha(v)$, $\alpha'(v) \geq 0$ to describe the technological side of the process of capital accumulation (the function i is of the same type as in section 1). Summing up, the revised model consists of the following four equations

$$\hat{u} = f(v) - \alpha(v) - (1-\eta(u))\lambda(u)[Au-1] \qquad (23)$$

$$\hat{v} = i(v)[g^* + c(x-1)] - g^* + \hat{x} \qquad (24)$$

$$g^*(v) = \alpha(v) + \beta \qquad (25)$$

$$x = \frac{c - g^*}{c - \sigma[(1-t)(1-u) + \mu(v-v^*)]} \qquad (26)$$

Inspecting (24) in comparison to fig. 2 shows that the invariant domain $(o,1)^2$ of this figure need no longer have this property, due to the occurence of $\hat{x}$ in eq. (24) [which makes the i(·)-

reaction insufficient for this purpose]. Problems for the condition $\dot{v} < 0$ at $v = 1$ arise to the right of the $\dot{u} = 0$-isocline, cf. again fig. 2 . Though we expect that this will not endanger the stability of the dynamics (23), (24) within the range of economically meaningful values of u, v, we nevertheless should add as side condition [8)]

$$v \leq 1 \tag{27}$$

to the above model to exclude any problems of the above mentioned type. We again make the assumptions of section 1 and assume in addition: $A = 1+a = 1/u^*$, $(f-\alpha)'(v) > 0$, $c > g^*$ sufficiently [9)] large and $\alpha(v) \equiv \alpha$ near v^*. The steady-state hence is the same as before $v^* = f^{-1}(\alpha) = \frac{\alpha+\gamma}{\rho}$ [if $f(v) = \rho v - \gamma$] , $u^* = 1 - g/[\sigma(1-t)]$. Logarithmic differentiation of (26) gives

$$\hat{x} = \frac{g^{*\prime}(v)}{c-g^*(v)}\dot{v} + \frac{\sigma[\mu\dot{v} - (1-t)\dot{u}]}{c - \sigma[(1-t)(1-u) + \mu(v-v^*)]} \tag{28}$$

Inserting (26),(28) into (24) leads to:

$$\hat{v} = \left[1 - \frac{g^{*\prime}(v)v}{c-g^*(v)} + \frac{\sigma\cdot\mu\cdot v}{c - \sigma[(1-t)(1-u)+\mu(v-v^*)]}\right]^{-1}$$

$$\cdot\left\{ i(v)\left(g^*(v) + c\left(\frac{c-g^*(v)}{c-\sigma[(1-t)(1-u)+\mu(v-v^*)]} - 1\right)\right)\right.$$

$$\left. -g^*(v) - \frac{\sigma(1-t)u}{c-\sigma[(1-t)(1-u)+\mu(v-v^*)]}\hat{u}\right\}$$

The assumption on $\alpha(v)$ near v^* and the fact that $i(v^*) = 1$ then imply for the basic case $\mu = 0$:

$$\hat{v} = c\left(\frac{c-g^*(v)}{c-\sigma(1-t)(1-u)} - 1\right) - \frac{\sigma(1-t)}{c-\sigma(1-t)(1-u)}\cdot\hat{u}$$

$$= h(u,v) - k(u)\cdot\hat{u} \quad \text{with} \quad h_u < 0,\ h_v \leq 0 \text{ and } k'(u) < 0.$$

The above calculation now allows the investigation of the stability of the steady state u^*, v^*

$$J_{11} = -(1-\eta(u^*))\hat{p}'(u^*)u^* > 0$$
$$J_{12} = (f-\alpha)'(v^*)u^* > 0$$
$$J_{21} = [h_u(u^*,v^*) - k(u^*)\cdot J_{11}]v^*$$
$$J_{22} = [h_v(u^*,v^*) - k(u^*)J_{12}]v^*$$

This gives for the steady state u^*, v^*:

$$\text{trace } J = \underbrace{-\ (1-\eta(u^*))}_{+}\underbrace{\hat{p}'(u^*)}_{+}u^* + \underbrace{[-k(u^*)J_{12}]v^*}_{-} \gtreqless 0$$

and

$$\det J = J_{11}J_{22} - J_{12}J_{21}$$
$$= -J_{11}k(\cdot)J_{12}u^*v^* - (h_uJ_{12} - k(\cdot)J_{12}J_{11})u^*v^*$$
$$= -\underset{+}{J_{12}}\underset{-}{h_u}\,u^*v^* > 0.$$

Whether the equilibrium u^*,v^* is stable or not consequently solely depends on the strength of the destabilizing term

$-(1-\eta(u^*))\hat{p}'(u^*)u^*$ in the trace of J. In striking contrast to the results obtained in section 3 this modified model is thus locally instable if and only if the conflict about income distribution is sufficiently intense [$\eta(u^*) > 1$ and $\lambda(u^*)$ sufficiently large, for example]. And when this situation prevails, we here again get the limit cycle result of section 1 and have consequently removed from the private sector's behaviour the implausible type of instability investigated in section 3.

Due to (27) an immediate application of the Poincaré-Bendixon theorem to prove this last assertion is, however, not possible. Yet, computer simulations show: [10]:

a.) the private sector now exhibits again the Marxian type of reproducibility (see section 1) despite the presence of an independent investment function.

b.) the results of section 2 on the non-equivalence of classical and Keynesian fiscal policy carry over to this new situation where problems of effective demand, of capacity utilization, etc. are now present.

This indicates that the analysis of section 3 is not conclusive and that its symmetry assertion may rest on shakier grounds than the opposite claim.

5. Footnotes:

*) The author thanks E.Wolfstetter for helpful suggestions and comments.

1) This assumption can be justified through a simple reinterpretation of the standard one-good model by assuming that it includes luxury goods as well. In a sufficiently small neighbourhood of full employment capitalists then simply prefer to use their inputs as (nontransferable) luxury goods instead of investing them (irreversibly) into capital formation.

2) $A = 1 + a \geq 1/u^*$.

3) We here assume $\eta'(u^*)=0$ and can thus neglect the term $\eta'(u^*)\hat{p}(u^*)u^*$ in J_{11}

4) These assumptions (which imply local instability) do not yet look very convincing. Further destabilizers can, however, easily be introduced,e.g., by means of production lags, wage-drift lags etc., yet are difficult to treat from a purely qualitative point of view.

5) All following simulations were performed with "TUTSIM" (of Meerman Automation, Neede, Holland and Applied i , Palo Alto, USA).

6) The hypothesis underlying the reaction function i(v) is that capitalists stop physical investment when the full employment ceiling is reached, but they, of course, find no problem in further investing in government bonds.

7) Note here, that large values of 'μ' may give rise to two (and) more intersections between $\dot{u}=0$ and $\dot{v}=0$ and also to a further equilibrium on the v-axis (if $\mu > 0$ becomes so large that the intersection of the $\dot{v}=0$-isocline with the u-axis is removed).

8.) Such a side condition is investigated in detail in the unpublished Ph.D.Thesis of C. Groth (in Danish), University of Copenhagen.

9.) To make this more precise demands an investigation of the positivity of $c - \sigma[(1-t)(1-u) + \mu(v-v^*)]$ for all economically meaningful values of u,v. Note here, that this positivity implies an unstable reaction of capacity use x and a stable reaction of prices p to demand/supply imbalances $I \neq S$ (see (20)). Eq.(20) therefore has to be interpreted with some care. I thank Peter Skott for making me aware of this point.

10.) An extended version of this section (which includes such simulations) can be obtained on request from the author.

References

Flaschel, P.(1984): Some stability properties of Goodwin's growth cycle model. A critical elaboration. *Zeitschrift für Nationalökonomie 44*, 63 - 69 .

Flaschel, P. and M. Krüger (1984): Endogenous aspirations in a model of cyclical grwoth. *Ricerche Economiche 38*, 598 - 612 .

Hirsch, M. and S. Smale (1974): *Differential Equations, Dynamical Systems and Linear Algebra*. London: Academic Press .

Marx, K. (1954): *Capital, Volume 1*. London: Lawrence and Wishart.

Phillips, A.W. (1961): A simple model of employment, money and prices in a growing economy. *Economica 73* , 360-370.

Wolfstetter, E. (1977): *Wert, Profitrate und Beschäftigung*. Franfurt a.M.: Campus Verlag.

Wolfstetter, E. (1982): Fiscal policy and the classical growth cycle. *Zeitschrift für Nationalökonomie 42*, 375 - 393.

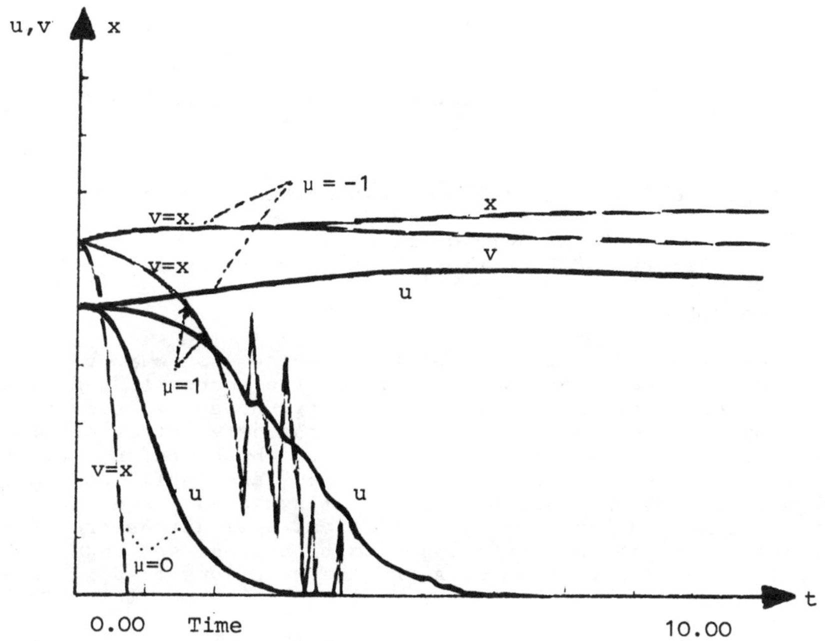

Fig. 5: A comparison of classical and Keynesian fiscal policies for economically meaningful (small) values of μ .

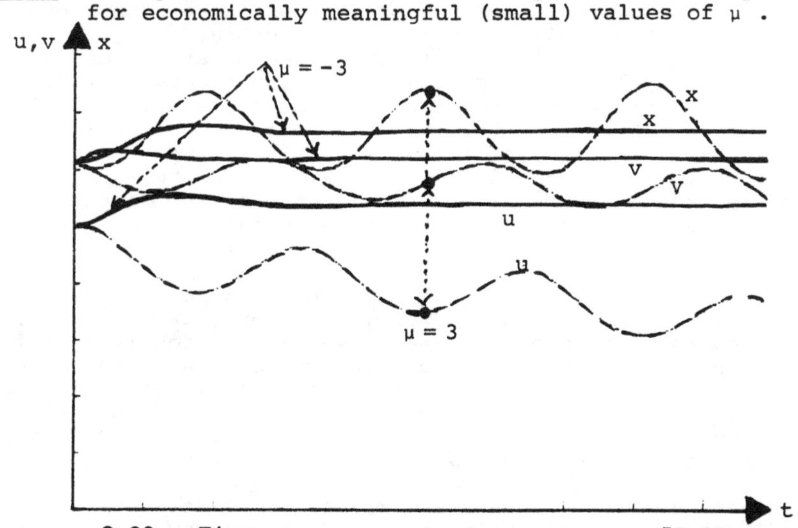

Fig. 6: A comparison of both policies for 'large' values of μ (note that the above value μ=3 is still insufficient to make the model stable).

"The question is whether Left or Right makes a difference."

Kurt W. Rothschild (1986, p.361)

INTERACTION OF AN ECONOMIC AND A POLITICAL CYCLE

Jörg Glombowski
Faculty of Economics, Tilburg University

1. Introduction[1]

Whatever the definition of political economics may be, it will contain the interaction between the economic and the political sphere. This formal similarity, however, is superseded by differences arising from the characteristics of the alternative approaches to the field. These refer, among others, to the ways in which "economics" and "politics" are delimited, which actors are distinguished and which assumptions about economic and political motivations are put forward. Also the historical and institutional settings may be described in different ways and the relations of dominance between the "economic" and the "political" be put in different perspectives.

In standard (functionalist) Marxist accounts the "state" is imputed the central tasks of preserving the economic order and supplying the indispensable infrastructure. Its policies would be related to the promotion of accumulation and the provision of political legitimation to the economic system. These functions are derived from system "necessities" while little attention is paid to party competition and individual motivations. (Well-known examples for this line of thought are Kalecki 1943 and Boddy/Crotty 1975.) As the state is primarily thought of as a stabilizer of the capitalist system, its role is not very much appreciated - to put it mildly.

The paradigm of quantitative economic policy rests on functionalist foundations as well. However, the state is understood as a promotor of general welfare, which is a noble task. The state's objective function is taken as given, i.e. it is neither derived from an explicit analysis of politics nor is it considered to change with economic and political circumstances. We can refer to Tinbergen 1956 or Fox/Sengupta/Thorbecke 1966 as examples for this line of thought.

The neoclassical "economic theory of politics" pays more attention to political motivations of voters, governments and bureaucrats. (Cf. for instance Bernholz 1975, Borooah/Van der Ploeg 1983, Van den Doel 1978, Frey 1977, Hibbs/Fassbender 1981, Van Winden 1981.) Thus, changes in the economic situation may affect political choices. Making the lonely "homo economicus" enjoy the company of a twin brother, the "homo politicus", seems to be a useful extension of neoclassical reasoning. Nevertheless, the results of this generalization seems to suffer from the same types of shortcomings as neoclassical economics proper.

In the remainder of the paper we shall dicuss a model of the interaction between economics and politics, which draws from various paradigmatic sources. There does not seem to be a clear-cut way to label it either Marxist or neoclassical (or whatsoever). Yet, as will soon become obvious, there exist some links with these bodies of thought.

The model focusses on economic and political determinants of employment. The economic part is a version of Goodwin's profit squeeze cycle (Goodwin 1972), which bears some relation to Marxist business cycle theory. The political element enters the economic stage as private sector wage rate changes will not only be related to private employment, but also to state employment, which is created in order to raise the overall degree of employment.

The political system is a democratic one and made up of two political parties, a government and the voters. While the parties are taken to represent objectives of collectivities, voters are regarded - at least in part - as individual utility maximizers. This mixture of assumptions is shown to give rise to political election cycles: As unemployment becomes "too high", voters turn to the left; if, on the contrary, the costs of fighting unemployment rise and the threat of unemployment fades, voters shift to the right. These political vote shifts eventually bring about switches in majorities and, consequently, policies.

The outcome of the interaction between the capitalist economy and its politics will be discussed in the final section of the paper. The intermediate paragraphs deal with the economic and political building blocks of the model in some detail.

2. The Economic Subsystem

Our model will be formulated in discrete time (t=0,1,2,..). We assume that labour productivity (y) increases with a constant rate of growth (m) per period, i.e.

$$y(t+1) = y(t)(1+m) \tag{2.1}$$

In addition, technical progress is taken to be Harrod-neutral, which means that the output-capital-ratio (σ) remains constant over time:

$$Y(t) = \sigma K(t) \tag{2.2}$$

In (2.2) Y denotes (net) output of the capitalist sector and K its stock of (fixed) capital. While the available capital stock is assumed to be fully used throughout, labour demand may fall short of supply. The latter is denoted by A and is assumed to increase with the constant growth rate n,

$$A(t+1) = A(t)(1+n) \tag{2.3}$$

Labour demand in the private (=capitalist) sector (L_p) is obtained by dividing net output by labour productivity (thereby neglecting labour hoarding):

$$L_p(t) = Y(t)/y(t) \tag{2.4}$$

We denote the ratio of private employment to labour supply by β_p,

$$\beta_p(t) = L_p(t)/A(t) \tag{2.5}$$

and refer to this variable as the "private degree of employment".

A constant share (α) of profits will be accumulated every period. Making use of the profit rate, r(t), we can express this classical assumption by

$$K(t+1)/K(t) = 1+\alpha r(t) \tag{2.6}$$

The profit rate itself can be written as

$$r(t) = \{Y(t)-w(t)L_p(t)\}/K(t), \tag{2.7}$$

where w(t) is the wage rate per year per worker in the capitalist sector and the term in brackets denotes profits. Profits are therefore considered as residual income, i.e. the surplus of net income above wages.

The final asumption of the economic sub-model is its most important one. We assume that the growth rate of wages depends positively on the private degree of employment as well as on the "degree of state employment" (β_s). This entails a modification of Goodwin's assumption as he abstracted from state employment. For the sake of simplicity we use a linear specification of the general assumption:

$$\{w(t+1)-w(t)\}/w(t) = -a_1+a_2\beta_p(t+1)+a_3\beta_s(t+1) \tag{2.8}$$

where $a_1 > 0$, $a_2 > a_3 \geq 0$.

The degree of state employment is defined analogously to that of capitalist employment, i.e. as the number of state employees divided by total labour supply. The choice of the time structure is motivated by the desire to keep our results comparable to those of Goodwin's model, which was formulated in continuous time: If $a_3 = 0$, we would fall back to a discrete-time version of Goodwin's continuous model; but then it would be the dependence on $\beta_p(t+1)$ rather than on $\beta_p(t)$ which would produce essentially the same results as Goodwin's model (i.e. oscillations of constant amplitudes in the degree of private employment and the wage share).

The economic part of the model would be self-sufficient and could be treated separately, were it not for the fact that it is linked to the political system by taking the influence of state employment on private wage formation into account. This provides, in our present simple model, the only link by which the political system affects the economic one and precludes its isolated solution.

3. The Political Subsystem

We assume a capitalist democracy with only two parties. The left one will adopt a full employment policy whenever it comes to rule. More specifically, whenever the private degree of employment falls short of 1, the state degree of employment will be increased by b_l (>0) per period until full employment ($\beta_p+\beta_l=1$) is established. The instrument of left-wing employment policy is conceived in an extremely simple way. Government agencies initiate employment projects in the state sector, the costs of which are covered by (additional) taxes on wages in the private sector. Costs of state employment other than wage costs are ignored, as is any contribution to marketable output the state employed might provide. Whenever full employment is achieved by employment programmes and the private sector tries to expand, the left government will set state sector workers free for private industry. Taking this attitude into account the left state employment policy can be described by

left employment policy: (3.1)

$$\beta_s(t) = \min[\beta_s(t-1)+b_l, 1-\beta_p(t)]$$

The right-wing party does not approve of full employment policies by creating state employment. Whenever it comes to power it will reduce the degree of state employment by b_r (>0) per period until it becomes zero:

right employment policy: (3.2)

$$\beta_s(t) = \max[\beta_s(t-1)-b_r, 0]$$

Of course, "zero" could be substituted by some positive lower limit. In the same way strict full employment could be replaced by a "high" degree of employment as a left government's objective. The right government would claim that the decrease of public employment contributes to the growth of private employment by reducing the tax burden. In order to analyse this claim properly, one would have to introduce profit taxes and let investment activity depend on net profits. This matter is not pursued in the present paper but will be taken into account in a more elaborate future version of the model.

Which kind of regime operates, is decided by elections which are assumed to be held every θ years. For simulation we need a rule to determine election years. The following function will do:

"election year selector": (3.3)

$$t/\theta - \mathrm{int}(t/\theta) \begin{cases} = 0 \Rightarrow \text{t is an election year} \\ = 0 \Rightarrow \text{t is no election year} \end{cases}$$

"int" means the operation of rounding off a number to the next smaller integer, e.g. int(5,95) = int(5,02) = 5. Elections are assumed to be held at the <u>end</u> of election years with the new government starting its term at the beginning of the next year. Under a strictly proportional system of parliamentary representation, simple majority will decide which party is going to form the government. Let $L(t)$ be the potential or actual left vote share at the end of period t. Then we can formulate the following rule:

"regime selector": (3.4)

$$L(t) \begin{cases} > 0{,}5 \Rightarrow \text{left government till the end of } t+\theta \\ \leq 0{,}5 \Rightarrow \text{right government till the end of } t+\theta \end{cases}$$

Notice that we have - arbitrarily - assumed that a right regime will come to office in the case of an exact balance of votes and seats. Alternatively, one could specify additional rules for this particular case: the incumbent government stays in power, elections will have to be repeated, the decision is made by throwing a coin or by taking penalties.

Finally and most important, vote behaviour has to be modelled. In doing so, we abstract from all non-economic factors influencing voting decisions, while with respect to economic circumstances we only take two variables into account. The first one is the degree of unemployment, $g_1(t)$, the level of which is thought to measure the voters' fears of loosing their jobs or the difficulties to find other ones. The second variable refers to the tax burden associated with state employment programmes, which we denote by $g_2(t)$. High levels of unemployment and low levels of the tax burden work in favour of the left-wing party, while low unemployment rates and high tax burdens cause (net) shifts of votes to the right. Again, we choose a linear function to express these assumptions, "vote shift function": (3.5)

$$L(t)-L(t-1) = b_1 g_1(t) - b_2 g_2(t)$$

$$b_1, b_2 > 0$$

Note that according to (3.5) the voters' reactions are assumed to be lagged, as L(t) was defined to be the left-wing vote share at the end of period t. One should admit that there could be strictly ideologically oriented voters who would never shift from one party to the other, irrespective of the economic situation. Therefore, a more complete formulation would have to take upper and lower bounds to the voting shares into account. Moreover, our vote shift function is a hypothesis on net aggregate preference shifts. It does not exclude the existence of individual voters who swim against the stream.[a]

The overall degree of unemployment, $g_1(t)$, can be written as

$$g_1(t) = 1-ß_p(t)-ß_s(t) \quad (3.6)$$

The expression for the tax burden is a little bit more complicated and depends, among other things, on the specification of the social security system. We assume that the unemployed receive a certain percentage g_u of the private sector wage rate. Their income (w_u) is thus given by

$$w_u(t) = g_u w(t) \quad (3.7)$$

Likewise, the income of state employees (w_s) is linked to private sector wages by a fixed percentage g_s:

$$w_s(t) = g_s w(t) \quad (3.8)$$

The following restrictions are imposed

$$1 > g_s > g_u > 0 \quad (3.9)$$

which means that the state employees receive a lower income than private sector workers but a higher one than the unemployed.

There are, of course, a whole range of possibilities to finance unemployment benefits and government employment programmes. Here we stipulate that total expenditures are fully covered by taxes (or contributions) to be paid exclusively by private sector workers. (Of course, one could take various schemes of profit taxes into account, compare for instance Glombowski/Krüger 1984a. Here we abstract from capitalists' contributions for the sake of simplicity.) As we assumed g_s and g_u to be fixed, the tax burden has to be adjusted to financial need every period. Total expenditures are given by

$$\beta_s(t)A(t)w_s(t) + g_1(t)A(t)w_u(t),$$

where the first term denotes state employees' income and the second refers to unemployment benefits. Taking into account what has been said about finance, the sum of both terms has to be equal to total tax payments (in the absence of other taxes and other kinds of government expenditure). Dividing total taxes by the sum of private sector wages, i.e. by $w(t)\beta_p(t)A(t)$, we obtain the following expression for the (relative) tax burden:

$$g_2(t) = g_s\beta_s(t)/\beta_p(t)+g_u[1-\beta_p(t)-\beta_s(t)]/\beta_p(t) \qquad (3.10)$$

Knowing the tax ratio on wages, one can also determine net private sector wages. As unemployment benefits and the wages of state employees are linked to gross instead of net wages, the question comes up, whether those incomes could possibly catch up or even exceed net private sector wages. Under our present assumptions this eventuality cannot strictly be ruled out, although it will only occur in extreme cases.[3] In a more elaborated version of the present model, we will link transfer incomes to net private sector wages.

Inserting (3.6) and (3.10) into (3.5), leads to

$$L(t)-L(t-1) = b_1[1-\beta_p(t)-\beta_s(t)]- \qquad (3.11)$$

$$-b_2\{g_s\beta_s(t)/\beta_p+g_u[1-\beta_p(t)-\beta_s(t)]/\beta_p(t)\}$$

Thereby the vote shift function becomes an explicit function of both, private and state employment.[4]

With regard to state employment we can show unambiguously that, other things being equal, a rise of state employment will decrease the vote shift to the left and will eventually reverse it into a shift to the right. This happens because additional government employment programmes reduce job insecurity while at the same time they raise the tax burden.

Changes in private employment have more complicated consequences. Isolated increases in private employment reduce job insecurity as well as the tax burden, and, therefore, have contradictory effects on the vote shift. We can show that

$$\partial[L(t)-L(t-1)]/\partial\beta_p(t) \geq 0 \iff \qquad (3.12)$$

$$\beta_p^2(t) \geq [g_u+(g_s-g_u)\beta_s(t)]b_2/b_1$$

holds. A closer consideration of the likely numerical magnitudes of the parameters involved allows the conclusion that in economically relevant ranges of private employment the vote shift to the left will decrease and eventually be reversed. Only if state employment were zero, the expansion of private employment towards full employment would fail to render the vote shift negative. In this limiting case, i.e. $\beta_p(t)=1$ and $\beta_s(t)=0$, it would become zero for all parameter constellations.

A graphical representation of the vote shift function in the β_p,β_s-plane can be constructed as follows. Let c be any preassigned constant vote shift, then from (3.11) we find

$$\beta_s(t) = \frac{[1-\beta_p(t)][b_1\beta_p(t)-b_2 g_u]-c\beta_p(t)}{b_2(g_s-g_u) + b_1\beta_p(t)} \qquad (3.13)$$

to be the geometrical loçus of all β_s,β_p-combinations giving rise to c. In figure 1 the heavy drawn curve belongs to $c = L(t)-L(t-1) = 0$, while above (below) this "neutral" curve the left party looses (wins) votes, the more so the greater the distance of points to the neutral curve. (See the **appendix** for a more rigorous dicussion of (3.13).) As both components of employment are non-negative and their sum cannot exceed 1, the set of feasible points forms a triangle (like OAB in figure 1).

The vote shift function completes our model. Yet it still remains to be solved.

4. Solutions to the Model

It will be convenient for the further considerations to reduce the number of variables and equations. Actually, the economic part of the model can be "summarized" by two dynamic equations.

The first one will express the determinants of change in the degree of capitalist employment. Starting from (2.5), (2.4) and (2.2), we obtain

$$\beta_p(t) = \sigma K(t)/[y(t)A(t)] \qquad (4.1)$$

The growth factor of private employment can then be stated by making use of assumptions (2.1) and (2.3), i.e.

$$\frac{\beta_p(t+1)}{\beta_p(t)} = \frac{K(t+1)/K(t)}{(1+m)(1+n)} \qquad (4.2)$$

Next, we define the (private) wage share, $\mu(t)$, by

$$\mu(t) = w(t)L_p(t)/Y(t) = w(t)/y(t) \qquad (4.3)$$

and write the growth factor of capital as a function of it. From (2.6), (2.7) and (4.3) we obtain

FIGURE 1: GRAPHICAL REPRESENTATION OF THE VOTE-SHIFT FUNCTION

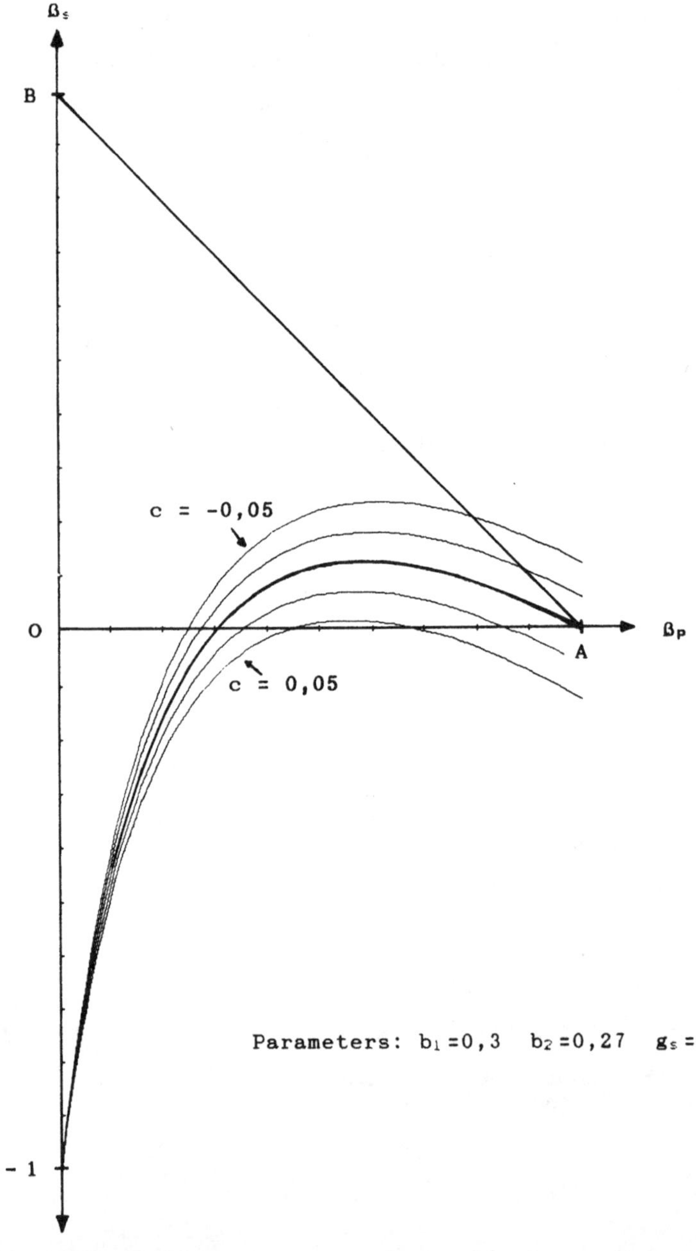

Parameters: $b_1=0,3$ $b_2=0,27$ $g_s=2/3$ $g_u=1/3$

$$K(t+1)/K(t) = 1+[1-\mu(t)]\alpha\sigma \qquad (4.4)$$

Therefore, the growth factor of private employment also becomes a function of the wage share, as can be seen be substituting (4.4) into (4.2), which yields

$$\frac{\beta_p(t+1)}{\beta_p(t)} = \frac{1+[1-\mu(t)]\alpha\sigma}{(1+m)(1+n)} \qquad (4.5)$$

The second basic equation shows the dynamic behaviour of the wage share to depend on our two employment variables. (4.3) implies

$$\frac{\mu(t+1)}{\mu(t)} = \frac{w(t+1)}{w(t)} \frac{y(t)}{y(t+1)} \qquad (4.6)$$

Using (2.1) and (2.8) we find

$$\frac{\mu(t+1)}{\mu(t)} = \frac{1-a_1+a_2\beta_p(t+1)+a_3\beta_s(t+1)}{1+m} \qquad (4.7)$$

Assume for the moment that a_3 were zero. Then our two "summarizing" equations (4.5) and (4.7) would make up a self-sufficient dynamic economic system in the wage share and the private degree of employment (cf. Glombowski/Krüger 1984b). This system would show a regular cyclical oscillation around the equilibrium values

$$\beta_p^* = (m+a_1)/a_2$$
$$\mu^* = 1-[(1+m)(1+n)-1]/(\alpha\sigma) \qquad (4.8)$$

in μ,β-space. However, we did assume that wage rate increases in the private sector depend to some degree, indicated by the size of a_3, on state employment, since increases of the latter back unions in their bargaining with employers. Therefore, the political subsystem cannot be ignored in the description of the economy.

The political system as summarized by the vote shift function (3.11), the state employment function (3.1) or (3.2), the election year and the regime selectors, would form a self-sufficient system giving rise to an electoral cycle, if private employment remained constant in a suitable range of unemployment.[5] Of course, this cannot be taken for granted here, because private employment varies endogenously.

Since the economic and the political subsystems are interlocked, precluding the solution of one subsystem before the other, we have to look for solutions to the system as a whole. As its parts contain cyclical tendencies, we may expect the whole (dynamically coupled) system to do so a fortiori.

Let us first check whether any long-run equilibrium does exist for the system as a whole. In the present context it seems appropriate to define an equilibrium as a process in time, in which both, the private and the state degree of employment, remain constant and in which no change in political power occurs. By implication, there may be two (types of) equilibria, supporting either a permanent right-wing or a permanent left-wing majority. Let us call the former one a "free enterprise" type of equilibrium and the latter one an "interventionist" type of equilibrium.

Obviously, a free enterprise equilibrium would imply state employment to be identically zero. For this case (4.8) would give the appropriate equilibrium value of private employment; not because a_3 were zero, but since $ß_s$ would vanish. We should expect that the equilibrium rate were less than full employment. On the other hand, an interventionist equilibrium is only conceivable as a full employment one, which implies that it has to satisfy

$$ß_{sl}^* = 1 - ß_{pl}^* , \qquad (4.9)$$

where the additional subscript "l" points to the left majority. While the equilibrium value of the wage share would not differ from that belonging to the free enterprise equilibrium, the potential (unique) interventionist equilibrium value of private employment would read

$$ß_{pl}^* = (m+a_1-a_3)/(a_2-a_3) \qquad (4.10)$$

as can be derived from (4.9) and by putting $\mu(t+1)$ equal to $\mu(t)$ in (4.7).

Provided the existence of a free enterprise equilibrium with less than full employment, the equilibrium degree of private employment would even be lower in a hypothetical interventionist equilibrium. Moreover, the net wage of private sector workers would be lower in the interventionist case because more (state employed rather than unemployed) persons would have to be paid a higher percentage of gross private sector wages.

We can show, however, that under our assumptions neither a free-enterprise nor an interventionist equilibrium is likely to persist for political reasons. Our hypothetical equilibria are only partial economic ones. It turns out that they fail to be politically supported in the long run.

For the interventionist case this is easily proven by substituting $ß_{sl}^*$ and $ß_{pl}^*$ into our vote shift funcion (3.11). While the term representing job insecurity vanishes, costs of maintaining full employment remain positive. Consequently, sticking to that partial economic equilibrium will cause a constant decline of the left vote share which must eventually lead to the fall of any left-wing government.

The situation is a little bit different with respect to the free enterprise equilibrium. A sufficient condition for a right-wing government to stay in power is given by

$$L^*(t)-L^*(t-1) = b_1(1-ß_p^*)-b_2 g_u(1-ß_p^*)/ß_p^* \leq 0 \qquad (4.11)$$

which boils down to

$$ß_p^* = (m+a_1)/a_2 \leq (b_1/b_2) g_u \qquad (4.12)$$

Clearly, this condition is not met unless unemployment benefits (measured by g_u) are fixed at a suffiently high level. As numerical calculations show, the parameters b_1 and b_2 should not differ much in size for the vote shift function to make sense. With b_1/b_2 around 1, however, we see that g_u has to take on unresonably high values in order to guarantee the existence of a free enterprise equilibrium. (Notice that we did not discuss necessary conditions.)

The political unsustainability of the partial economic equilibria can also be demonstrated graphically. In figure 2, the points R and L correspond to a free-enterprise and an interventionist equilibrium, respectively. Since L lies above the locus for c=0, the left-wing party will continue to loose votes. When the right-wing party finally takes over, it will reduce state employment, i.e. it will move the actual positon of the economy to the south of L. On the other hand, if R lies to the right of the intersection point D, which corresponds to the restriction (4.12), and to the left of A, which corresponds to private full employment, then it lies below the neutral vote shift locus. Hence, the right-wing vote share is gradually reduced. After having regained political power, the left-wing party steps up state employment, i.e. it moves the actual position from R in northern direction.[6]

We now turn our attention to the question whether our integrated politico-economic system will exhibit cyclical features or not. As we are not aware of general mathematical methods to decide this question for the type of model under consideration, we resort to numerical simulation. Only a single numerical example is presented, the results of which are summarized by figure 3. Along the horizontal axis time is measured, while the vertical axis measures various relevant variables as percentages. These are (from below): the degree of state employment, the left vote share, the degree of private employment, and the overall degree of employment.

FIGURE 2: THE UNSUSTAINABILITY OF LONG-RUN EQUILIBRIA

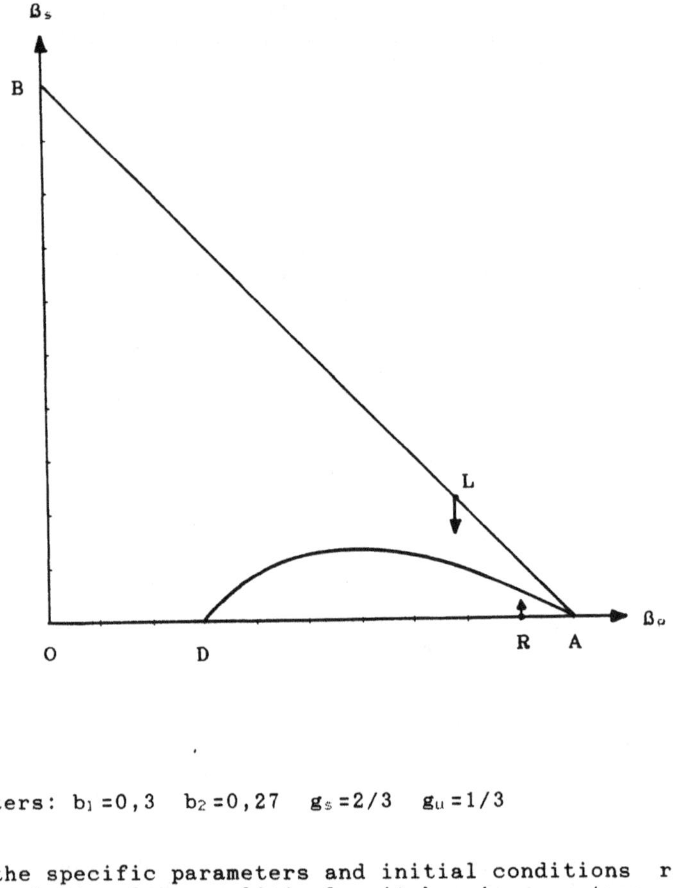

Parameters: $b_1=0{,}3$ $b_2=0{,}27$ $g_s=2/3$ $g_u=1/3$

Under the specific parameters and initial conditions reported, we obtain eleven political switches in a century, which trigger off increases and decreases of state employment, respectively. The overall behaviour of the variables in time appears to follow an irregular cyclical pattern without leaving admissible intervals of variables. There are substantial lapses of time in which state employment adds significantly to private employment. While full employment is not achieved throughout, employment policy is not ineffective either. In assessing the improvement we must, however, keep in mind that private employment in general is lower under left-wing regimes than it would be otherwise.

FIGURE 3: TIME PATHS OF RELEVANT VARIABLES FOR A NUMERICAL EXAMPLE

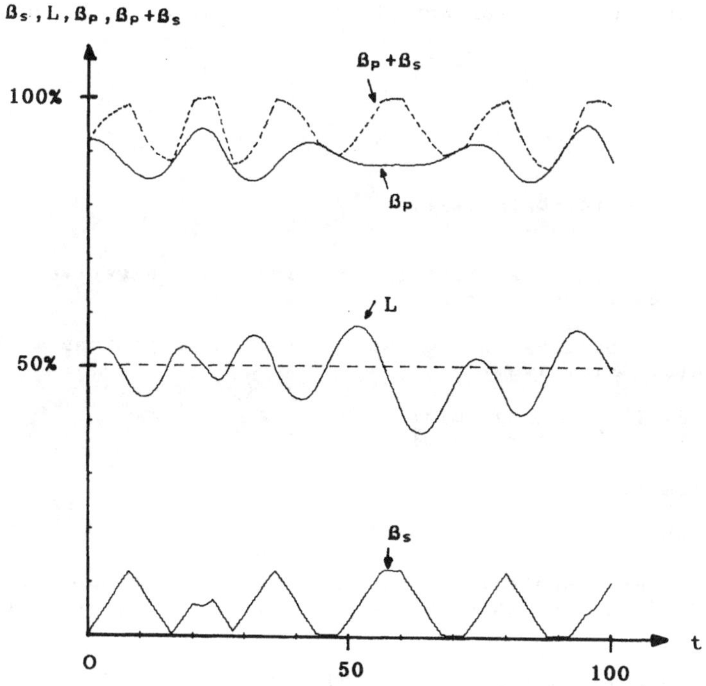

Initial conditions:
β_p=0,92 β_s=0 μ=0,7 L=0,52

Parameters:
a_1=0,86 a_2=1 a_3=0,2 b_1=0,3 b_2=0,27 b_l=0,015 b_r=0,015
α=0,336 σ=0,5 m=0,04 n=0,01 g_s=2/3 g_u=1/3

Our numerical example is not a unique case. Bounded cyclical motions with irregular periods and amplitudes occur for a large range of parameters. However, as one would expect, there are also cases in which variables take on irrelevant values, such as negative vote shares or private overemployment. These cases point to the necessity of introducing further non-linearities and restrictions on variables. It goes without saying that the basic model presented here offers plenty room for modifications and more realistic assumptions - at the cost of simplicity.

APPENDIX: PROPERTIES OF THE VOTE SHIFT FUNCTION (3.13)

Function (3.13) has the following general properties:
For $\beta_P \in [0,1]$ the numerator is strictly positive due to (3.9). The family of curves obtained by variations of the shift parameter c all intersect the vertical axis in the same point, i.e.

$$\lim_{\beta_P \to 0} \beta_S = -g_U/(g_S-g_U) < 0$$

Moreover, we have

$$\lim_{\beta_P \to 1} \beta_S = -c/[b_1+b_2(g_S-g_U)]$$

which shows that this limit is positive for negative c, negative for positive c, or zero if c is zero.

The curve corresponding to c=0 has got two intersections with the horizontal axis, i.e.

$$(c=0,\ \beta_P=1) \quad \Rightarrow \quad \beta_S=1$$
$$(c=0,\ \beta_P=b_2 g_U/b_1) \quad \Rightarrow \quad \beta_S=1$$

Under the restriction

$$b_2 g_U < b_1$$

this particular "neutral" curve must have a maximum with respect to β_S in the interval $\beta_P \in (b_2 g_U/b_1,\ 1)$ due to its continuity.

Curves belonging to a positive (negative) value of c lie below (above) the neutral one in the interval $\beta_P \in (0,\ 1]$; this means that the intervals defined by their intersections with the β_P-axis increase (shrink) with falling (rising) values of c. As long as

$$0 < c < b_2 g_U + b_1 g_S/(g_S-g_U)$$

holds, there will be intersection points. For suffiently low values of c (large in absolute amount), the maxima of the corresponding curves will appear only for private degree of employments above 1.

FOOTNOTES

1) Critical comments by Marco Wilke are gratefully acknowledged.
2) Our assumptions differ from those of other voting models in a number of respects. For instance, we do not take the inflation rate into account which has been frequently done in

models which employ the Phillips-curve as their economic background. Next, our parties are neither conceived of as vote maximizers nor as satisficers; rather they are committed to political goals. We employ a simple macro-formulation of the vote shift, while other models try to go back to individual decision making. Moreover, we consider the <u>levels</u> of the variables to explain <u>changes</u> in voting behaviour; in the literature, most often levels of economic variables are associated with levels of votes, while also links between changes in economic variables and levels of votes can be found. Finally, we do not find it worthwhile to test models like the present one empirically, as this wrongly suggests, according to our opinion, that the abstractions made in the construction of the model are of a negligible charcacter.

3) One can show that the income of the state employed (and, a fortiori, that of the unemployed) cannot exceed net private sector wages, as long as the private degree of employment exceeds the percentage of gross private sector wages obtained by the state employed, i.e. as long as $\beta_p > g_s$ holds.

4) We admit deliberately that our vote shift function is open not only to material but also to logical objections. As the relative tax burden defined by (3.10) addresses the position of private sector workers only, the vote shift function incorporating this variable should - strictly speaking - refer to the voting behaviour of this particular group only. In vote shift functions for the unemployed as well as for the state employed the relative tax burden would have to be neglected. The same is true for the capitalists, although the feedback from state employment programmes on wage dynamics could affect their political attitudes. For the sake of simplicity we neglect the differential impact of taxes on the various groups of the population assuming that the voting behaviour of wage earners in the private sector dominates the political outcomes.

5) For the sake of brevity, I refer to Glombowski 1987. It is shown there that political cycles are also obtained for a large variety of exogenously given time profiles of private employment.

6) There is one special case in which a right or a left majority could be maintained forever, i.e. if $\beta_p^* = 1$. In this case there would be neither a threat of loosing one's job nor would there be any costs associated with unemployment or state employment. Therefore, there would be no vote shifts at all. The party which happened to be in power when the equilibrium was reached would monopolize it forever. Note that with continuous private full employment our only political issue would disappear. Thus, this case would deprive our model of whatever significance it possibly might have. Therefore, it has to be ruled out.

REFERENCES

Bernholz, P. (1975): *Grundlagen der politischen Ökonomie* , *2nd vol*. Tübingen: UTB.

Boddy,R. and J.Crotty (1975): "Class Conflict and Macro-Policy: The Political Business Cycle", *Review of Radical Political Economics* , 7, 1-19.

Borooah, V.K. and F. van der Ploeg (1983): *Political Aspects of the Economy* . Cambridge: Cambridge University Press.

Van den Doel,J. (1978): *Democratie en welvaartstheorie. 2nd print*. Alphen aan de Rijn: Samson.

Fox, K.A. and J.K. Sengupta and E. Thorbecke (1966): *The Theory of Quantitative Economic Policy*. Amsterdam: North-Holland.

Frey, B.S. (1977): *Moderne politische Ökonomie. Die Beziehungen zwischen Wirtschaft und Politik*. München and Zürich: Piper.

Glombowski, J. (1987): "Arbeitslosigkeit trotz Beschäftigungspolitik", in: Rahmann, B. and O. Roloff (eds.): *Beschäftigungspolitik zwischen Steuerwiderstand und Ausgabenwachstum*. Regensburg: Transfer Verlag, 1-16.

Glombowski,J. and M. Krüger (1984a): "Unemployment Insurance and Cyclical Growth", in: Goodwin,R.M. and M. Krüger and A. Vercelli (eds.): *Nonlinear Models of Fluctuating Growth*. Berlin e.a.: Springer, 25-46.

Glombowski,J. and M. Krüger (1984b): "Profit-Squeeze und Fall der Profitrate als Elemente eines integrierten Überakkumulationsansatzes". *Prokla*, 57, 40-54.

Goodwin, R.M. (1972): "A Growth Cycle", in: Hunt, E.K. and J. G. Schwartz (eds.): *A Critique of Economic Theory*. Harmondsworth: Penguin, 442-449.

Hibbs, D.A.jr. and H. Fassbender (1981): *Contemporary Political Economy*. Amsterdam: North-Holland.

Kalecki, M. (1971): "Political Aspects of Full Employment", in: idem: *Selected Essays on the Dynamics of the Capitalist Economy*. Cambridge: Cambridge University Press, 138-145.

Rothschild, K.W. (1986): "'Left' and 'Right' in 'Federal Europe'", *Kyklos*, 39, 359-376.

Tinbergen, J. (1956): *Economic Policy. Principles and Design*. Amsterdam: North-Holland.

Van Winden, F. (1981): *On the Interaction between State and Private Sector. A Study in Political Economy*. Den Haag: Plasmans.

B. Problems of Micro- and Macrodynamics

OPTIMAL PERSISTENT OSCILLATIONS AND ADJUSTMENT COSTS

Gustav Feichtinger, Technical University of Vienna

1. Introduction

Recently there has been a surge of interest in endogenous cycles that arise in autonomous infinite-time optimal control models. It is known that in a continuous time framework an optimal state trajectory is monotone if there is only one state variable (see Hartl, 1987). Thus, we need at least two state variables to construct an example of cycles in an continuous time optimal control model; see Benhabib and Nishimura (1979), Medio (1987ab), Feichtinger and Sorger (1986). In a discrete time framework, however, it is easier to provide examples where cyclical paths are optimal (see Benhabib and Nishimura, 1988, for a good survey).

The purpose of the present paper is to construct a model which generates an endogenous cycle as optimal solution. By endogenous we mean that the oscillating behaviour is neither caused by exogenous fluctuations nor by suitable boundary conditions. More specifically, we consider a deterministic, autonomous, continuous time optimal control model with an infinite planning horizon.

The paper is organized as follows. In Section 2 the model is presented in a general setting. The following sections contain three different applications of the model; namely optimal employment, R & D, and pollution.

2. A Class of Non-concave Two-state Control Models

In order to define the model the following notation is needed:

$x(t)$ = first state variable measuring the *actual* level of a variable at time t

$y(t)$ = second state variable measuring the '*normal*' or '*expected*' level of the above variable at time t.

$y(t)$ is defined as an exponentially weighted average of past levels of $x(.)$:

$$y(t) = \gamma\int_{-\infty}^{t} \exp\{-\gamma(t-s)\}x(s)ds. \qquad (1)$$

Here the parameter $\gamma > 0$ represents the relative weight given to more recent actual levels. By differentiating (1) with respect to time t we get the differential equation

$$\dot{y}(t) = \gamma[x(t) - y(t)] \tag{1a}$$

describing the adjustment of the expected level to the current actual level. The parameter γ measures the speedness of the adjustment process.

A similar adjustment dynamics is considered, e.g., in Schmalensee (1973), Ryder and Heal (1973), Long and Siebert (1985), Steindl et al. (1986), and Luhmer et al. (1988).

The development of the actual stock x(t) is governed by the following differential equation

$$\dot{x}(t) = u(t) - \delta x(t), \tag{2}$$

where u(t) is the control variable, $u \in \mathbb{R}$, and δ is a constant decay rate.

To define the objective functional we use the following notation:

f(x) = return resulting from x

c(u) = adjustment costs, i.e. costs for adjusting the state variable x by the control u

h(x, y) = reward/penalty function depending both on the actual and the normal level of the state variable.

The following assumptions are made

$$f(0) = 0, \ f'(x) \gtreqless 0 \text{ for } x \lesseqgtr \bar{x} > 0, \ f''(x) < 0 \tag{3}$$

$$c(0) = 0, \ c'(u) \gtreqless 0 \text{ for } u \gtreqless 0, \ c''(u) > 0. \tag{4}$$

For simplicity we restrict ourselves to the following linear reward/penalty function:

$$h(x, y) = \begin{cases} \alpha(x-y) & \text{for } x \geq y \\ \beta(x-y) & \text{for } x \leq y, \end{cases} \tag{5}$$

where α, β are positive constants. To open the possibility for cycles we assume that

$$0 < \beta < \alpha. \tag{6}$$

This means that an excess of the actual level over the normal one implies a reward which is marginally higher than the penalty implied by a negative difference of the both levels.

Finally, let r be a positive discount rate assumed to be constant.

The intertemporal decision problem of the firm is to maximize the present value of the profit stream over an infinite time horizon:

$$\max \int_0^{\infty} e^{-rt}[f(x(t) - c(u(t)) + h(x(t),y(t))]dt \qquad (7)$$

subject to the state equations (1a) and (2), and given initial states

$$x(0) = x_0, \; y(0) = y_0. \qquad (8)$$

In the following the time arguments are omitted.

The optimal control problem has two state variables, x and y, and one control, u. Since the objective functional is non- differentiable with respect to the state x and y for x = y, the decision problem faced by the firm is *not* a standard optimal control problem. To solve it a generalized maximum principle has to be used (cf. Clarke, 1983, Hartl and Sethi, 1984).

Consider the current-value Hamiltonian

$$H = f(x) - c(u) + h(x,y) + \lambda\gamma(x-y) + \mu(u-\delta x),$$

where λ is the adjoint variable belonging to y and μ corresponding to the state variable x.

The necessary optimality conditions are

$$u = \arg_u \max H, \text{ i.e. } \mu = c'(u) \qquad (M)$$

$$(\dot\lambda-r\lambda, \dot\mu-r\mu)\in\partial_{(y,x)}H = (\lambda\gamma, -f'(x)-\lambda\gamma+\mu\delta) - \partial_{(y,x)}h(x,y), \qquad (A)$$

where

$$\partial_{(y,x)}h(x,y) = \begin{cases} \{(-\beta, +\beta)\} \\ \{(-\eta, \eta) \mid \beta \le \eta \le \alpha\} \\ \{(-\alpha, +\alpha)\} \end{cases} \text{ for } x \begin{Bmatrix} < \\ = \\ > \end{Bmatrix} y.$$

The four-dimensional state-costate space if divided by the hyperplane x = y along which the right-hand side of the adjoint equation (A) is discontinuous. $\partial_{(y,x)}h(x,y)$ denotes the super-gradient of h(x,y) with respect to (y,x) (see, e.g., Clarke, 1983). From the maximizing condition and the adjoint equation we obtain a four-dimensional canonical system of differential equations.

Since an analytical solution of the canonical system is impossible, a *numerical* analysis has to be carried out to solve the system. For this we have to specify the functions f and c as well as the occuring model parameters.

In order to find a cyclical solution of the canonical system the nonlinear boundary value problem solver COLSYS has been applied (see, e.g., Ascher et al., 1978, Steindl, 1981, and Steindl et al., 1986). It turns out that for certain regions of the parameter space a closed orbit exist. In figs. 1 and 2 the projections of this *persistent oscillations* into the (y,x) and (x,u) planes an depicted as solid lines. Moreover some trajectories that converge to the *limit cycle* are sketched as dashed curves. Apparently the convergence is very fast, so that after one period it almost coincides with the closed orbit (see figs. 1 and 2).

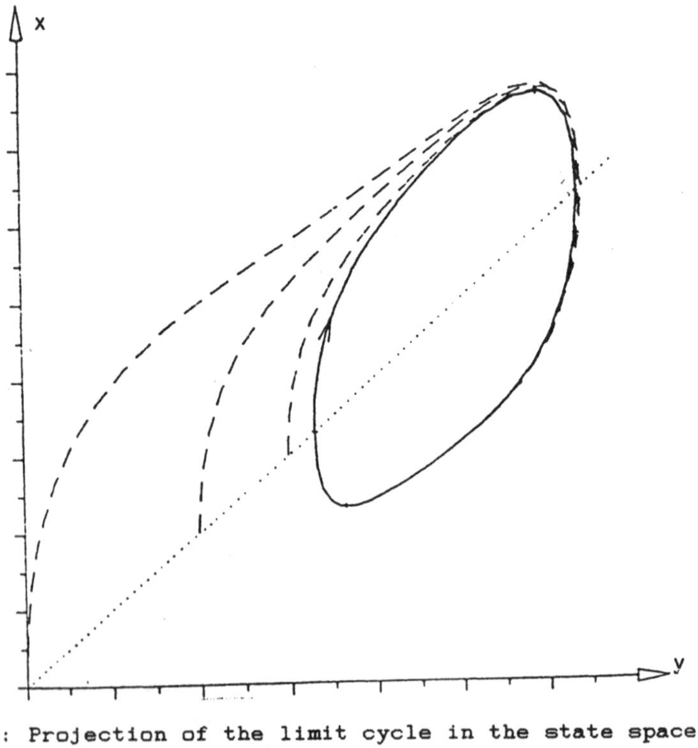

Figure 1: Projection of the limit cycle in the state space.

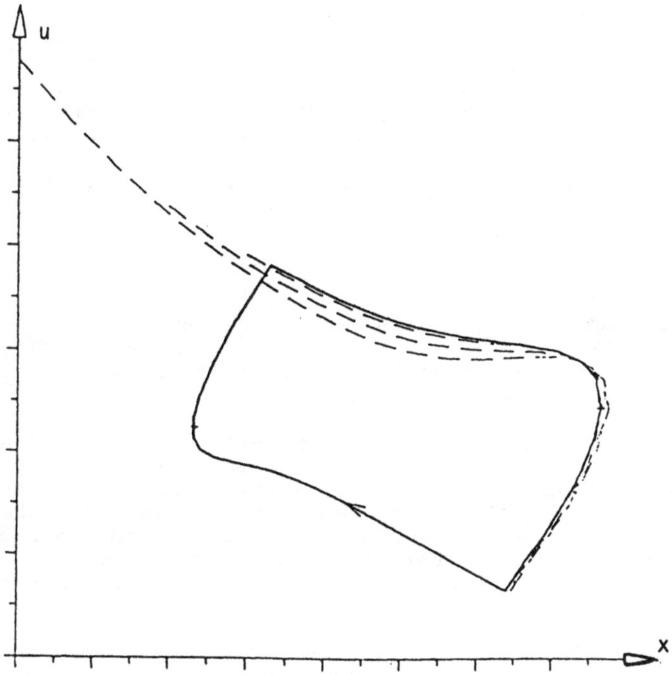

Figure 2: Projection of the limit cycle in the (x,u) phase plane.

To determine the solutions depicted in figs. 1 and 2 the following specifications has been made (source: Steindl et al., 1986):

$$f(x) = \sqrt{x} - x, \quad c(u) = u^2/2$$

$$r = 0.1, \quad \gamma = 0.5, \quad \delta = 0.1, \quad \alpha = 1, \quad \beta = 0.5.$$

Thus, $u = \mu$ because of (M), and the canonical system becomes

$$\dot{y} = 0.5(x-y) \qquad \dot{\lambda} = (r+\gamma)\lambda + \eta$$

$$\dot{x} = \mu - 0.1x \qquad \dot{\mu} = (r+\delta)\mu - 1/(2\sqrt{x}) + 1 - 0.5\lambda - \eta,$$

where

$$\eta = \left\{ \begin{matrix} \alpha \\ \beta \end{matrix} \right. \text{ for } x \left\{ \begin{matrix} > \\ < \end{matrix} \right\} y.$$

The important conclusion is that we have given a numerical proof of the existence and stability of a persistent oscillatory solution (limit cycle) in a family of two-state optimal control problems. Note, however, that the usual *sufficient* optimality conditions are not satisfied since the Hamiltonian is not concave in (x,y) (compare (5) and (6)).

In the rest of the paper several interpretations of this model are given.

3. Hiring and Firing (Steindl et al., 1986)

We consider a profit-maximizing firm which produces a homogenous good using only labour as production factor.

Interpreting the state variable, $x(t)$, as number of workers *actually* employed by the firm at time t, and the control variable, $u(t)$, as hiring rate (for $u > 0$) and firing rate (for $u < 0$), the state equation (2) describes the dynamics of the stock of employees. The constant δ is interpreted as voluntary quit rate. Defining the firm's *normal* employment level, $y(t)$, by (1), equation (1a) describes the adjustment process of y to x.

Furthermore, let us denote by $\varphi(x)$ the production function depending only on labour x (capital is assumed to be constant), and by w the constant wage rate. Then

$$f(x) = \varphi(x) - wx \qquad (9)$$

measures the return of the firm, where (3) corresponds to the usual assumptions made for production functions.

The costs for labour adjustment (training and integration costs, layoff costs etc.) are denoted by $c(u)$. According to (4) this cost function is assumed to be convex (see, e.g., Salop, 1973).

The purpose of the model is to study the impact of financial penalties or subsidies imposed by the government on the firm for its deviation from the 'normal' employment level. Following Long and Siebert (1985) the government imposes an asymmetric incentive scheme in sense that the marginal subsidy for an employment surplus exceeds the marginal penalty for a deficit of employees. More specifically, the incentive scheme is given by (5) and (6).

In Steindl et al. (1986) it is illustrated that under certain parameter constellations the government's good intentions result in periodic hiring and firing. The economic explanation for this cyclical behaviour is as follows (cf. also

Long and Siebert, 1985): At first, the firm increases its level of employment to obtain the subsidy. But as the subsidy vanishes if there is no change in unemployment level, the firm will reduce its employment level incurring a penalty in order to increase it again. The limit cycle represents a tradeoff between profits by production for which a stationary medium level would be optimal and the gains from the premium/tax scheme. In the case of low marginal subsidy and high marginal penalty a periodic solution turns out to be suboptimal. However, in the opposite case, i.e. if (6) is satisfied, an optimal limit cycle may occur (see figs. 1 and 2). The conclusion is that incentive schemes of the proposed form imply cyclical hiring and firing rates which is clearly not desirable for obvious reasons.

The next interpretation of the general model of section 2 refers to opitmal R & D strategies of a firm.

4. Research and Development (Feichtinger and Sorger, 1988)

L'Ecuyer et al. (1985) have studied the effect of an incremental tax incentive scheme on optimal R & D expenditurers of a profit-maximizing firm. Using discrete dynamic programming the authors are able to characterize the optimal policy for a one-period-memory and a piecewise linear incentive scheme.

The model presented in section 2 can be interpreted to extend the above problem to a continuous time many-period memory model (cf. Feichtinger and Sorger, 1988, and Haurie and Hollander, 1988) for a related approach.

Let us consider a profit maximizing firm. Denote by $x(t)$ its (actual) intensity of research and development at time t, and by $\varphi(x)$ the return resulting from x. Denoting by w the unit costs of R & D, the firm's return is given by (9).

To encourage the R & D process the authority imposes an incentive scheme (premiums and taxes). This scheme depends on the current as well as on the past levels of R & D. The past history of R & D is reduced to the weighted average (1). Thus, $y(t)$ measures the expected (normal) intensity of R & D, and (1a) again describes the adjustment dynamics. The positive parameter γ describes the memory of the authority. A high (low) value of γ corresponds to a short (good) memory.

The incentive scheme is assumed to be piecewise linear. More specifically, the premium/tax function is given by (5) and (6).

It seems reasonable to suppose that altering the level of R & D induces adjustment costs. Denoting by

$$\dot{x} = u \qquad (10)$$

the rate of change of R & D, and by c(u) the adjustment costs, the firm faces the following optimal control problem: Determine the time path u such that the discounted net profit stream (7) is maximized under the dynamic constraints (1a) and (10), the initial conditions (8), and the assumptions (3), (4), (5), (6) and (9).

Using the TPBV problem solver COLSYS it has been shown that a limit cycle exists under various parameter constellations (see Feichtinger and Sorger, 1988). Thus, evaluating the necessary optimality conditions it turns out that under certain circumstance the asymmetric (non-concave) incentive scheme( ) imply oscillating research and development expenditures.

Finally, we sketch a model which generates production-pollution cycles under a nonconcave incentive scheme. To fit this model into the general framework presented in section 2, the latter must be slightly changed.

5. Production and Pollution (Feichtinger, 1986)

This model combines in some sense two approaches of environmental economy. Feichtinger and Luptacik (1987) have studied a pollution control problem for a profit maximizing firm. They defined an 'actual' emission rate, e, which is influenced both by the production and the abatement activities. Moreover, they consider also a 'normal' (standard) emission level, $\bar{e}$, assumed to be constant, which is compared with e. If e exceeds $\bar{e}$, penalty costs are charged to the firm, while for e smaller than $\bar{e}$, the representative firm obtains a financial reward. In the present model the following extensions are made: first, the emission rate e is replaced by an (actual) stock of pollution x. Second, rather than to consider a constant given, 'normal' pollution level $\bar{x}$, the environmental standard *(normal pollution level)*, y, is defined by (1) as a weighted average of past pollution levels. This yields the known adjustment dynamics

$$\dot{y} = \gamma(x-y); \; y(0) = y_0. \qquad (11)$$

The *actual* level of pollution, x, is governed by the following differential equations (see also Forster, 1977):

$$\dot{x} = g(u) - \delta x, \quad x(0) = x_0. \tag{12}$$

Here δ is the natural cleaning rate, and u denotes the rate of production of the firm. It might appear strange to take into consideration a pollution level for an individual firm. However, x could be the current amount of slag in a foundry or the imissions of a firm into a lake which is used only by this firm.

It seems reasonable to restrict the production rate to a finite nonnegative interval, i.e.

$$u \in [0, \bar{u}]. \tag{13}$$

The functional g(u) determines the contribution of pollution generated by the production rate u. According to Forster (1977) we assume that g(u) is convex with

$$g'(u) > 0, \quad g''(u) > 0. \tag{14a}$$

Moreover, there exists a pollution-neutral level of production, $u_0 \in (0, \bar{u})$ such that

$$g(u) \lesseqgtr 0 \text{ for } u \lesseqgtr u_0. \tag{14b}$$

Here it is assumed that low level of production make it possible to carry out abatement activities by using partly idle produciton capacities.

The level of pollution which is accumulated by g(u) is assumed to be nonnegative:

$$x \geq 0. \tag{15}$$

The instantaneous utility rate of the firm is now given by

$$\pi(u) - d(x) - k(x, y), \tag{16}$$

where $\pi(u)$ denotes the profit rate of the firm producing with rate u, and d(x) is the disutility (damage) induced by the pollution level x. The pollution control authority fixes the penalty/premium function k(x,y) depending on the excess of actual and normal pollution level. The following assumptions are made:

$$\pi(u) > 0, \quad \pi'(u) > 0, \quad \pi''(u) < 0 \text{ for } u > 0 \tag{17}$$

$$d(x) > 0, \quad d'(x) > 0, \quad d''(x) > 0 \text{ for } x > 0. \tag{18}$$

As before we restrict ourselves to *linear* penalty functions where the marginal penalty is lower than the marginal reward, i.e.

$$k(x-y) = \begin{cases} \alpha(x-y) & \text{for } x > y \\ \beta(x-y) & \text{for } x < y \end{cases} \tag{19}$$

with

$$0 < \alpha < \beta. \tag{20}$$

Thus, if the actual pollution exceeds the normal level, the firm is penalized increasingly in accordance to the difference between both levels. On the other hand, the firm is rewarded proportionally to y-x provided this difference is positive.

Note that this approach can be realized by assuming that a representative firm has a certain ancount of emission 'rights'. The firm can use these emission rights itself or sell them to other firms in form of emission trading, i.e. there is a market of emission certificates.

Combining these pieces we get the following two-state optimal control model:

$$\max_u \int_0^\infty e^{-rt}[\pi(u) - d(x) - k(x,y)]dt$$

subject to the state equations and initial condition (11) and (12). Moreover, the control and the state variable is restricted by (13) and (14), respectively.

The conditions (14), (17), (18), (19) and (20) imply that the present model fits also into the scheme described in section 2. To see this it should be noted that the pollution level has a negative adjoint variable (shadow price).

The main result is that a non-concave incentive scheme (19), (20) may imply a cyclical production-pollution tradeoff. We conclude that the effect of a certain institutional arrangement which aims at the protection of the environment may run counter to its good intention. Note that this behaviour is similar to the results of the previous sections. A useful incentive scheme should not generate cycles of the pollution level, instead it should be used to clean up the environment at a reasonably economic level.

6. Concluding Remarks

In this paper a family of two-state continuous time optimal models has been considered, where one state variable is an exponentially weighted average of past levels of the other state variable. Charging convex costs for the underlying adjustment process and imposing a non-concave reward/penalty scheme to the difference of the state variables it has been shown numerically that the 'optimal' solutions converge to a limit cycle. Thus, the main result of the paper is a *numerical* proof for the existence and stability of a limit cycle for a class of non-concave two state- optimal control models. One other way to prove the existence *analytically* is to apply the Hopf bifurcation theorem. There are a few papers in economic literature where the Hopf technique is applied to prove the existence of persistent cycles analytically; see Benhabib and Nishimura (1979), Medio (1987ab), and Feichtinger and Sorger (1986). A complete analysis of this type, however, is generally quite cumbersome.

Three applications arising in the dynamics of the firm have been provided. More specifically, we have studied the optimization of hiring and firing, R & D expenditures, and production and pollution over an infinite time horizon for a profit-maximizing firm. For a more detailed economic explanation of the regimes in the persistent cycles we refer to the cited literature. More than that oscillatory behaviour plays a role as optimal solution in several other models of operations research and economics; see, e.g., Ryder and Heal (1973), Sorger (1986), Colonius (1988), Luhmer et al. (1988), Sieveking (1988).

References

ASCHER, U., J. CHRISTIANSEN, AND R.D. RUSSEL (1978): "A Collocation Solver for Mixed Order Systems of Boundary Value Problems", *Mathematics of Computation*, 33, 659-679.

BENHABIB, J., AND K. NISHIMURA (1979): "The Hopf Bifurcation and the Existence and Stability of Closed Orbits in Multisector Models of Optimal Economic Growth", *Journal of Economic Theory*, 20, 421-444.

BENHABIB, J., AND K. NISHIMURA (1988): "On Endogenous Cycles in Discrete Time Optimal Growth Models, in *Optimal Control Theory and Economic Analysis 3*, ed. by G. Feichtinger. Amsterdam: North-Holland.

CLARKE, F.H. (1983): *Optimization and Nonsmooth Analysis*. New York: Wiley.

COLONIUS, F, (1988): "Poincaré-Bendixson Theory for Control Problems with Continuous Optimal Controls", in *Optimal Control Theory and Economic Analysis 3*, ed. by G. Feichtinger. Amsterdam: North-Holland.

FEICHTINGER, G., AND M. LUPTACIK (1987): "Optimal Production and Abatement Policies of a Firm", *European Journal of Operational Research*, 29, 274-285.

FEICHTINGER, G., AND G. SORGER (1986): "Optimal Oscillations in Control Models: How Can Constant Demand Lead to Cyclical Production?", *Operations Research Letters*, 5, 277-281.

FEICHTINGER, G., AND G. SORGER (1988): "Periodic Research and Development", in *Optimal Control Theory and Economic Analysis 3*, ed. by G. Feichtinger. Amsterdam: North-Holland.

FORSTER, B.A. (1977): "On a One State Variable Optimal Control Problem: Consumption-Pollution Trade-Offs", in *Applications of Control Theory to Economic Analysis*, ed. by J.D. Pitchford, and S.J. Turnovsky. Amsterdam: North-Holland.

HARTL, R.F. (1987): "A Simple Proof of the Monotonicity of the State Trajectories in Autonomous Control Problems", *Journal of Economic Theory*, 41, 211-215.

HARTL, R.F., AND S.P. SETHI (1984): "Optimal Control Problems with Differential Inclusions: Sufficiency Conditions and an Application to a Production-Inventory Model", *Optimal Control Applications and Methods*, 5, 289-307.

HAURIE, A., AND A. HOLLANDER (1988): "A Note on Incremental Incentives in Continuous Time", in *Optimal Control Theory and Economic Analysis 3*, ed. by G. Feichtinger. Amsterdam: North- Holland.

LONG, N.V., AND H. SIEBERT (1985): "Lay-off Restraints, Employment Subsidies, and the Demand for Labour", in *Optimal Control Theory and Economic Analysis 2*, ed. by G. Feichtinger. Amsterdam: North-Holland.

LUHMER, A, A. STEINDL, G. FEICHTINGER, R.F. HARTL, AND G. SORGER (1988): "ADPULS in Continuous Time", *European Journal of Operational Research*, 34, 171-177.

MEDIO, A. (1987a): "Oscillations in Optimal Growth Models". To appear in *Journal Economic Behaviour and Organization*.

MEDIO, A. (1987b): "Complex Behaviour in Multisector Growth Models", Paper presented at the Third Viennese Workshop on Optimal Control Theory and Economic Analysis, May 20-22, 1987.

RYDER, H.E., Jr., AND G.M. HEAL (1973): Optimal Growth with Intertemporally Dependent Preferences", *Review Economic Studies*, 121, 1-31.

SALOP, S.C. (1973): "Wage Differentials in a Dynamic Theory of the Firm", *Journal of Economic Theory*, 6, 321-344.

SCHMALENSEE, J.R. (1972): *The Economics of Advertising*. Amsterdam: North-Holland.

SIEVEKING, M. (1988): "Present Values of One Dimensional Autonomous Resources", in *Optimal Control Theory and Economic Analysis 3*, ed. by G. Feichtinger. Amsterdam: North-Holland.

SORGER, G. (1986): "Two-Dimensional Autonomous Optimal Control Problems with a 0-1 State", Forschungsbericht Nr. 94, Institut für Ökonometrie und Operations Research, Technische Universität Wien.

STEINDL, A. (1981): "COLSYS: Ein Kollokationsverfahren zur Lösung von Randwertproblemen bei Systemen gewöhnlicher Differential- gleichungen", Diplomarbeit, Technische Universität Wien.

STEINDL, A., G. FEICHTINGER, R.F. HARTL, AND G. SORGER (1986): "On the Optimality of Cyclical Employment Policies: A Numerical Investigation", *Journal of Economic Dynamics and Control*, 10, 457-466.

MONOPOLISTIC PRICE-SETTING BEHAVIOUR IN A DISEQUILIBRIUM MODEL

Ulrich K. Schittko*
and
Bernhard Eckwert**

Contents

* Universität Regensburg, Institut f. Volkswirtschaftslehre einschl. Ökonometrie, Universitätsstraße 31, 8400 Regensburg

** Universität Augsburg, Wirtschafts- u. Sozialwissenschaftliche Fakultät, Memmingerstraße 14, 8900 Augsburg

Monopolistic Price-Setting Behaviour in a Disequilibrium Model

§1 Introduction and outline of the dynamic concept

For more than two decades economists are interested in the stability of non-Walrasian systems. Among others Böhm [1978], van den Heuvel [1983], Honkapohja/Ito [1982] investigate stability issues within a disequilibrium framework. Honkapohja/Ito [1980] study inventory dynamics in a non-Walrasian model while Ito [1980] and Picard [1983] introduce capital accumulation into a disequilibrium approach.

In this paper we address the dynamic properties of a non-Walrasian economic system with monopolistic price setting behaviour. Technically formulated we provide an analysis of a dynamic system with a time-discontinuous vector field. To motivate our investigation we would like to illustrate first under which conditions a dynamical system possesses a discontinuous vector field. Secondly, in which economic contexts systems with discontinuous vector fields show up and thirdly, which new economic problems arise from the discontinuities of a vector field.

For the differential equation system

$$\dot{x} = g(x) \quad , x \in \mathbb{R}^n \quad , g = (g_1, \cdots, g_n) \tag{1}$$

the mapping $g : \mathbb{R}^n \to \mathbb{R}^n$ is called a vector field.

To illustrate our point let $Q \subset \mathbb{R}^n$ be a region. The region Q is subdivided by a curve S into the domains Q^- and Q^+:

$$S := \{ x \in Q \mid s(x) = 0 \}$$

$$Q^+ := \{ x \in Q \mid s(x) > 0 \}$$

$$Q^- := \{ x \in Q \mid s(x) < 0 \} \quad , s : Q \to \mathbb{R}$$

This situation is demonstrated in the following Figure 1

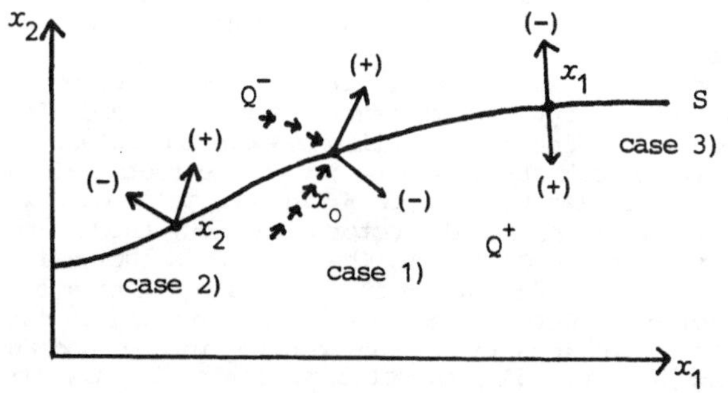

Figure 1

Here we have drawn a vector field in the following manner. To each point x in the plane we assign the vector $g(x)$. $g(x)$ is visualized as an arrow based at x and pointing to $x+g(x)$, i.e. we assign to x the directed line segment from x to $x+g(x)$. In x_0 the mapping g does not possess a limit and the vector directions point into different subregions if we approach x_0

from Q^- and Q^+. We conjecture that our economy, if started in x_0, will move along the separating surface S. Indeed, the Filippov-solution, which we employ in this paper, behaves this way. In case 2) the Fillipov-solution traverses the separating curve from Q^+ to Q^-, whereas in case 3) the solution gets stuck on $s(x) = 0$.

Discontinuous vector fields typically arise from dynamic formulations of disequilibrium models. These models differ from Walrasian models in their stability behaviour. If a non-Walrasian system travels through the phase-space it will typically cross boundaries on which the differential equations describing its motion change in a discontinuous way. The dynamics of the system depend on the economic situation in the particularly given disequilibrium.

The points on the separating curve between two disequilibrium regimes (e.g. Q^- may denote an unemployment regime and Q^+ an overemployed regime) are in general discontinuity points. This is because the equilibria in Q^- and Q^+ have completely different characteristics and hence the law of motion which governs the dynamical variables changes abruptly if the system traverses the boundary between two disequilibrium regimes.

The discontinuity phenomenon has a simple economic justification if markets do not clear.

The economic agents face quantity constraints for their transactions which they take into account when they decide. These quantity constraints which are not invariant across the types of different temporary disequilibria may change in a discontinuous way causing an abnormal stability behaviour of the economic system.

Honkapohja and Ito [1983] have shown that a discontinuous dynamical system can be stable although it is unstable in all subregimes and vice versa.

This is a remarkable statement, which makes clear that in general it is by no means sufficient to investigate the dynamic behaviour of an economy in different disequilibrium regimes separately. Regime stability is neither necessary nor sufficient for overall stability. From this it is obvious that usual methods cannot be applied to economic systems which arise from disequilibrium models.

Referring to our paper [1988], we give an intuitive geometrical explanation for the Filippov-solution concept of differential equations here: In points where the vector field is continuous, the Filippov-solution coincides with the ususal solution. If we look at our figure 1 then the vector field is continuous in Q^+ and Q^- and discontinuous on S. In each point on S we can determine the limit of the vector field (vector direction) through the Q^+-region and the limit through the Q^--region. The Filippov-solution is defined in such a way that its time derivative in $x_0 \in S$ is equal to a convex combination of all limits of vector directions in x_0. Turning to our figure 1 we have two vector directions in x_0, one pointing into Q^+ and another pointing into Q^-. If $\phi^F(t)$ is a Filippov-solution with $\phi^F(t_0)=x_0$, its time derivative satisfies

$$\dot{\phi}^F(t_0)=\alpha g^+(\phi^F(t_0))+(1-\alpha)g^-(\phi^F(t_0)) \text{ for some } \alpha \in [0,1] \tag{2}$$

where e.g. $g^+(\phi^F(t_0)$ is the limit of vector directions through the Q^+-region in the point $\phi^F(t_0)$. In case 1), point x_0, the vector directions point into different regions. The time-derivative of the Filippov-solution as an convex combination of both vector directions lies on the dashed line between the arrows. In the special case depicted here according to Theorem 1 (Schittko/Eckwert [1988]) it is uniquely determined and is tangential to the curve S in point x_0. One can be even more specific. If the constellation of vector directions in x_0 would apply to any point x on the separating curve, the Filippov-solution does not leave the curve (compare Theorem 2, Schittko/Eckwert [1988]). In a point like x_1 with the limit of vector directions through $Q^+(Q^-)$ pointing into $Q^-(Q^+)$ and both arrows forming an angle of 180° the Filippov-solution comes to rest and x_1 is a stationary point. If the limits of the vector directions are as in x_2, then a Filippov-solution travels from x_2 into the Q^--region. Its time derivative lies on the dashed line between the arrows and is not uniquely determined.

§2 The Model

We consider a small open monetary economy, whose exports to the rest of the world (foreign country) are fixed at the level $\bar{m}_2$. The economy imports a production factor from abroad in amount m_1. For production of one unit output a constant amount of the imported factor is needed, while the marginal product of labour is decreasing. We assume that the fixed export level is such that

$$T := \bar{m}_2 - (1/\tau)m_1 = 0 \tag{3}$$

holds, if all workers are employed. In (3) T denotes the trade balance and τ the terms of trade. The labour supply is fixed at a constant level $\bar{l}$. The terms of trade are also fixed by monetary authorities and are constant throughout the analysis.

2.1 Specifications of the behavioural functions

Since the price of the imported factor is constant, the output- and labour demand decision depends exclusively on the real wage, if the production is not rationed in any market.

$$z=z(w/p);y=y(w/p), z \mathrel{\hat{=}} \text{labour demand},\ y \mathrel{\hat{=}} \text{goods supply},\ w/p \mathrel{\hat{=}} \text{real wage} \tag{4}$$

In case the goods supply is rationed (by the demand side) at the level $\hat{y}$, employment l is determined by the effective labour demand, $l = \tilde{z}(\hat{y})$. Analogously $y = \tilde{y}(\bar{l})$, if the labour demand is rationed at the full employment level $\bar{l}$.

Since labour and the imported factor are complementary in the production process, by (3) the trade balance is equilibrated in full employment situations and positive, if unemployment prevails.

The consumption demand c depends on the vector $(\frac{w}{p}l, M/p)$, where M is the money stock, held by the private sector. If $l<\bar{l}$ the effective consumption demand is denoted by $\tilde{c}$. The government demand g is exogenous and financed by a lump sum tax which is collected from the consumers. We assume that the profits of the production sector are distributed to the consumers <u>after transactions have taken place</u>, so that the consumption sector holds the total stock of money at each point of time.

A monetary authority converts the money flows, arising from a non-equilibrated trade balance, into the currency needed, so that there is no need for agents to hold foreign money. The money stock of our economy grows, if the trade balance is positive:

$$\dot{M} = pT, \tag{5}$$

respectively,

$$(\dot{M/p}) = T - \frac{\dot{p}}{p^2}M . \tag{6}$$

2.2 Equilibria of the model

Since in our model prices are not flexible enough to achieve market equilibrium at each point of time, disequilibria - i.e. temporary equilibria with quantity rationing - usually will occur. Following Malinvaud [1977], we distinguish three types of disequilibria:

a) A Keynesian equilibrium (K) is a solution $(\hat{y},l)$ of the equation system

$$\hat{y} - \tilde{c}(\tfrac{w}{p}l, M/p) - \bar{m}_2 - g = 0 \;;\; z(\hat{y}) - l = 0 . \tag{7}$$

b) A classical equilibrium (C) is a solution $(\bar{c},l)$ of the equation system

$$y(w/p) - \bar{c} - \bar{m}_2 - g = 0 \;;\; z(w/p) - l = 0 . \tag{8}$$

c) A repressed inflation equilibrium (R) is a solution $(\bar{c},l)$ of the equation system

$$\tilde{y}(\bar{l}) - \bar{c} - \bar{m}_2 - g = 0 \;;\; \bar{l} - l = 0 . \tag{9}$$

In (7) - (9) we used a $\sim$ to characterize behavioural functions, depending on quantity constraints. We drop the $\sim$ if the quantity constraints are replaced by equilibrium conditions.
Apart from the equilibria listed above one or both markets can be cleared. The latter case is the Walrasian equilibrium, which is assumed to be unique for a given real wealth M/p.

2.3 The separating surfaces between the disequilibrium regimes

The set of all vectors (w,p,M/p), which lead to a K-, C-, R- equilibrium, will be called K-, C-, R-regime, whereby the separating surfaces between two regimes belong to either regime. It is verified immediately by the continuity of all behavioural functions in a given disequilibrium regime that the regions $\mathring{K}$, $\mathring{C}$, $\mathring{R}$, which remain if the separating curves are subtracted from the disequilibrium regimes, are relatively open in $\mathbb{R}^3_+$.
The separating surface between the K- and C-regime, S_{KC}, is the solution set of

$$s_{KC}(x) = y(w/p) - c(\tfrac{w}{p}z(w/p), M/p) - \bar{m}_2 - g = 0; s_{KC}: \mathbb{R}^3_{++} \to \mathbb{R}; x = (w,p,M/p) \in \mathbb{R}^3_{++} \tag{10}$$

Analogously, the separating surfaces between the regimes K and R, S_{KR}, respectively C and R, S_{CR}, are the solution sets of (11) and (12), respectively.

$$s_{KR}(x) := y(\bar{l}) - c(\tfrac{w}{p}\bar{l}, M/p) - \bar{m}_2 - g = 0 \;;\; s_{KR}: \mathbb{R}^3_{++} \to \mathbb{R} \;;\; x = (w,p,M/p) \in \mathbb{R}^3_{++} \tag{11}$$

$$s_{CR}(x) = z(w/p) - \bar{I} = 0\ ;\ s_{CR} : \mathbb{R}^3_{++} \to \mathbb{R}\ ;\ x = (w,p,M/p) \in \mathbb{R}^3_{++} \tag{12}$$

The derivatives of the functions s_{KC}, s_{KR} and s_{CR} are collected in (13):

$$\frac{\partial s_{KC}}{\partial x}(x) = \Big[\Big(\underbrace{-y'(w/p)\frac{w}{p^2}}_{(+)}\underbrace{- c_I I_p + c_{M/p}\frac{M}{p^2}}_{(+)}\Big), \underbrace{\Big(\underbrace{y'(w/p)\frac{1}{p}}_{(-)} \underbrace{- c_I I_w}_{(\stackrel{?}{-})}\Big)}_{(-)}, \Big(\underbrace{-c_{M/p}}_{(-)}\Big)\Big] \quad \text{(a)}$$

$$\frac{\partial s_{KR}}{\partial x}(x) = \Big[\underbrace{-c_I I_p + c_{M/p}\frac{M}{p^2}}_{(+)}, \underbrace{-c_I \bar{I}}_{(-)}, \underbrace{-c_{M/p}}_{(-)}\Big] \quad \text{(b) (13)}$$

$$\frac{\partial s_{CR}}{\partial x}(x) = \Big[\underbrace{-z'(w/p)\frac{w}{p^2}}_{(+)}, \underbrace{z'(w/p)\frac{1}{p}}_{(-)}, 0\Big] \quad \text{(c)}$$

$I \mathrel{\hat{=}}$ real current labour income.

§3 Dynamics

In this section we analyse the global stability behaviour of our economy using the theorems stated in Schittko/Eckwert [1988].

3.1 Monopolistic price setting behaviour with linked price-wage dynamics

The dynamics we are concerned with here are defined for each disequilibrium regime separately

a) $\overset{\circ}{K}$-regime

In a situation of excess supply on both the goods and labour market the production sector (representative producer) behaves like a monopolist. He maximizes his profits, knowing (or expecting) that his sales depend on the price demanded.

$$\max\ \pi = p\zeta(x) - wf^{-1}(\zeta(x))$$

f = production function; $x := (w,p,M/p)$; ζ = expected conjectural demand function

In producer equilibrium we have

$$\frac{\partial \pi}{\partial p} = \zeta(x) + p\frac{\partial \zeta}{\partial p}(x) - wf^{-1'}(\zeta(x))\,\frac{\partial \zeta}{\partial p}(x) = 0$$

respectively

$$\left.\frac{\partial p}{\partial w}\right|_{\pi=\pi_{max}} = - \frac{\overset{(+?)}{\frac{\partial^2\zeta(x)}{\partial p\partial w}}\overset{(+)}{\left[p-wf^{-1'}\zeta(x)\right]}+\overset{(+)}{\frac{\partial\zeta(x)}{\partial w}}-\overset{(+)}{\frac{\partial\zeta(x)}{\partial p}}\overset{(+)}{\left[f^{-1'}(\zeta(x))+wf^{-1''}(\zeta(x))\frac{\partial\zeta(x)}{\partial w}\right]}}{\underset{(-?)}{\frac{\partial^2\zeta}{\partial p^2}(x)}\underset{(+)}{\left[p-wf^{-1'}(\zeta(x))\right]}+\underset{(-)}{\frac{\partial\zeta(x)}{\partial p}}\underset{(+)}{\left[2-wf^{-1''}(\zeta(x))\frac{\partial\zeta(x)}{\partial p}\right]}} > 0 .$$

Thus $p'(w) > 0$. Now the dynamics in the K-regime can be formalized as

$$\dot{p} = p'(w)\dot{w}, p'(w) > 0; \dot{w} = k_1\left[f^{-1}(\hat{y}(x))-\bar{l}\right], k_1>0; (\dot{M/p}) = T(\hat{y}(x)) - \frac{\dot{p}}{p^2}M . \tag{14}$$

The last equation was already derived in (6). The price is set monopolistically and is closely linked to the wage rate. The wage rate in turn reacts to disequilibria in the labour market. Finally the dynamics of the real money stock depend on the trade balance and the price dynamics. The output level $\hat{y}$ is endogenously determined in a Keynesian equilibrium and therefore depends on the vector x.

b) $\overset{\circ}{C}$-regime

We assume that in C-equilibria the wage rate is related to the price level by $\dot{w} = \gamma_1\dot{p}$, while the price reacts to disequilibria in the goods market.

$$\dot{p} = -k_2\left[y(w,p) - c(x) - \bar{m}_2 - g\right], k_2 > 0; \dot{w} = \gamma_1\dot{p}; (\dot{M/p}) = T(y(w/p)) - \frac{\dot{p}}{p^2}M . \tag{15}$$

$c(x)$ is not the consumption demand function but represents the effective consumption demand in the C-equilibrium, which is realized at the parameter constellation x.

c) $\overset{\circ}{R}$-regime

In the $\overset{\circ}{R}$-regime we assume that the price dynamics are related to the wage dynamics by $\dot{p} = \gamma_2\dot{w}$, while the wage rate reacts to disequilibria in the labour market.

$$\dot{p} = \gamma_2\dot{w}; \; w = k_3\left[z(w/p) - \bar{l}\right]; \; (\dot{M/p}) = T(y(\bar{l})) - \frac{\dot{p}}{p^2}M . \tag{16}$$

<u>3.1.1 regime-stability</u>

Consider the differential equation system (14) which describes the dynamics of our economy in the K-region. Obviously there are more than just one stationary point, since $\dot{w} = 0$ implies $\dot{p} = (\dot{M/p}) = 0$. In particular, all stationary points are Walrasian equilibria since $\dot{w} = 0$ implies a labour market equilibrium as well as $\hat{y} = y(\bar{l})$. In a Keynesian equilibrium the latter condition holds only if the goods market is equilibrated.

Honkapohja and Ito [1983] introduced the notion of "regime stability". The notion' "regime stability" is used in the hypothetical situation where a system of differential equations for one regime is extended, given that it is extendable, to an entire domain and that the equilibrium is asymptotically stable in such a situation'. Since the stationary points of our differential equation systems are not unique in general we call a disequilibrium regime stable (more exactly quasi globally stable) if the extended system (extended to $\mathbb{R}^3_{++}$) converges to an element of W, where W denotes the set of all Walrasian equilibria.

We first check the extendability of system (14) to $\mathbb{R}^3_{++}$: $\hat{y}$ denotes the activity level of our economy which equals the demand for home consumption goods in all K-equilibria, i.e. $\hat{y} = \tilde{c} + \bar{m}_2 + g$. This relationship can be extended in an obvious manner to all disequilibrium regimes by defining $\tilde{\hat{y}} := c + \bar{m}_2 + g$; c is the consumption demand in an arbitrary disequilibrium situation. Especially we have $\tilde{\hat{y}} = \hat{y}$ in all K-equilibria and no discontinuity in the vector field of the extended system (which is defined on $\mathbb{R}^3_{++}$) arises. In the sequel we omit the $\sim$ on $\hat{y}$ in the extended system.

In order to show stability we first prove that no trajectory of the extended system leaves the range of definition of $\hat{y}$, i.e. no trajectory leaves the domain $\mathbb{R}^3_{++}$ (c and by that $\hat{y}$ is not defined outside of $\mathbb{R}^3_{+}$), so that a solution is well defined. Suppose p approaches zero from above, i.e. $p \to 0+$. If the price level is sufficiently low, consumption demand c becomes so large that $\hat{y} > f(\bar{I})$, from which $\dot{w} > 0$ and $\dot{p} > 0$ follows. Thus p remains positive throughout the motion. Also (M/p) cannot become negative: M = 0 implies $(\dot{M/p}) = T(\hat{y})$; but if the money holdings are zero a negative trade balance cannot be financed, so $T(\hat{y}) \geq 0$ must hold. In general $T(\hat{y}) > 0$ will be satisfied if M = 0 (which we assume) and by continuity we have $T(\hat{y}) > 0$ for sufficiently small (but positive) M. Concerning the wage rate we examine the function $\left(\frac{\partial}{\partial t}(\dot{w}^2)\right)$:

$$\frac{\partial}{\partial t}(\dot{w}^2) = 2k_1\dot{w}\left[f^{-1'}(\hat{y})\left(\hat{y}_w\dot{w} + \hat{y}_p\dot{p} + y_{M/p}(\dot{M/p})\right)\right] =$$

$$= \underbrace{2k_1\dot{w}^2 f^{-1'}}_{(\geq 0)}\underset{(+)}{(\hat{y})}\left(\hat{y}_w + \underbrace{p_k'(w)\hat{y}_p}_{(-)}\right) + \underbrace{2k_1\dot{w}(\dot{M/p})}_{(\leq 0)}\underset{(+)}{\hat{y}_{M/p}} \quad (\leq 0)$$

The factor $2k_1\dot{w}(\dot{M/p})$ in the last term is not grater than zero: $\dot{w} < 0$ implies $\hat{y} < f(\bar{I})$ and $\dot{p} < 0$, so that $T(\hat{y}) > 0$ and $\left(-\frac{\dot{p}}{p^2}M\right) > 0$ from which $(\dot{M/p}) > 0$ follows. If $\dot{w} > 0$ we conclude $\dot{p} > 0$ and $\hat{y} > f(\bar{I})$ which leads to $T(\hat{y}) < 0$ and $\left(-\frac{\dot{p}}{p^2}M\right) < 0$; thus $(\dot{M/p}) < 0$.

We assume $\hat{y}_w + p_k'(w)\hat{y}_p < 0$ and obtain a negative time derivative of $\dot{w}^2$ if the

labour market is not equilibrated. Thus $\dot{w}$ converges <u>monotonically</u> to zero and $w \geq 0$ holds throughout the motion.
The asymptotic stability of the extended version of (14) is now easy to show. $\dot{w} \to 0$ implies $\dot{p} \to 0$ and $\hat{y} \to y(\bar{l})$. By continuity we conclude $T(\hat{y}) \to 0$ respectively $\dot{(M/p)} \to 0$. Since $\hat{y} \to y(\bar{l})$ the stationary point of any trajectory, which stays in the $\mathring{K}$-region forever, must be a Walrasian equilibrium. So far we know that the $\mathring{K}$-region is regime stable.
The system (15) need not be extended to see that any trajectory in the $\mathring{C}$-regime does not leave the range of definition $\mathbb{R}^3_{++}$ of the behavioural functions. Clearly $\dot{w} \geq 0$ and $\dot{p} \geq 0$ holds in any C-equilibrium. Furthermore $y(w/p) \leq y(\bar{l})$ if $(w,p,M/p) \in \mathring{C}$, so that $M = 0$ implies $\dot{(M/p)} > 0$. The time derivative of $\dot{p}$ is

$$\frac{\partial}{\partial t}(\dot{p}) = -k_2\Big\{\underbrace{\underset{(-)}{y'(w/p)}\frac{w}{p^2}\dot{p}\underset{(-)}{\left(\varepsilon^C_{w,p}-1\right)}}_{(+)}-\Big[\underbrace{\dot{p}\Big(\underset{(-)}{c_p}+\underset{(+)}{\underline{c_w}\,\underline{w'_c}(p)}\Big)}_{(-)}+\underset{(+)}{c_{M/p}}\underset{(?)}{\dot{(M/p)}}\Big]\Big\}, \tag{17}$$

$\varepsilon^C_{w,p} = w'(p)\,\frac{p}{w} \mathrel{\widehat{=}}$ price elasticity of nominal wages in the $\mathring{C}$-regime.

In (17) we adopted the assumption $(c_p + c_w w'_c(p)) < 0$. The sign of (17) cannot be determined in general, since $\dot{(M/p)}$ can be positive. But we can argue as follows: In the C-regime (w/p) decreases monotonically because $\varepsilon^C_{w,p} < 1$ and $\dot{p} > 0$ and does not converge in the interior of the C-regime since $\dot{p}$ does not converge there. Therefore, along any trajectory that does not leave the C-regime, $y(w,p)$ converges monotonically to $y(\bar{l})$ and hence $T(y(w/p))$ approaches zero. It follows that either $\dot{(M/p)}$ goes to zero or there exists t^* such that $\dot{(M/p)} \leq 0$ for $t > t^*$. For all $t > t^*$ (respectively $\dot{(M/p)}$ sufficiently small) $\partial/\partial t(\dot{p})$ is negative (compare (17)) and $\dot{p}$ converges to zero. Since also $T(y(w/p))$ converges to zero, $\dot{(M/p)} \to 0$ and $\dot{w} \to 0$ results for all trajectories that do not leave the C-regime. The asymptotic equilibrium for such a trajectory must be Walrasian, because $y(w/p) \to y(\bar{l})$ and $\dot{p} \to 0$ imply a labour- and goods market equilibrium.

We finally turn to the differential equation system (16). w and p cannot become negative as long as the economy is located in the $\mathring{R}$-region, because $\dot{w} \geq 0$, $\dot{p} \geq 0$ holds in any temporary R-equilibrium. To show $(M/p) > 0$ we first note that along any trajectory $(\dot{p}/p)$ is bounded from above as long as the trajectory stays in the $\mathring{R}$-regime; this assertion follows immediately from the fact that $\frac{\partial}{\partial t}(\dot{w})$ is negative (w/p increases because $\varepsilon^R_{p,w} < 1$) as long as $\dot{w} > 0$. Suppose a trajectory starts at time t_0 in $(w_0,p_0,M_0/p_0) \in \mathring{R}$ and $\dot{p}/p \leq P$ throughout the motion. It follows $\dot{(M/p)} \geq -P(M/p)$ so that $M/p \geq (M_0/p_0)\exp(-P(t-t_0))$ holds and we conclude $M/p > 0$ if repressed inflation prevails in our economy.
So far we have proved that all behavioural functions in (16) are well defined along any trajectory in the $\mathring{R}$-region. We have already argued above that $\frac{\partial}{\partial t}(\dot{w})$ is negative as long as $\dot{w} > 0$. Thus $\dot{w}$ is monotonically decreasing and

it is easily verified that $\dot{w}$ cannot converge to any point but zero. So we conclude $\dot{w} \to 0 \Rightarrow \dot{p} \to 0 \Rightarrow (\dot{M/p})_o \to 0$, and the stationary point of any trajectory that does not leave the $\mathring{R}$-region is a Walrasian equilibrium.

In this section we have shown, that any trajectory which starts in one of the three disequilibrium regimes either converges to a Walrasian equilibrium of leaves the region after finite time.

3.1.2 The dynamic behaviour of the economy near points of discontinuity

In section 2.3 we have constructed the separating surfaces (boundaries) between the disequilibrium regimes. We will show in a moment that the vector field of our dynamical system is in general discontinuous on these boundaries. A trajectory near a boundary can (a) stay in that region where it is located, (b) travel across the boundary or (c) slide along the separating surface. If it slides along a boundary and has a continuous time derivative discontinuities can normally arise when two boundaries intersect. For our model, however, the set of intersection points of two boundaries contains exclusively Walrasian equilibria, where the dynamics formulated in 3.1 and - as we will see - also a sliding Filippov-solution come to rest. These rest points are approached only asymptotically so that a trajectory which slides on a boundary will never reach another separating surface and thus will never exhibit any discontinuities in the vector field. The stability behaviour of a sliding trajectory can be analyzed using standard methods for dynamical systems without discontinuities.

In all points, that do not belong to any boundary the vector field of our dynamical system is continuous and the time derivative of the Filippov-solution is equal to the dynamics, specified for the respective disequilibrium regime.
In order to apply the theorems established in Schittko/Eckwert [1988] we have to assess the signs of the scalar products

$$\left\langle \frac{\partial s_{ij}}{\partial x}(x_o), g^{(\pm)}(x_o) \right\rangle \quad ; \qquad i,j = K,C,R; \quad i \neq j \; .$$

3.1.2.1 The S_{KC}-boundary

We start with the separating surface, defined by $s_{KC}(x) = 0$. Let $x_o \in S_{KC} \setminus W$ ($W \hat{=}$ set of Walrasian equilibria).

$$\left\langle \frac{\partial s_{KC}}{\partial x}(x_o), g^+(x_o) \right\rangle = \underbrace{\underset{(+)}{\frac{\partial s_{KC}}{\partial p}(x_o)} \lim_{\substack{x \to x_o \\ K}} \dot{p}}_{(-)} + \underbrace{\underset{(-)}{\frac{\partial s_{KC}}{\partial w}(x_o)} \lim_{\substack{x \to x_o \\ K}} \dot{w}}_{(-)} + \underbrace{\underset{(-)}{\frac{\partial s_{KC}}{\partial (M/p)}(x_o)} \lim_{\substack{x \to x_o \\ K}} (\dot{M/p})}_{(+)} \qquad (18)$$

with

$$\lim_{\substack{x \to x_o \\ K}} \dot{p} = \lim_{\substack{x \to x_o \\ K}} \frac{p}{w}\left(\varepsilon^K_{p,w}\right) k_1\left(f^{-1}(\hat{y}) - \bar{l}\right) < 0 \, ; \lim_{\substack{x \to x_o \\ K}} \dot{w} = \lim_{\substack{x \to x_o \\ K}} k_1\left(f^{-1}(\hat{y}) - \bar{l}\right) < 0$$

$$\lim_{\substack{x\to x_0\\ K}} (\dot{M/p}) = \lim_{\substack{x\to x_0\\ K}} \left(T(\hat{y})\ \frac{\dot{p}}{p^2}M\right) > 0\ .$$

$\varepsilon^K_{p,w}$ denotes the wage elasticity of prices in the $\overset{\circ}{K}$-regime. (18) can be rewritten as

$$\left\langle \frac{\partial s_{KC}}{\partial x}(x_0), g^+(x_0)\right\rangle = \underbrace{\frac{\partial s_{KC}}{\partial (M/p)}(x_0) \lim_{\substack{x\to x_0\\ K}} (\dot{M/p})}_{(-)} + \lim_{\substack{x\to x_0\\ K}} \underset{(-)}{\dot{w}}\underbrace{\left[\underset{(-)}{\frac{\partial s_{KC}}{\partial w}(x_0)} + \underbrace{\frac{p}{w}\left(\varepsilon^K_{p,w}\right)\frac{\partial s_{KC}}{\partial p}(x_0)}_{(+)}\right]}_{(+)} < 0 \quad (19)$$

The limit of the last bracket is assumed to be positive, because in a Keynesian equilibrium (p/w) is relatively large and price changes have real wealth effects (compare (13(a)) in addition to the direct and the labour income induced goods market effects (which also show up if the wage rate changes).
If the point x_0 is approached through the $\overset{\circ}{C}$-regime we obtain different limits for $\dot{p}$, $\dot{w}$ and $(\dot{M/p})$, so that the vector field is discontinuous in x_0:

$$\lim_{\substack{x\to x_0\\ C}} \dot{p} = -\lim_{\substack{x\to x_0\\ C}} k_2\left(y(w/p) - c(x) - \bar{m}_2 - g\right) = 0\ ;\ \lim_{\substack{x\to x_0\\ C}} \dot{w} = \lim_{\substack{x\to x_0\\ C}} \frac{w}{p}\left(\varepsilon^C_{w,p}\right)\dot{p} = 0$$

$$\lim_{\substack{x\to x_0\\ C}} (\dot{M/p}) = \lim_{\substack{x\to x_0\\ C}} T\left(y(w/p)\right) > 0$$

The inner product in (20)

$$\left\langle \frac{\partial s_{KC}}{\partial x}(x_0), g^-(x_0)\right\rangle = \underset{(-)}{\frac{\partial s_{KC}}{\partial (M/p)}(x_0)} \cdot \underbrace{T\left(y(w_0/p_0)\right)}_{(+)} < 0 \qquad (20)$$

is negative and has the same sign as (19). By definition the time derivative of a Filippov-solution in x_0 is a convex combination of $g^+(x_0)$ and $g^-(x_0)$:

$$g^F(x_0) = \alpha g^+(x_0) + (1-\alpha)g^-(x_0)\quad ,\quad 0 \le \alpha \le 1\ .$$

For any Filippov-solution with time derivative of the above type it holds that

$$\left\langle \frac{\partial s_{KC}}{\partial x}(x_0), g^F(x_0)\right\rangle = \alpha\left\langle \frac{\partial s_{KC}}{\partial x}(x_0), g^+(x_0)\right\rangle + (1-\alpha)\cdot\left\langle \frac{\partial s_{KC}}{\partial x}(x_0), g^-(x_0)\right\rangle < 0\ .$$

By Theorem 2 in Schittko/Eckwert [1988] our economy will travel from x_0 into the $\overset{\circ}{C}$-regime (in particular it does not slide along the boundary).

Furthermore, the economy will <u>not</u> approach the boundary S_{KC} coming from the C-regime (compare Theorem 3 in Schittko/Eckwert [1988]).

3.1.2.2 The S_{KR}-boundary

Turning to the S_{KR}-boundary we calculate for $x_o \in S_{KR} \setminus W$:

$$\lim_{\substack{x \to x_o \\ K}} \dot{p} = \lim_{\substack{x \to x_o \\ K}} \frac{p}{w}\left(\varepsilon^K_{p,w}\right) k_1 \left(f^{-1}(\hat{y}) - \bar{l}\right) = 0; \lim_{\substack{x \to x_o \\ K}} \dot{w} = \lim_{\substack{x \to x_o \\ K}} \frac{w}{p} \frac{1}{\varepsilon^K_{p,w}} \dot{p} = 0$$

$$\lim_{\substack{x \to x_o \\ K}} (\dot{M/p}) = \lim_{\substack{x \to x_o \\ K}} \left(T(\hat{y}) - \frac{\dot{p}}{p^2} M\right) = 0$$

Thus $g^+(x_o) = (0,0,0)$ and the inner product $\left\langle \frac{\partial s_{KR}}{\partial x}(x_o), g^+(x_o) \right\rangle$ vanishes. However, as we shall see in the sequel condition (6) of Corollary 1 in Schittko/Eckwert [1988] is satisfied under weak assumptions: In the interior of the K-regime

$$\left\langle \frac{\partial s_{KR}}{\partial x}(x), (\dot{p}, \dot{w}, (\dot{M/p})) \right\rangle = \underset{(-)}{\dot{w}} \Bigg[\underset{(-)}{\frac{\partial s_{KR}}{\partial w}(x)} + \underbrace{\frac{p}{w}\left(\varepsilon^K_{p,w}\right)\frac{\partial s_{KR}}{\partial p}(x)}_{(+)} - \tag{21}$$

$$- \underbrace{\frac{M}{p} \frac{\varepsilon^K_{p,w}}{w} \frac{\partial s_{KR}}{\partial (M/p)}(x) \Bigg]}_{(+)} + \underbrace{\frac{\partial s_{KR}}{\partial (M/p)}(x) T(\hat{y})}_{(-)} < 0 .$$

So far occasionally we have assumed the consumption demand to react sufficiently more sensitive to price changes than to w-changes. If we maintain this assumption the sum of the first two terms in the bracket above will be positive (compare (13b)) and the sign of (21) negative for all $x \in \overset{\circ}{K}$.

Again x_o is a point of discontinuity: If x_o is approached through the $\overset{\circ}{R}$-regime the limits of $\dot{p}, \dot{w}$ and $(\dot{M/p})$ are not all zero:

$$\lim_{\substack{x \to x_o \\ R}} \dot{p} = \lim_{\substack{x \to x_o \\ R}} \frac{p}{w}\left(\varepsilon^R_{p,w}\right)\dot{w} > 0; \lim_{\substack{x \to x_o \\ R}} \dot{w} = \lim_{\substack{x \to x_o \\ R}} k_3 \left[z(w/p) - \bar{l}\right] > 0; \lim_{\substack{x \to x_o \\ R}} (\dot{M/p}) = -\lim_{\substack{x \to x_o \\ R}} \frac{\dot{p}}{p^2} M < 0.$$

The inner product

$$\left\langle \frac{\partial s_{KR}}{\partial x}(x_0), g^-(x_0) \right\rangle = \lim_{x \to x_0 \atop R} \Big\{ \underset{(+)}{\dot{w}} \Big[\underset{(+)}{p'_R(w)} \Big(\underset{(+)}{\frac{\partial s_{KR}}{\partial p}(x_0)} - \underbrace{\frac{M}{p^2} \frac{\partial s_{KR}}{\partial (M/p)}(x_0)}_{(+)} \Big) + \underset{(-)}{\frac{\partial s_{KR}}{\partial w}(x_0)} \Big] \Big\} > 0 \quad (22)$$

is positive under our assumption that price changes have sufficiently stronger effects on consumption demand than wage changes (compare (13b)).

For the boundary S_{KR} the conditions of Corollary 1 in Schittko/Eckwert [1988] are satisfied and the time derivative of any Filippov-solution in $x_0 \in S_{KR}$ can be calculated according to Theorem 1 in Schittko/Eckwert [1988]: $g^F(x_0) = \alpha g^-(x_0)$ since $g^+(x_0) = (0,0,0)$. α can be determined from the equation

$$0 \overset{!}{=} \left\langle \frac{\partial s_{KR}}{\partial x}(x_0), g^F(x_0) \right\rangle = \alpha \left\langle \frac{\partial s_{KR}}{\partial x}(x_0), g^-(x_0) \right\rangle .$$

Provided that the regularity condition of Theorem 1 in Schittko/Eckwert [1988] is satisfied we have $\alpha = 0$ and

$$g^F(x_0) = (0,0,0) \; ; \quad (23)$$

i.e. the Filippov-solution comes to rest and all points on S_{KR} are stationary. Thus the set of long run equilibria for our economy is larger than the set of Walrasian equilibria; the latter one is not attractive, i.e. there is no need for a trajectory (in the sense of Filippov) to approach asymptotically a Walrasian equilibrium.

3.1.2.3 The S_{CR}-boundary

If $x_0 \in S_{CR} \setminus W$, we proceed in a straightforward manner and calculate

$$\lim_{x \to x_0 \atop C} \dot{p} = \lim_{x \to x_0 \atop C} -k_2 \left[y(w/p) - c(x) - \bar{m}_2 - g \right] > 0 \; ; \; \lim_{x \to x_0 \atop C} \dot{w} = \lim_{x \to x_0 \atop C} \frac{w}{p} \left(\varepsilon^C_{w,p} \right) \dot{p} > 0 \quad .$$

The limit of $(M/p)^{\cdot}$ under approach to x_0 is not needed because the third component of $\frac{\partial s_{CR}}{\partial x}(x_0)$ is zero. For the inner product

$$\left\langle \frac{\partial s_{CR}}{\partial x}(x_0), g^-(x_0) \right\rangle = \lim_{x \to x_0 \atop C} \underset{(+)}{\dot{p}} \; z'(w/p) \frac{w}{p^2} \underbrace{\left(\varepsilon^C_{w,p} - 1 \right)}_{(-)} > 0$$

we get a positive sign. Since $\lim\limits_{\substack{x \to x_0 \\ R}} \dot{p} = \lim\limits_{\substack{x \to x_0 \\ R}} \frac{p}{w}\left(\varepsilon^R_{p,w}\right)\dot{w} = 0$ we have

$\left\langle \frac{\partial s_{CR}}{\partial x}(x_0), g^+(x_0) \right\rangle = 0$ and Theorem 1 (Schittko/Eckwert [1988]) is not directly appliquable. However the conditions of Corollary 1 (Schittko/Eckwert [1988]) are satisfied, because $\left\langle \frac{\partial s_{CR}}{\partial x}(x_0), g(x_0) \right\rangle = \frac{\dot{w}}{p} z'(w/p)\left(1-\varepsilon^R_{p,w}\right) < 0$ holds for all $x \in \mathring{R}$ (recall that $s_{CR}(x) > 0 \Leftrightarrow x \in \mathring{R}$-regime). By Corollary 1 and Theorem 1 (Schittko/Eckwert [1988]) all Filippov-solutions passing through x_0 have the same time derivative in x_0 and $g^F(x_0) = (0,0,0)$ follows immediately, if we assume the regularity condition of Theorem 1 (Schittko/Eckwert [1988]) to hold. We conclude that a Filippov-solution comes to rest on the boundary between the $\mathring{C}$- and $\mathring{R}$-regime. All points on S_{CR} are stationary.

3.1.3 Conclusions for the dynamic behaviour of the economy

We have shown in 3.1.1 that any trajectory which do not leave the regime where it was started converges to a Walrasian equilibrium; in this sense the regimes are stable. Now the vector field is discontinuous on all boundaries between two regimes and we have to analyze the different regimes separately. Consider a trajectory $\Phi^F(t)$ that starts in a point of the $\mathring{C}$-regime. In 3.1.2.1 we showed that Φ^F will not approach a point on the boundary S_{KC} and in particular will not travel into the $\mathring{K}$-regime. On the other hand all points on the separating surface between the $\mathring{C}$- and $\mathring{R}$-regime are stationary (compare Section 3.1.2.3), so that the economy cannot leave the C-regime. It either converges to an equilibrium on S_{CR} or (by regime stability) it approaches asymptotically a Walrasian equilibrium. *In the long run equilibrium all workers are employed (labour market equilibrium) and the goods market is equilibrated or in excess demand if the economy starts in the $\mathring{C}$-regime.*

Suppose Φ^F starts in a K-equilibrium. In 3.1.2.2 we showed that the trajectory cannot travel into the $\mathring{R}$-regime, because all points on S_{KR} are stationary. In case Φ^F does not leave the $\mathring{K}$-regime it either approaches a point on S_{KR} or (by regime stability) it converges to a Walrasian equilibrium. *If the asymptotic equilibrium of our economy lies on S_{KR}, excess demand on the labour market combined with a goods market equilibrium will prevail in the long run.* The trajectory can leave the $\mathring{K}$-region only in direction of the $\mathring{C}$-regime; if so, *a labour market equilibrium combined with an equilibrium or an excess demand on the goods market results in the long run,* as was stated above.

Finally a trajectory which starts in a R-equilibrium cannot leave the R-regime. It can either approach a stationary point on the S_{KR}-respectively S_{CR}-boundary or converge to a Walrasian equilibrium. *In the long run the labour- as well as the goods market will either be equilibrated or in excess demand.*

In Figure 2 the dynamic behaviour of our economy is illustrated graphically, where we have projected the 3-dimensional phase space to a 2-dimensional figure.

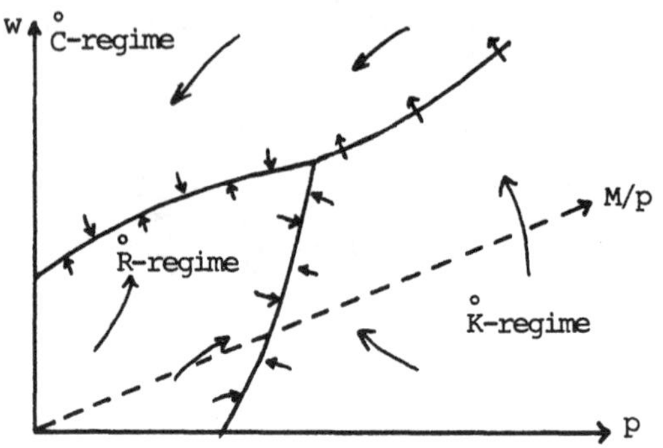

Figure 2

Conclusion

In this paper we have analyzed the dynamic behaviour of an economy within a disequilibrium context using the technical concept of a Filippov-solution to differential equations with discontinuous vector fields. If commodity prices are set monopolistically the economy may converge to a long run equilibrium, in which the goods- or the labour market (or both) do not clear. The final (long run) position of the economy depends sensitively on the point of the phase space where the economy was started.

References

1. Böhm, V.: 'Disequilibrium Dynamics in a simple macroeconomic model', Journal of Economic Theory 17 (1978), 179-199.

2. van den Heuvel, P.: 'The Stability of a Macroeconomic System with Quantity Constraints', Springer-Verlag, Berlin etc.,(1983).

3. Honkapohja, S./Ito, T.: 'Inventory Dynamics in a Simple Disequilibrium Macroeconomic Model', Scandinavian Journal of Economics, Vol. 82, No. 2, (1980), p. 184-198.

4. Honkapohja, S./Ito, T.: 'Disequilibrium Dynamics with Monetarist Price Expectations', Economics Letters 9, (1982), p. 69-75.

5. Honkapohja, S.Ito, T.: 'Stability with Regime Switching', JET 29, (1983), p. 29-48.

6. Ito, T.: 'Disequilibrium Growth Theory', Journal of Economic Theory 23, (1980), p. 380-409.

7. Malinvaud, E.: 'The Theory of Unemployment Reconsidered', Basil Blackwell, Oxford, (1977).

8. Picard, P.: 'Inflation and Growth in a Disequilibrium Macroeconomic Model', Journal of Economic Theory 30, (1983), p. 266-295.

9. Schittko, U.K./Eckwert, B.: 'Disequilibrium Dynamics', Scandinavian Journal of Economics, (1988), forthcoming.

Formation and Diffusion of Expectations

by **Hermann Schnabl**, Universität Stuttgart

1. Introduction

Both categories of models of expectations formation, the extrapolative as well as the so-called Rational expectations (Nerlove (**14**)) are based on the ground of an indivualistic concept: Each agent develops his own expectations independently of all others. This viewpoint is totally consistent, as all expectations are homogenous.

We do not intend to give a broad critique on this subject here. The assumptions have already been criticized in various ways in the literature (cf.Barro and Fischer (2), Laidler (**10**), Hey (6), Shiller (20)).

Although being very different in their starting points, both kinds of expectation theory can be criticized on two characteristics, which they have in common:

- The adaptive reactions on behalf of a changed informational basis do not need "any time" at all, they happen to take place exactly at the juncture of two periods (Machinek (**12**))

- The adaptive reactions of all agents are **homogenous** and **synchronised.**

On the second point, we have to admit that the lack of homogeneity has already been recognized and taken into consideration by Conlisk (**4**) as well as Haltiwanger and Waldman (5) introducing heterogeneity of expectations between agents. However the assumption of a synchronism of reactions remains to be still in use.

Paying attention to the possibility of heterogeneity of expectations, a new deficit of the theory comes to view which seemed not to exist before: The actual formation of expectations in the **social environment** (Schlotthauer (**18**, p.147)). Taking into account the aim of optimization, each agent would have to include the others' expectations into his computations (Ribhegge (**17**, p.101)).

2. Basics of a model of expectations formation

The formation of expectations happens to take place only if there is **uncertainty** about the future. Its function is primarily to **reduce** the feelings of uncertainty and to deliver a fixed point of reference for planning as a kind of surrogate for the lacking knowledge. Of course, the individual tries to **make use of existing information** in order to establish this surrogate.

For this purpose, it is very interesting to check the quality of the informational basis used in this process. It consists of:

a) Situative information (received by perception)

b) Information from the past (adding to the content of memory)

c) Information processing after the **actual inner model of the world**

Elements a to c show all characteristics of **imperfection**:

a) speaking with psychology, every perception is distorted, i.e. it does not give a "true" copy of reality but reconstructs it from "selected" subsets of perceptions. As far as the situation the agent is in, is relevant for the formation of expectations it is not based on reality itself but on a cognitive copy of it that contains greater or lesser alterations.

b) stored information had to be perceived once in the past and is therefore of no better quality than described above. Moreover, it is likely to suffer more distortion with time because memory failures may add some more drop outs.

c) The processing of perceived and remembered information is done after the rules given by an "inner model of the world" (Steinbuch (**21**, p.137)), which consists of hypotheses constructed after perceived and stored information and made consistent by argument and comparison. This model shows a huge variety between persons due to differing experiences and education.

Thus, not only highly individual sources of information input but also information processing lead to **diverging individual expectations.**

Examining the result of the process described above which is used in order to individually form expectations, it is clear that not only expectations differ amongst agents, but also differences between individuals may become disfunctional for the purpose of that process if the individual becomes aware of the fact that it has expectations that diverge from those of another. If this happens, the individually formed expectation does no more what it should do: reduce the feeling of uncertainty. The individual will then feel what the psychologist calls a "cognitive dissonance" and is likely to alter his/her expectations in a way that is more consonant with the expectations of others. Thus the integration of an individual expectation into the pattern of existing social expectations turns the possible disfunctional dissonance into an efficient reduction of uncertainty.

These effects are well known in social psychology. Hofstätter (7, p.78f) for example, sees the reduction of uncertainty by **social acknowledgement** to be fundamental. Allport even considers the individual to "mirror prevailing opinions" (Allport (1, p.376)). Especially in the context of the economy the social environment and the experience made under its influence play an important role (Katona (9)).

It is not even necessary to have what is called group pressure, well known from social psychological experiments, in order to induce an individual to adapt his opinion to an averaged group opinion. The fact alone that the individual is **uncertain** how to estimate future prospects, increases the propensity to join the supposed opinion of the group (Hofstätter (7, p.94f)). Behind an apparent conformity of other expectations, the individual presumes a higher degree of certainty of informations to exist. Lavington (11, p.33) compared the agents in this situation to skaters on a frozen lake who increase their reliance on the load capacity of the ice according to the rising number of skaters on the ice. With rational expectations they would behave in the opposite direction.

> "...the aggregate effect is not simply a summation of individual expectations. People take each other's expectations into account and change their own expectations accordingly." (van Raaij (16, p. 394)).

From these quotes, it should be clear that the formation of expectations cannot only be viewed on an individualistic basis but has to be seen also as influenced by the individual's contacts with its **social environment** - this also implies TV and press as the eminent forms of **mass communication** - which have to be part of the subject of forming expectations in order to get a meaningful theory about expectations.

As it would be too ambitious to try a complete theory of expectations formation, the following arguments focus only on the second phase, the social interaction, taking individually formed expectations for given and therefore neglecting this step.

3. The model

Expectations of individuals are essentially influenced by the expectations of other individuals they are in contact with. This relationship can be taken as an immediate starting point of our model of expectations formation, thus delivering these assumptions:

- influence of individuals is mutual
- it happens in the course of communication

The overall result of these rules was checked using a simulation model which copies the process of mutual affection in a large enough population (256 "persons" or even more, if simulation time gives no restrictions). The basic concept is given by the following elements:

The strength of an expectational attitude of a single individual was reflected by a variable E_i, $i=1,\ldots,n$, where the range of values was restricted to the interval (-100,100). A negative E_i indicated the **rejection** of a certain expectation. Values around zero were signalling irresolution (no distinct expectation at all). Moreover, a certain threshold was fixed where an expectational attitude was assumed to turn to be distinct or known. In most simulations this threshold was fixed at 35 or 40.

If an individual meets another, they are supposed to enter discourse, i.e. communication occurs which will influence the expectational attitudes of both as a result of simply talking together. This process of change of expectational attitudes is sketched by equation (1) for a single communication act.

$$E_i^{+} = E_i + (1-\beta_i)*E_j*f \qquad (1)$$

where β_i is a constant, reflecting the individual tendency to stick to his or her attitude and f a general adaption-coefficient which has the task to "size" the alteration term into a reasonable range. In simulation runs, it was assumed that in every period t, t=1,..,35, each individual i, i=1,..,256, had communication with several neighbours j and f was usually fixed to 0.1.

The strength value of an indivual expectation could lie in one of the following 4 intervals:

(-100,-35(; (-35,0(; (0,35(; (35,100)

Therefore, principally 16 possible situations of communication exist as fig. 1 shows:

E_i \ E_j	--	-	+	++
--	>>	>>	><	<>
-	>>	>>	0	><
+	><	0	>>	>>
++	<>	><	>>	>>

Fig. 1 Table of all combinations in communication

Each of the 16 cases was inspected for the plausibility of the communications results. It turned out that in 8 of the 16 given cases (same sign of attitude-strength) equation (1) (i.e. mutual acknowledgement and **confirmation**, that means fostering of the existing attitude, >>) produced a very acceptable result. The same equation (1) described a mutual **approach** of expectational attitudes (><), in 4 cases (differing signs of E, but only one above threshold), which also made very much sense. Only in the last 4 cases was the rule of equation (1) abandoned (same interval but different sign of E): If both were **below the threshold**, then no change (0) was assumed because of lacking impulse on the side of both communicators. If both were **above the threshold**, it was assumed that the confrontation of the differing attitudes resulted in a further polarization of the already existing expectational attitudes (<>).

This was achieved by simply changing the sign of the alteration term in equation (1). This assumption however, was restricted to the condition that both individuals had a coefficient of persistence $\beta > 0.5$. The behaviour of polarization not only may claim plausibility, but proved to exist in experimental situations of empirical communications research (Schneider (19)).

Fig. 2 shows a typical result of the dynamic simulation of the above model for 16*16 = 256 "individuals" over 35 periods. Starting with a nearly normal distributed set of basic expectational attitudes E_i, with the bulk of "persons" below threshold (marked "+" if pro- and "-" if contra-expectations are given) i.e. with latent expectations and only a few "conscious" keepers of expectations (pro =● or contra = o) the process of

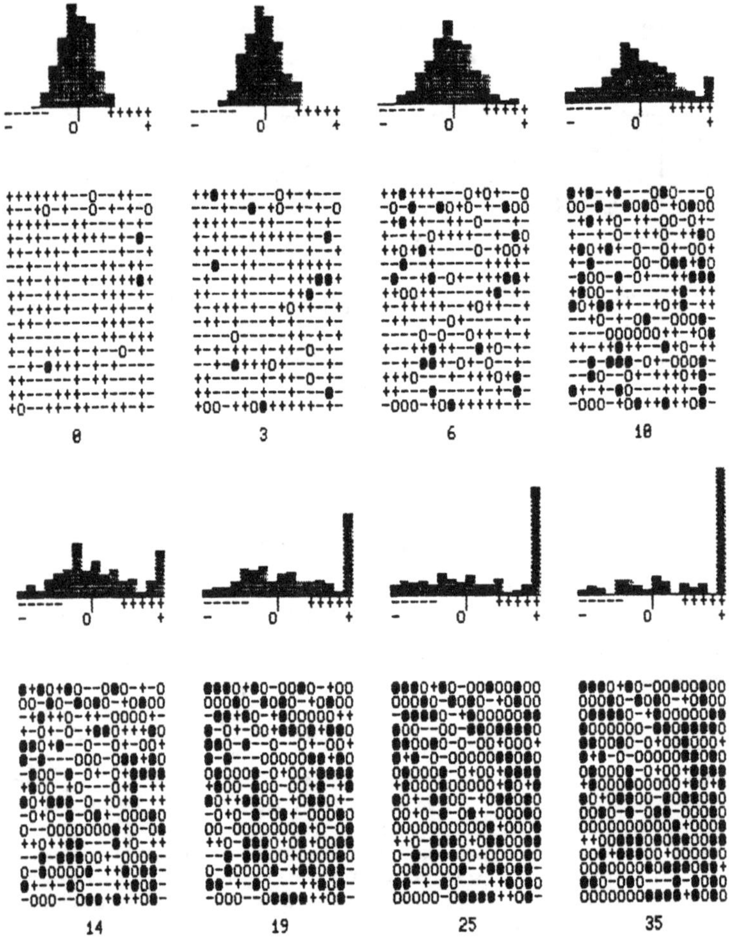

a)

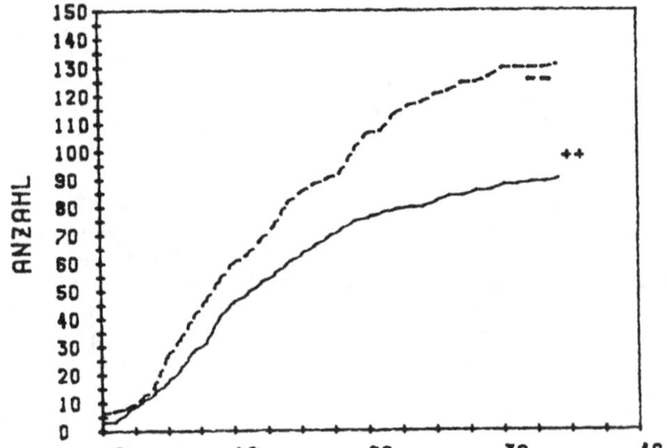

Fig. 2 a,b
Standardexample of
a simulation result
● = ++
o = --

b)

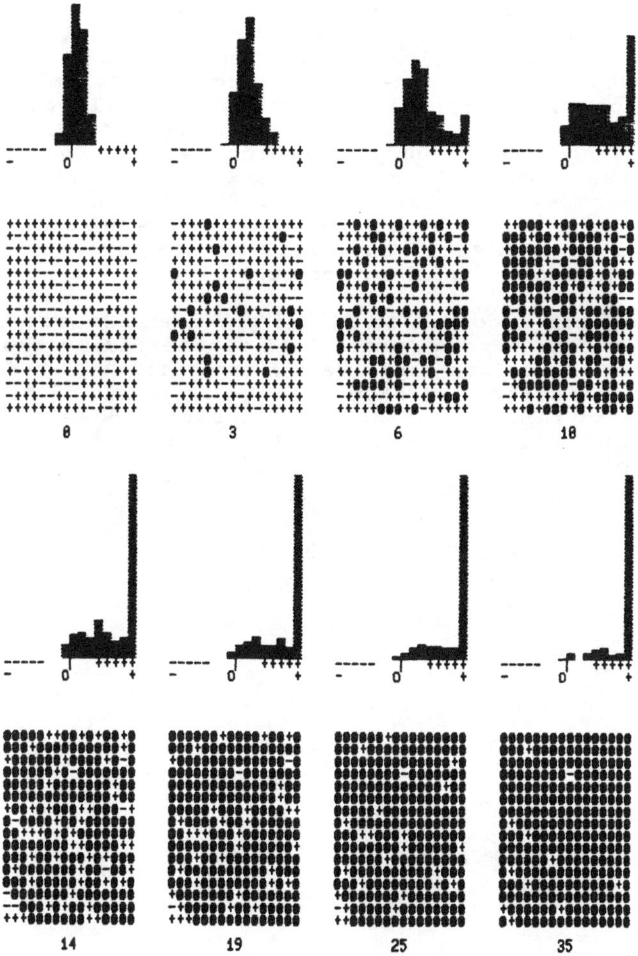

a)

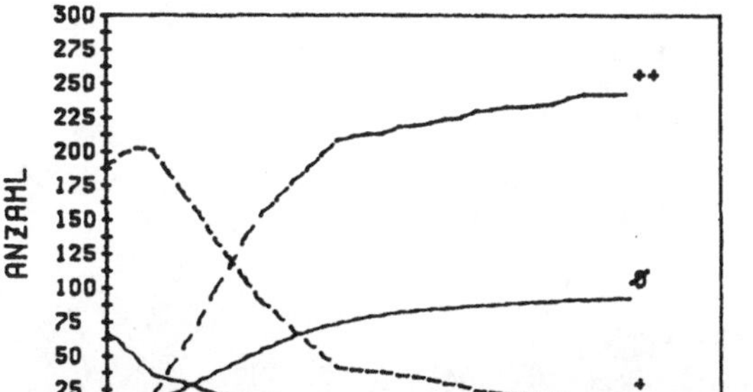

b)

Fig. 3 a,b
Another simulation result: dominance of one group (++)
● = ++

expectations diffusion goes on, appearing in the sequence of topological plots as well as in the corresponding distributions of strengths of expectation plotted above. This result is due primarily to the **reinforcement-effect** which stems from the confirmation by a congenial individual, as well as the fact that this situation covers statistically about 50% of all cases. Showing some latency, the process - given a critical threshold was passed - booms until a phenomenon of saturation takes place. The second example (fig. 3) shows that one of both possible expectational attitudes can assume absolute dominance if there is an even small asymmetry between pro- and contra-groups in the beginning of the process.

The development recorded in fig. 3 gives an example that the dominance of the strong "pros" (++) can be fed by the latent positive (+) as well as negative (-) expectations, where a symmetric "polarisation" of expectations as in fig. 2 didn't get any chance to rise as the ground for such a development was occupied right from the beginning. This course also shows up in the curve of mean values of all E_i (Ø).

This pattern, which is a result of interactions on the microanalytical level, resembles the curve of infection as is well known from other sciences such as biology or even marketing. The behaviour manifests itself as a **diffusion process of expectations** at the macroanalytic level, the basics of which are defined by interactive communication at the **microanalytical level.** Thus, the formation of expectations is at least in part an epidemical process.

This fact gives us the possibility of replacing the modelling of the interactions at the microanalytic level - which has to cover a lot of details in the simulation - by a **macroanalytic** formulation via the well known diffusion processes (cf. Bass (3) or for a survey Kaas (**8**)). Especially in the range of marketing, the **Bass-model** is regarded as standard for the scientific research on diffusion. However, in 1845 the same basic structure was developed by **Verhulst**, in order to describe the growth of an animal population (Peitgen/Richter (**15**, p. 23)). The basic idea is the **infection** resp. the transfer of something (bacteria, quality, or even an idea or an opinion) from that part of the population that is already "infected" to other, not yet infected members of that population . The velocity of the transfer (which corresponds to the increase in percentage of infection per period) is proportional to three quantitities:

- a coupling (=transfer-)constant k
- the percentage of the infected part p and
- the not yet infected "pool" still to be infected, (1-p)

In the so-called **Verhulst-equation** p_{t+1} is a function of these factors as well as of the value p_t

$$p_{t+1} = p_t + k*p_t*(1-p_t) \qquad (2)$$

We can use the Verhulst-equation in order to describe the development of expectations diffusion. p_t would then be the percentage of a population i.e. people, with a well defined expectation at a given time t, as e.g. "the price level will remain constant" or "we will have a period of boom". The simple structure of (2) considerably facilitates the analysis as compared to the very detailed simulation model. Moreover, the Verhulst-equation offers the advantage of having been researched thoroughly by experimental mathematics (May (**13**)). Therefore the application of this structure even in the field of economic expectations could turn out to be fruitful.

4. Dynamic properties

Fig.4 shows a graphical plot of an iteration of equation (2) for alternative coupling-coefficients

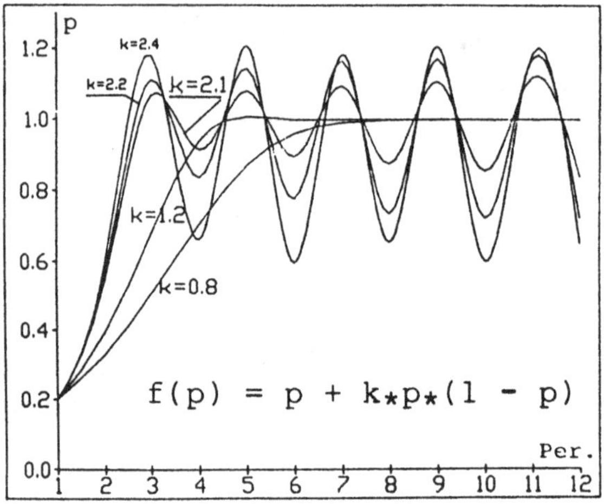

Fig.4 Iterations of the Verhulst-equation for different k-rates

If the coupling coefficient is less than 1, the result of the iteration in time is a logistic curve with saturation level 1. For coupling coefficients k greater than 1, we get oscillations which evolve increasing amplitudes beyond k = 2. A 3-dimensional plot (fig. 5) showing the time periods on the x-axis, the coupling coefficients from 0 to 2.5 on the y-axis and p on the z-axis gives a good overlook.

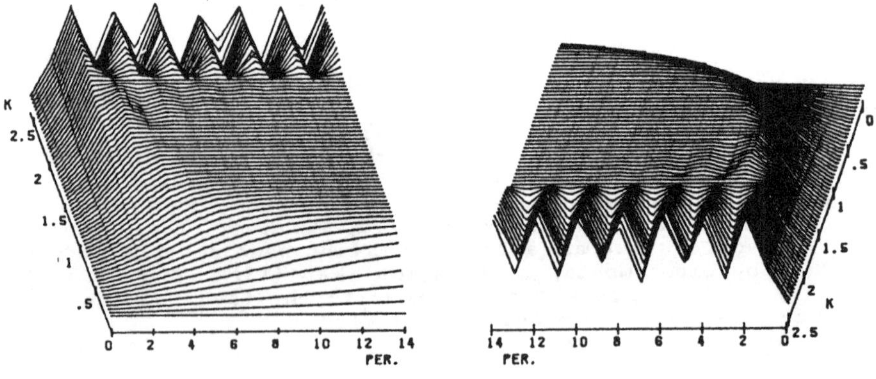

Fig. 5 3-D-Plot of the diffusion behaviour

While the pattern of a logistic curve with saturation level of 100 % was rather predictable in the case of the oscillations for k greater 1, especially those for k greater 2 are a surprising feature of the model which need to be examined closer.

Fig. 6 shows the **Feigenbaum-diagram** of the system for a range of the coupling coefficient k between 1.8 and 3.0. It represents the final result of p after a sequence of iterations (100 - 200) for different values of k where the first 100 iterations are not plotted due to their provisional character on their course to the final attractor.

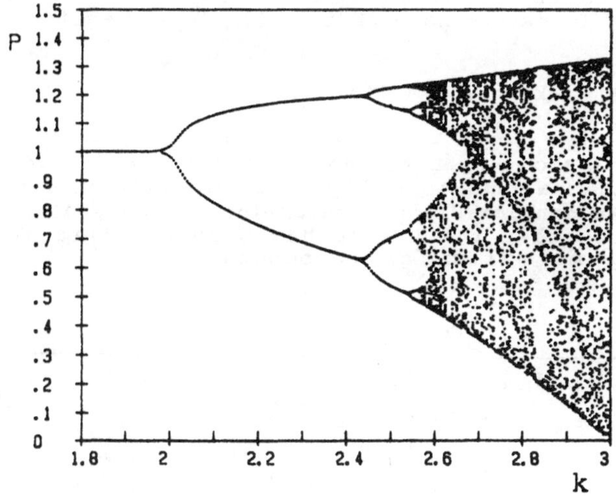

Fig. 6 The Feigenbaum-diagram of the Verhulst-equation

Although we know from fig. 4 that the model already produces moderate oscillations for k greater than 1, iterations always end up with p=1, no matter how long the iteration endures. This reaction appears to be **stable** and actually **independent** of the starting value of p_0 for all k values less than 2. In this range, the value of p=1 is the **attractor** of the system as the result is always the same, irrespective of the starting point.

If k=2 or slightly more, the attractor does not consist of one single line but of two; if k is 2.45, they double to 4 and at k=2.544 to 8 branches and so forth. Fig. 6 shows for k greater than 2.544, more and more bifurcations occur similiar to those formed before. This feature of self-similarity is due to the recourse which equation (2) takes with respect to its own results. (The gaps between the clouds of points are not mistakes, but so-called periodical windows).

Behind the feature of continued bifurcation, stands the dimension of "chaotic behaviour": Which one of the many possible points that will be reached in a certain situation cannot be predicted. Thus, the Feigenbaum-graph shows the transition from **order** (i.e. stability of behaviour) to **chaos.**

This transition can be made more understandable by a thorough analysis of equation (2). For this purpose, we draw the so-called phase diagram of the system by plotting consecutive values of p in a graph with the axes p and f(p) (cf. fig. 7). This graph shows points on the angle bisection for the "order phase" (f(p) = p) but jumps to different points of the parabola in the "chaos phase" depending on the k-value of the equation. These parabolas can directly be derived from the Verhulst-equation by transforming equation (2). Thus we get:

$$f(p) = -k*p^2 + (k+1)*p \qquad (3)$$

This is a parabola open to the p-axis, the width of which varies with k. If we additionally plot the angle bisection of the quadrant, we can immediately determine by means of a graphical iteration whether the system has only one (fig. 7a) or two (fig. 7b) or more (fig. 7c) end-points.

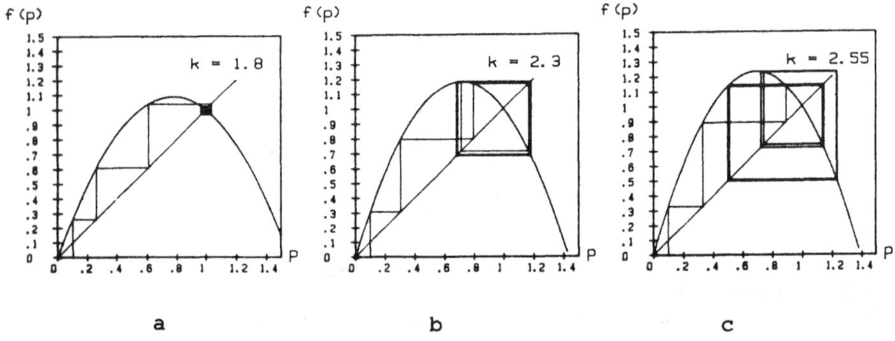

Fig. 7 Graphical iteration of the p-values

Starting with p_o (In the example p_o was always set to 0.1) we find the corresponding point f(p) on the parabola. Since after the recoursive formula of equation (2) for the next period f(p) replaces the old p we picture this fact by drawing a horizontal line till the angle bisection (where f(p)=p). For this **new** value of p, we find the corresponding f(p) by drawing again the vertical line through the new p till the parabola and so on.

5. The interpretation of the model-structure

In paragraph 4., we discussed the system theoretical implications of the Verhulst-equation. But the question how to interpret the behaviour of this system in economic terms is a problem of its own. Since the structure is formed recoursively it is indispensible that each value p_t be realized, in order to produce the following value p_{t+1}. Since in reality no more than 100% of the population can have the same expectation, the appearance of a chaos-phase would then be impossible, which would mean a more trivial result of the underlying structure.

Of course, this does not mean that we are unable to obtain any valuable economically interpretable results from the "normal" diffusion state characterized by coupling coefficients belonging to the order phase. However, a theory that misses the onset of chaos, would without doubt, be somewhat castrated or deficient as compared to the creativity a complete theory could produce.

The restrictions sketched above can however, be avoided by a simple asumption which well meets reality. As everybody knows, even during an epidemic of influenza some people are not infected. They are just "immune" towards the virus. Its the same thing with communication. The reason for this are certain barriers existing that have to be overcome in order to change one's mind. There has to happen " a little bit more than usual" to jump over this barrier. If we assume that in a given group ca. 20% are immune, the above Verhulst-equation (2) would change to

$$f^*(p) = p + k*p*(0.8 - p) \qquad (4)$$

the graphical presentation of which is shown in fig. 8

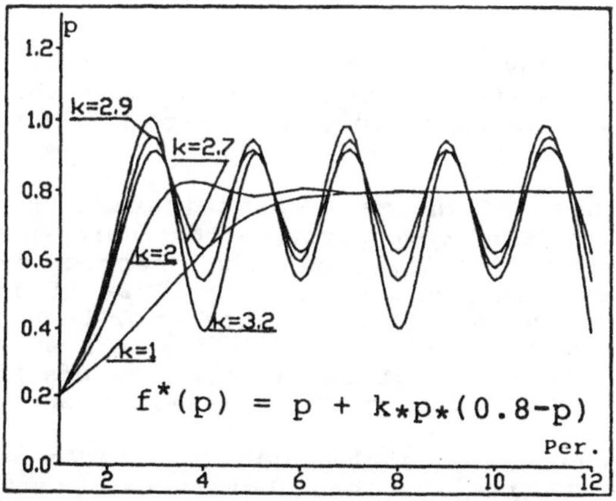

Fig. 8 Verhulst-iteration with 80%-barrier

As a consequence of the introduction of this limit, the percentage of infection does no more transgress 100 % of the whole population. For this equation, the downward movements are more distinct than they were in the original Verhulst-equation. Since this new equation shows the same behaviour as the original one, there is no need for further representation because it only would be a repetition of figures just shown above, except for the shift of the critical coupling coefficients, as can be seen from fig.8.

6. Conclusions

The Verhulst equation in the modified form

$$f^*(p) = p + k*p*(0.8-p)$$

is able to describe the process of expectations formation on the basis of an underlying diffusion within two regimes:

- the coupling coefficient is either in a range where "order" is the result. This can even yield an oscillation between two or even some more points, since we have a nonlinear system.

- or the coupling coefficient takes values that produce "chaos" in the formation of expectations

For reasons of simplicity we will talk in the following of "stable" and "chaotic" coupling coefficients.

For the regime of stable coupling coefficients that may reflect "normal" conditions of communication, the described behaviour of expectations formation could be a basis for endogenizing (heterogenous) expectations into economic theory. This would better integrate the factor "time" which is an essential variable of the process and thus help to ameliorate the modelling of the dynamics of economic processes that involve cognitive or expectational features.

Beyond the range of "normal" communication the above model showed how intensified communication that corresponds to higher coupling coefficients, may lead to a "chaotic" dynamic of the underlying diffusion process. They could be of importance for special markets in special situations (e.g. bonds or shares at the stock exchange) as well as for situations sensitive to the development of business cycles, if expectations become crucial for the volume of consumer goods or investments.

The great advantage of such a model to be developed on the basis of the dynamics sketched above would be that we would need only **one** structure in order to describe both kinds of processes.

REFERENCES

(1) ALLPORT,G.W.:Persönlichkeit. Struktur,Entwicklung und Erfassung der menschlichen Eigenart. Meisenheim am Glan, 1959

(2) BARRO,R.J. AND FISCHER,S.: Recent Devolopments in Monetary Theory,Journ. Monet. Econ. 2(1976),pp.133-167

(3) BASS,F.M.:A New Product Growth Model for Consumer Durables Management Science 15(1969),pp.215-227

(4) CONLISK,J.: Costly Optimizers Versus Cheap Imitators, Journal of Economic Behavior and Organization 1(1980) pp.275-293

(5) HALTIWANGER,J.,WALDMAN,M.: Rational Expectations and the Limits of Rationality: An Analysis of Heterogeneity. Amer. Ec. Review 75(1985), pp. 326-340

(6) HEY,J.D.: Economics in Disequilibrium, Oxford, 1981

(7) HOFSTÄTTER,P.: Gruppendynamik. Die Kritik der Massenpsychologie, Hamburg, 1957

(8) KAAS,K.: Diffusion und Marketing, Stuttgart, 1973

(9) KATONA,G.: Psychological Analysis of Business Decisions and Expectations, in: Amer. Ec. Review 36(1946), p.51ff

(10) LAIDLER,D.: Money and money income: An Essay on the 'Transmission Mechanism', Journ. Monet. Econ. 4(1978), pp.151-191

(11) LAVINGTON,F.: The Trade Cycle. An account of the causes producing rythmical changes in the activity of business. London, 1925

(12) MACHINEK,P.: Behandlung und Erkenntniswert der Erwartungen in der Wirtschaftstheorie. Berlin, 1968

(13) MAY,R.M: Simple mathematical models with very complicated dynamics. Nature 261(1976) pp.459-467

(14) NERLOVE,M.: Expectations, Plans and Realisations in Theory and Practice, Econometrica 51(1983), pp.1251 -1279

(15) PEITGEN,H.O., RICHTER,P.H.: The Beauty of Fractals. Berlin a.o.,1986

(16) VAN RAAIJ,W.F.: Micro and macro economic psychology. in: Journ. Ec. Psych.5 (1984), pp.385-401

(17) RIBHEGGE,H.: Grenzen der Theorie rationaler Erwartungen. Tübingen, 1987

(18) SCHLOTTHAUER,K.H.: Inflationserwartungen. Hamburg, 1981

(19) SCHNEIDER,FR.: Journalismus als Vermittlung oder als Proporzritual. Spectrum der Wissenschaft, March 1985, p.32f

(20) SHILLER,R.J.: Rational expectations and the dynamic structure of macroeconomic models. Journ. Monet. Econ. 4 (1978), pp.1-44

(21) STEINBUCH,K.: Automat und Mensch. 4th Ed. Berlin, 1971

TECHNOLOGICAL CHANGE AND INCREASING RETURNS

Frank C. Englmann

Universität Tübingen

1. Introduction

In recent years increasing returns phenomena have been recognized as central to several areas of economics (trade theory, industrial organization, economic development, and economic history). In what follows we shall investigate the importance of increasing returns phenomena in the economics of technological change.

In the theory of the choice of technique one usually assumes that one new technique is added to the book of blueprints whose actual efficiency equals its potential value right from the point in time of its first introduction. Then, applying methods of comparative **statics**, it is determined under which conditions it is profitable to introduce the new technology. In this way such characteristic increasing returns phenomena like **learning by doing** [2] and **learning by using** [8] are neglected.

In the present paper we shall assume that **Schumpeterian entrepreneurs** are introducing **two** new technologies. Both of them are characterized by learning by doing, i.e. their efficiency increases, starting from a given initial value, as a function of cumulated investment into the respective technology. One can imagine that the two technologies are two forms of one basic innovation like direct current and alternating current in the case of electricity [3].

Here the question arises whether always that technology is finally chosen which is characterized by the highest potential of efficiciency. This question can only be answered if the diffusion process of the new technologies is explicitely considered in a **dynamic** model. Taking up ideas earlier proposed by Prigogine, Allen [7] and Silverberg [9] we treat the introduction of new technologies by Schumpeterian entrepreneurs as fluctuations in Goodwin's "Growth Cycle" [5].

To each technology (the old one and the two new ones) corresponds one sector in which one good is produced by means of itself and labour. The goods which are produced in the various sectors differ from each other only if they are used as capital goods, not if they are consumed. In this sense we are considering process innovations. In the following we shall see that the final outcome of the diffusion process is heavily depending on the assumptions concerning factor substitutability, saving and imitation.

The remaining of the paper is organized as follows. In section 2 the so-called putty-putty case is considered, where the sectoral production functions are of the Cobb-Douglas type. In section 3 the same analysis is carried out for the so-called clay-clay case, where the sectoral production functions are of the Leontief type. Whereas imitation is excluded in section

2 and 3, its effects are analysed in section 4. As the sections 2 and 4 will show, the technology characterized by the highest potential of efficiency will always "win" in the end, if the sectoral production functions are of the Cobb–Douglas type. Therefore, we shall shortly raise the question in section 5, under which assumptions concerning the investment behaviour of the respective Schumpeter–entrepreneurs this conclusion is no longer true.

2. Putty–putty: the model

The three production functions are given by eq. (1):

$$R_i = a_i L_i^{b_i} K_i^{1-b_i}, \qquad a_i > 0,\ 0 < b_i < 1, \quad i=1,2,3 \qquad (1).$$

Index 1 refers to the old technology, index 2 to that new technology which is charactarized by the lower potential of efficiency relative to the other new technology (index 3). R denotes net output, L labour employed, K capital employed, a the actual level of technological efficiency, b the elasticity of production of labour.

The labour supply L^s increases at the natural rate of growth n

$$L^s = L^s(0)\, e^{nt}, \qquad L^s(0),\ n > 0 \qquad (2).$$

The rate of change of the real wage w is determined by means of a linearized bargaining equation:

$$\hat{w} = -m + 1 \cdot v, \qquad 1 > m > 0 \qquad (3),$$

where v denotes the rate of total employment which is defined as

$$v = (L_1 + L_2 + L_3) \,/\, L^s \qquad (4).$$

Eq. (3) determines the equilibrium of employment at which the real wage is constant as

$$v^* = m \,/\, 1 \qquad (5).$$

If the rate of total employment exceeds v^*, the real wage increases and vice versa.

Assuming that the commodities are competitively supplied and labour is competitively demanded by each single firm the amount of labour employed in each sector is determined by the condition of profit maximization:

$$w = \frac{\partial R_i}{\partial L_i} = a_i\, b_i\, (K_i/L_i)^{1-b_i}, \qquad i=1,2,3 \qquad (6).$$

Neglecting imitation at this stage of the argument we assume that the capitalists invest all the savings into the sector in which they are already operating. Thus we have:

$$\dot{K}_i = s_c\, (1 - b_i)\, R_i, \qquad i=1,2,3 \qquad (7)$$

where s_c denotes the capitalists' propensity to save.

We assume that in the two new sectors the levels of technological efficiency, starting from a level which lies above the actual and potential level of technological efficiency in the old sector, increase as a function of cumulated investment in the respective sector up to their potential level:

$$a_i = a_i(K_i), \qquad da_i/dK_i \geq 0, \qquad i=1,2,3 \qquad (8),$$

where $da_1/dK_1 = 0$. Defining the sectoral capital-labour supply ratios as

$$k_i = K_i \,/\, L^s, \qquad i=1,2,3 \qquad (9)$$

and the sectoral rates of employment as

$$v_i = L_i \,/\, L^s, \qquad i=1,2,3 \qquad (10)$$

we can finally find the following system of differential equations describing the time paths of the sectoral capital-labour supply ratios, of the real wage and of the labour supply:

$$\dot{k}_i = k_i \left\{ s_c (1 - b_i) \left[a_i (k_i L^s) \right]^{\frac{1}{1-b_i}} b_i^{\frac{b_i}{1-b_i}} w^{\frac{-b_i}{1-b_i}} - n \right\}, \qquad i=1,2,3 \qquad (11)$$

$$\dot{w} = w \left\{ -m + l \sum_{i=1}^{3} \left[a_i (k_i L^s) b_i \right]^{\frac{1}{1-b_i}} k_i w^{\frac{-1}{1-b_i}} \right\} \qquad (12)$$

$$\dot{L}^s = n L^s(0) e^{nt} \qquad (13).$$

Obviously there does not exist a stationary point in the system of differential equations (11) - (13), as dL^s/dt always remains positive. But eq. (13) can be neglected, if both k_2 and k_3 vanish. Thus, before the introduction of the new technologies the equilibrium values of the capital-labour supply ratio and of the wage rate are given by:

$$w_1^* = (n/s_c)^{\frac{-(1-b_1)}{b_1}} (1 - b_1)^{\frac{1-b_1}{b_1}} a_1^{\frac{1}{b_1}} b_1 \qquad (14)$$

$$k_1^* = (m/l) \left[(s_c/n) (1 - b_1) a_1 \right]^{\frac{1}{b_1}} \qquad (15).$$

As we assume that the two new technologies are more efficient already at the point in time of their first introduction, i.e. before any learning by doing takes place, we have

$$w_i' > w_1^*, \qquad i=2,3 \qquad (16),$$

where

$$w_i' = (n/s_c)^{\frac{-(1-b_i)}{b_i}} (1 - b_i)^{\frac{1-b_i}{b_i}} a_i'^{\frac{1}{b_i}} b_i, \quad i=2,3 \qquad (17)$$

and

$$a_i' = a_i(0), \qquad i=2,3 \qquad (18).$$

The evaluation of the Jacobian belonging to the system of differential equations (11) and (12) at the original equilibrium $w = w_1^*$, $k_1 = k_1^*$, $k_2 = k_3 = 0$ shows that this equilibrium becomes unstable, if condition (16) is fulfilled. Under this condition, initially, both the new technologies are successfully introduced, because they allow for a higher rate of profit than the incumbent technology at $w = w_1^*$. To put in another way: as in the neo-ricardian theory of the choice of technique a technology is introduced if it allows for a higher real wage at the equilibrium rate of profit (n/s_c).

The question whether a technology that is characterized by the highest potential of efficiency, in our case technology 3, can be crowded out of the market during the diffusion process by the other new technology, in our case technology 2, can be formulated in the following way: do there exist conditions which ensure that the stationary point ($w = w_2^*$, $k_1 = 0$, $k_2 = k_2^*$ and $k_3 = 0$) is locally stable?

w_2^* and k_2^* are determined in an analogous way as w_1^* and k_1^*:

$$w_2^* = (n/s_c)^{\frac{-(1-b_2)}{b_2}} (1 - b_2)^{\frac{1-b_2}{b_2}} a_{2m}^{\frac{1}{b_2}} b_2 \qquad (19)$$

$$k_2^* = (m/l) \left[(s_c/n)(1 - b_2) a_{2m}\right]^{\frac{1}{b_2}} \qquad (20).$$

a_{2m} denotes the maximum level of technological efficiency of technology 2. In order to answer the question concerning the stability of the stationary point ($w = w_2^*$, $k_1 = 0$, $k_2 = k_2^*$ und $k_3 = 0$) it is not enough to determine the sign of the eigenvalues of the Jacobian belonging to the eqs. (11) and (12). Eq. (13) also has to be taken into consideration.

In the model developed above the sectoral rates of accumulation are determined by the respective marginal productivities of capital multiplied by the capitalists' propensity to save. As is well known, in the case of a Cobb-Douglas production function, the marginal productivity of capital only goes to zero, if the real wage goes to infinity. Thus, at $w = w_2^*$, the capital stock in sector 3 increases, given the assumptions concerning the investment behaviour, if a small amount is introduced by Schumpeterian entrepreneurs. According to eq. (8) this leads to a steady increase of the actual level of technological efficiency a_3 up to its maximum level a_{3m}. By assumption we have

$$a_{3m} > a_{2m} \qquad (21).$$

In the system of differential equations (11) - (13) the equilibrium point ($w = w_3^*$, $k_1 = k_2 = 0$, $k_3 = k_3^*$) is the only attractor. Thus, for finite values of a_{2m} und a_{3m}, that technology always "wins" in the long run which is characterized by the highest potential of efficiency, if the production functions are of the Cobb-Douglas type.

In the next section we are going to ask the question, whether this conclusion remains unchanged if the production functions are of the Leontief type.

3. Clay-clay: the model

In this case we have the following production functions

$$R_i = \min \left[(1/c_i) K_i, d_i L_i \right], \qquad i=1,2,3 \qquad (22),$$

where c denotes the capital-output ratio and d the labour productivity. Assuming the sectoral capital stocks to be fully employed the corresponding levels of employment of labour are determined as follows:

$$L_i = \frac{1}{c_i d_i} K_i, \qquad i=1,2,3 \qquad (23).$$

Again we assume that the profits are reinvested into the same sector. Following Nelson and Winter [6] losses lead to a scrapping of capacities in that very sector:

$$\dot{K}_i = \begin{cases} K_i s_c (1 - w/d_i) / c_i & \text{if } w < d_i \quad i=1,2,3 \\ K_i (1 - w/d_i) / c_i & \text{if } w > d_i \quad i=1,2,3 \end{cases} \qquad (24).$$

As the capital-output ratio is more or less constant in the long run we assume that **learning by doing** just affects the sectoral labour productivities. Thus we have

$$d_i = d_i(K_i), \qquad dd_i/dK_i \geq 0, \qquad i=1,2,3 \qquad (25),$$

where $dd_1/dK_1 = 0$. Taking into account eqs. (9) and (10) we finally get the following system of differential equations:

$$\dot{k}_i = \begin{cases} k_i \left[s_c \left(1 - w/d_i\right) / c_i - n \right] & \text{if } w \leq d_i, \quad i=1,2,3 \\ k_i \left[\left(1 - w/d_i\right) / c_i - n \right] & \text{if } w > d_i, \quad i=1,2,3 \end{cases} \tag{26}$$

$$\dot{w} = w \left(- m + 1 \sum_{i=1}^{3} \frac{k_i}{d_i \, c_i} \right) \tag{27}$$

$$\dot{L}^s = n \, L^s(0) \, e^{nt} \tag{28}.$$

In order to make the putty-putty and the clay-clay case numerically comparable we determine the capital-labour ratios and the labour productivities as follows:

$$c_i = (1 - b_i) \cdot s_c / n \qquad i=1,2,3 \tag{29}$$

$$d_i = (1/c_i) \, (a_i \, c_i)^{\frac{1}{b_i}} \qquad i=1,2,3 \tag{30}.$$

Then, in both cases, w_i^*, w_i' and k_i^* ($i = 1,2,3$) have the same values. Therefore the original equilibrium ($w = w_1^*$, $k_1 = k_1^*$, $k_2 = k_3 = 0$) again becomes unstable, if the two new technologies are introduced by Schumpeterian entrepreneurs.

Again we assume that the third technology has the highest potential of efficiency. But now we add the assumptions that (a) the second technology has an early lead and (b) it allows for quicker learning processes. Again we are asking the question whether, given these assumptions, the second technology can crowd the third technology out of the market during the diffusion process. Again this amounts to asking the question whether the equilibrium ($w = w_2^*$, $k_1 = 0$, $k_2 = k_2^*$ und $k_3 = 0$) can be locally stable. Obviously this is the case, if

$$w_2^* > d_3' = d_3(0) \tag{31}.$$

Under this condition losses occur in the third sector, if the economy is in the neighbourhood of the equilibrium ($w = w_2^*$, $k_1 = 0$, $k_2 = k_2^*$, $k_3 = 0$). These losses lead to a scrapping of capacities and block learning processes in the third sector. The labour productivity remains at its initial value d_3'. The third capital good is crowded out of the market.

In the putty-putty case an analysis of the **local** stability was enough to establish the impossibility of the stability of the equilibrium ($w = w_2^*$,

$k_1 = 0$, $k_2 = k_2^*$, $k_3 = 0$). This is different in the clay-clay case. As we assume that both new technologies are successfully introduced initially, there does occur learning by doing also in the third sector. This makes a **global** stability analysis necessary.

Furthermore, the equilibrium ($w = w_3^*$, $k_1 = k_2 = 0$, $k_3 = k_3^*$) is locally stable, too. In the clay-clay case the locally stable equilibria are centers which are not *asymptotically* stable, whereas in the putty-putty case the locally stable equilibrium is also *asymptotically* stable. Thus, in the clay-clay case the whole $w - k_2$ plane, given $k_1 = k_3 = 0$, and the whole $w - k_3$ plane, given $k_1 = k_2$ 0, are attractors, if condition (31) is fulfilled. In the following the former will be called attractor 2, the latter attractor 3.

The locus

$$w = d_3(k_3 L^S) \qquad (32).$$

is of a special interest. When $w > d_3(k_3 L^S)$, capacities are scrapped and, thus, learning is blocked in sector 3, whereas accumulation and learning is going on in sector 3, when $w < d_3(k_3 L^S)$. As we assume that the economy is in the neighbourhood of the original equilibrium ($w = w_1^*$, $k_1 = k_1^*$, $k_2 = k_3 = 0$), when the two new technologies are introduced for the first time, initially we have $w < d_3(k_3 L^S)$ because of $w_1^* < d_3(0)$.

This means that the economy is in the domain of attraction of attractor 2 (domain 2), if and only if the initial values of the variables and the parameter values are such that the real wage exceeds the labour productivity of technology 3 during the diffusion process. For this to happen it is necessary that the learning process in sector 3 is sufficiently slow. This can be modeled by low values of dd_3/dK_3.

This leads us to the question how the time path of the real wage is determined in the model outlined above. The equilibrium wage rates w_i^* are determined by the equilibrium rate of profit (n/s_c) and the technological parameters c_i and d_i ($i = 1,2,3$). During the diffusion process the (average) real wage is driven up from w_1^* to w_2^* or w_3^* via the bargaining equation. Therefore, the time path of the real wage depends on the rates of diffusion of the new technologies.

As can be shown explicitly in the case of constant returns, the (relative) rate of diffusion of the currently most efficient technology (positively) depends on the capitalists' propensity to save and the differences between the actual level of labour productivity of this technology and that of the others. In the case of increasing returns these differences depend on the sectoral learning curves. Hence we get characteristic positive feed-back processes. An initial technological advantage leads to a higher rate of accumulation in that very sector which in turn leads to a further increase of productivity etc.

The lower the capitalists' propensity to save, the lower is cet. par. the initial equilibrium wage rate and the lower is the (relative) rate of diffusion of the currently most efficient technology. This implies that the real wage at a given point in time is the lower, the lower is s_c. Therefore it is less "probable" that the economy is in the domain 2 during the diffusion process. This is shown in the figures 1 and 2. [1]

Figure 1:

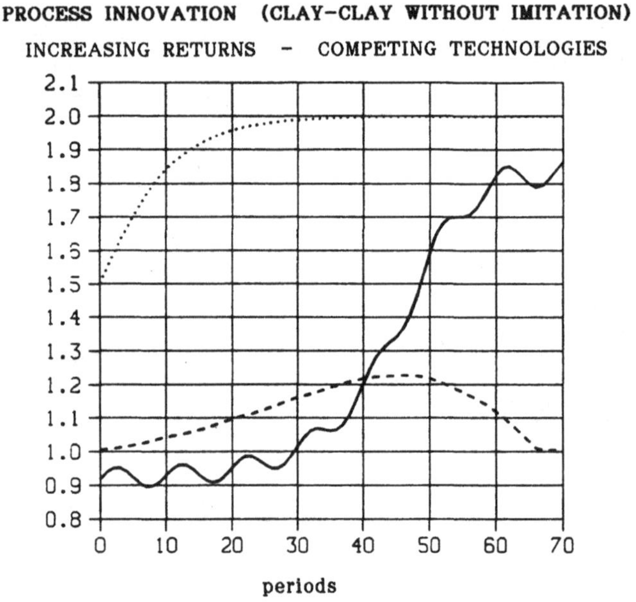

solid: real wage dotted: labour productivity in sector 2
dashed: labour productivity in sector 3
$n = 0.02$ $d_{20} = 1.50$ $d_{2m} = 2.0$ $l_2 = 1.0$ $v_{20} = 0.01$ $s_c = 1.00$
$d_{30} = 1.005$ $d_{3m} = 2.5$ $l_3 = 0.1$ $v_{30} = 0.001$

The parameter values of the simulation runs shown in the figures 1 and 2, just differ with respect to the value of the capitalists' propensity to save. In figure 1 we assume $s_c = 1.0$, in figure 2 $s_c = .5$. In the latter case technology 3 "wins" in the long run, as its actual level of labour productivity exceeds the corresponding value of technology 2 at the end of the time interval under consideration.

This is different in figure 1. Here technology 3 is crowded out of the market, as the economy is in the domain 2. In this case the selection process, mirrored by the (relative) rates of diffusion, is "too" quick relative to the learning process in sector 3. But the rates of diffusion and the time path of the real wage are also influenced by the technological coefficients of the old technology. Thus, contrary to the Arthur-models (see for an overview [1]), the old technolgy also plays a role in the determination of the final outcome of the selection process of the two new technologies.

Figure 2:

PROCESS INNOVATION (CLAY-CLAY WITHOUT IMITATION)

INCREASING RETURNS - COMPETING TECHNOLOGIES

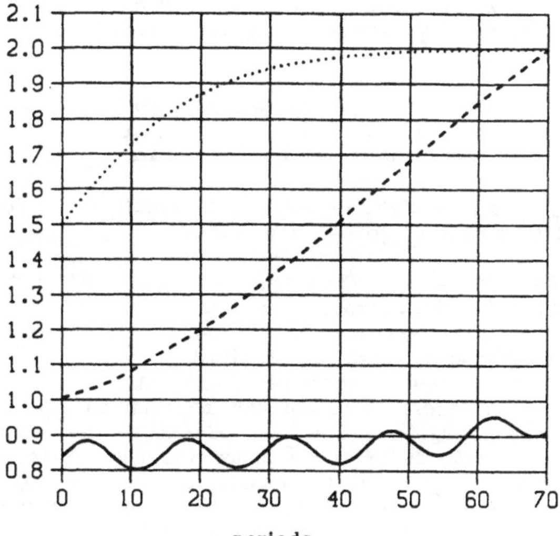

solid: real wage dotted: labour productivity in sector 2
dashed: labour productivity in sector 3
$n = 0.02$ $d_{2o} = 1.50$ $d_{2m} = 2.0$ $l_2 = 1.0$ $v_{2o} = 0.01$ $s_c = 0.50$
$d_{3o} = 1.005$ $d_{3m} = 2.5$ $l_3 = 0.1$ $v_{3o} = 0.001$

In this section we have derived the result that in the clay-clay case without imitation a lock-in to an inferior technology is *possible*. In the next section we shall ask the question, whether such a lock-in becomes more "probable", if imitation is taken into consideration.

4. IMITATION

Up to now we have assumed that only the Schumpeterian entrepreneurs who introduced the new technologies are investing into these two technologies throughout the diffusion process. This implies the unrealistic assumption that the capitalists using the old technology do not react, even if they are crowded out of the market.

In the case of **two** new technologies we have to make an assumption which one of the two is adopted by the capitalists operating in the old sector. We shall assume that they will choose that technology which is characterized by the highest *actual* level of productivity. This means that only the

Schumpeterian entrepreneurs using technology 3 are convinced of its superiority in the long run, as long as its actual productivity level lies below the one of technology 2. Furthermore we shall assume that the entreprises operating in a sector are realizing the same productivity level independently of the point in time of their entry. Thus, with respect to a given technology, there is no difference between internal and external learning (see on this distinction e.g. [4]).

As imitation means that the capitalists operating in sector 1 are investing a new capital good, we have to ask the question how this can be treated in the model outlined above. In principle there are two possibilities.

(a) A production process might be introduced which allows for the production of good 2 or 3 by means of good 1 and labour.

(b) In the modell developed above the new goods differ from the old one only with respect to their production characteristics, not with respect to their consumption characteristics. At the same time the consumption of the new goods both by the workers and capitalists constrains the accumulation of the new capacities. These consumer needs can equally well be satisfied by the old good. Therefore imitation can be modeled in the following way: The capitalists operating in the old sector supply the old good in order to satisfy the consumer needs of the workers and capitalists operating in that new sector whose technology is to be adopted. In this way they are receiving the corresponding amount of the new good which they are going to invest. As in [6] and [9] this implies the unrealistic assumption that the old capital stock can be "eaten up".

In the following we are taking up the alternative (b). Given the production processes this is the most extreme case of imitation which is possible. Nonetheless it does not change the conclusion that in the putty-putty case the technology characterized by the highest potential of efficiency is winning in the long run as long as the corresponding Schumpeterian entrepreneurs are just reinvesting their saved profits earned with that technology.

Assuming identical capital-output ratios we can derive the following system of differential equations in the clay-clay case, as long as $d_2 > d_3$:

$$\dot{k}_1 = \begin{cases} k_1 \left[s_C (1 - w/d_1) / c_1 - n \right] \\ - k_2 \left[1 - s_C (1 - w/d_2) / c_2 \right] & \text{if } k_1 > 0 \text{ and } w < d_1 \\ k_1 \left[(1 - w/d_1) / c_1 - n \right] \\ - k_2 \left[1 - s_C (1 - w/d_2) / c_2 \right] & \text{if } k_1 > 0 \text{ and } w > d_1 \\ 0 & \text{if } k_1 = 0 \end{cases} \qquad (33)$$

$$\dot{k}_2 = \begin{cases} k_2\,(1/c_2 - n) & \text{if } k_1 > 0 \text{ and } d_2 > d_3 > d_1 \\ k_2\,[\,s_c\,(1 - w/d_2)\,/\,c_2 - n\,] & \text{if } k_1 = 0 \text{ and } w < d_2 \\ k_2\,[(1 - w/d_2)\,/\,c_2 - n\,] & \text{if } k_1 = 0 \text{ and } w > d_2 \end{cases} \qquad (34)$$

$$\dot{k}_3 = \begin{cases} k_3\,[\,s_c\,(1 - w/d_3)\,/\,c_3 - n\,] & \text{if } k_1 = 0 \text{ and } w < d_3 \\ k_3\,[(1 - w/d_3)\,/\,c_3 - n\,] & \text{if } k_1 = 0 \text{ and } w > d_3 \end{cases} \qquad (35)$$

The differential equations for the real wage and the labour supply remain unchanged. Imitation obviously accelerates the diffusion process of that technology which is actually characterized by the highest level of technological efficiency.

Figure 3:

PROCESS INNOVATION (CLAY-CLAY WITHOUT IMITATION)

INCREASING RETURNS - COMPETING TECHNOLOGIES

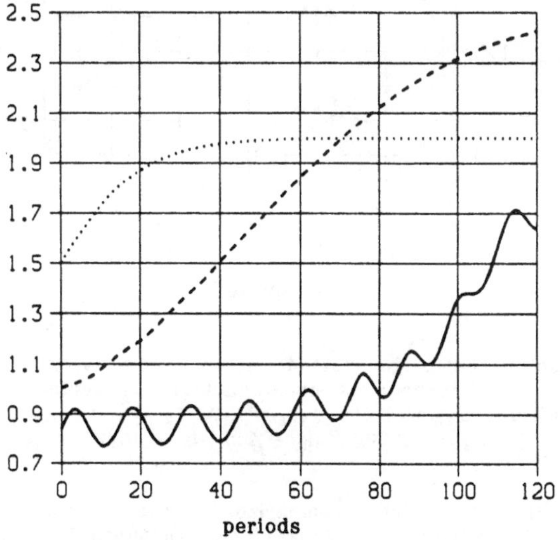

solid: real wage dotted: labour productivity in sector 2
dashed: labour productivity in sector 3
$n = 0.02$ $d_{2o} = 1.50$ $d_{2m} = 2.0$ $l_2 = 1.0$ $v_{2o} = 0.01$ $s_c = 0.50$
$d_{3o} = 1.005$ $d_{3m} = 2.5$ $l_3 = 0.22$ $v_{3o} = 0.01$

Even if the equilibria and the locus given by eq. (32) remain un changed, if imitation is taken into account, this is *not* true for the domains of attraction of the attractors 2 and 3. The real wage increases the faster, the higher is the rate of diffusion of one of the new technologies. By assumption, again technology 2 has an initial technological lead relative to technology 3. Then imitation leads to a higher rate of accumulation and a higher learning rate in the second sector. At the same the higher rate of increase of the wage rate hampers the accumulation and thus learning in the third sector. This widens the domain 2.

Figure 4: [2]

PROCESS INNOVATION (CLAY-CLAY WITH IMITATION)

INCREASING RETURNS - COMPETING TECHNOLOGIES

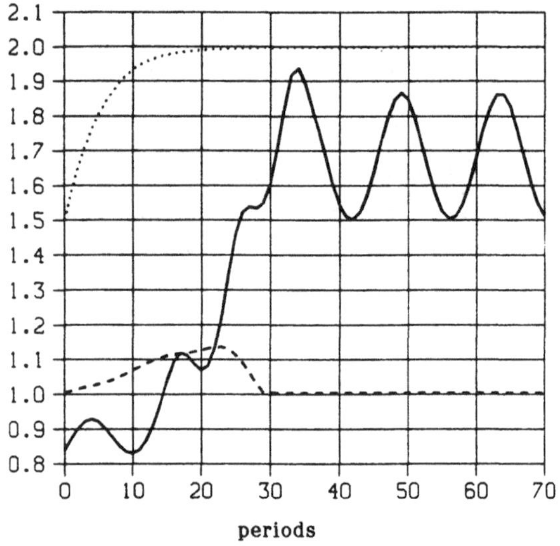

solid: real wage dotted: labour productivity in sector 2
dashed: labour productivity in sector 3
$n = 0.02$ $d_{20} = 1.50$ $d_{2m} = 2.0$ $l_2 = 1.0$ $v_{20} = 0.01$ $s_c = 0.50$
$d_{30} = 1.005$ $d_{3m} = 2.5$ $l_3 = 0.22$ $v_{30} = 0.01$

This is also shown by a comparison of the figures 3 and 4. The parameter values of the simulation runs plotted in these figures are identical. In figure 3 the economy is in the domain of attraction of attractor 3, whereas in figure 4 it is in the domain 2. In the latter case technology 3 is crowded out of the market, as the selection process is accelerated by imitation and the learning process in sector 3 is slowed down because of the decrease of the corresponding rate of accumulation.

5. CONCLUDING REMARKS

As outlined above, in the putty-putty case always that technology is winning in the long run which is characterized by the highest potential of technological efficiency. This result crucially depends on the assumption that the Schumpeterian entrepreneurs introducing technology 3 always reinvest a part of the profits earned in that sector, even if they forgo profits over a long time period.

This leads to the question whether their behaviour is rational. In principle, the present values of two investment projects are to be compared. If we assumed that all the economic agents had perfect foresight, only one new technology would be introduced. There would be no new **competing technologies.** At least some economic agents must not have perfect foresight. Let us assume the capitalists operating in sector 3 to have perfect foresight, the capitalists operating in the sectors 1 and 2 *not* to have perfect foresight. Thus we have a case of **asymmetric information.**

Perfect foresight of the Schumpeterian entrepreneurs potentially operating in sector 3 includes the knowledge about the other capitalists' future actions, especially with respect to imitation, as they influence the sectoral rates of profit and their distribution over time.

For the entrepreneurs potentially operating in sector 3 it is *not* reasonable first to adopt technology 2 and later on technology 3, as in the meanwhile no learning by doing would take place in sector 3. For them it is only rational either to adopt technology 2 once and for all or to adopt technology 3 once and for all. This does not exclude the possibility that the entrepreneurs initially operating in sector 2 and switching to technology 3 later on are earning higher profits. But this strategy is not open to (all) the entrepreneurs potentially operating in sector 3.

Obviously, the outcome of their decision process is heavily depending on their **technological expectations** with respect to the point in time of a further dominant innovation (technology 4, 5,..). The earlier they expect its introduction, the more probable becomes their choice of technology 2.

If we take into account **increasing returns to adoption** in the economics of technical change, there may exist more than one attractor in the corresponding dynamic system. In this case (and only in this case) it is necessary to carry out a dynamic analysis, as, otherwise, it cannot be determined which one of the **competing technologies** is finally chosen.

Given the assumptions specified in the paper more than one attractor only exist in the clay-clay case. Here the final outcome of the diffusion process is crucially depending on the capitalists' saving and imitation behaviour and on the properties of the sectoral learning curves. High rates of saving and imitation accelerate the selection process and increase the probability of a lock-in of a technology that is not characterized by the highest potential of technological efficiency, but has a technological lead in the early phase of the diffusion process.

6. FOOTNOTES

1 The time paths of the real wage and of the labour productivity in sector 3 should intersect in the maximum of the latter. The divergence in figure 1 is due to numerical inexactitudes of the NAG D02BBF subroutine. The assumption that d_3 declines when K_3 is decreasing could be modified by assuming that d_3 remains constant at the maximum level reached during the diffusion process. This would not alter the final outcome that technology 3 is crowded out of the market. In the figures l_2 and l_3 denote parameters of the respective learning curves.

2 Figure 4 does not show the extreme case of imitation corresponding to the eqs. (33)-(35), but a case of "epidemic" imitation where a growing part (x) of the consumption of the agents operating in sector 2 is satisfied by good 1. x is a logistic function depending on the difference between the rates of profit earned in sector 2 and sector 1.

7. REFERENCES

[1] ARTHUR, W.: "Competing Technologies: An Overview," mimeo, Stanford University, 1987.

[2] ARROW, K.: "The Economic Implications of Learning by Doing," *Review of Economic Studies*, 29, 155-173.

[3] DAVID, P.: "The Hero and the Herd in Technological History: Reflections on Thomas Edison and 'The Battle of the Systems'," mimeo, Stanford University, 1987.

[4] DOSI, G., ORSENIGO, L. AND G. SILVERBERG: "Innovation, Diversity and Diffusion: A Self-Organisation Model," paper presented to the *International Conference on Innovation Diffusion*. Venice, 1986.

[5] GOODWIN, R.: "A Growth Cycle," in *Socialism, Capitalism and Economic Growth*, ed. by C. Feinstein. Cambridge: Cambridge Univerity Press, 1967.

[6] NELSON, R. AND S. WINTER: *An Evolutionary Theory of Economic Change.* Cambridge, Mass.: The Belknap Press of Harvard University Press, 1982.

[7] PRIGOGINE, I.: "Order Through Fluctuation: Self - Organization and Social System," in *Evolution and Consciousness*, ed. by E. Jantsch and C. Waddington. Addison - Wesley, 1976.

[8] ROSENBERG, N.: *Inside the Black Box: Technology and Economics.* Cambridge: Cambridge University Press, 1982.

[9] SILVERBERG, G.: "Embodied Technical Progress in a Dynamic Economic Model: The Self - Organization Paradigm," in *Nonlinear Models of Fluctuating Growth*, ed. by R. Goodwin et al.. Berlin: Springer Verlag, 1984.

Micro-Macro-Modeling (MMM) of Schumpeterian Dynamics 1950-1984

by Gerhard Mensch |1| co-author of G. Haag and W. Weidlich, "The Schumpeter Clock", to appear 1988

Abstract

This survey article highlights mayor findings of a ten-year research agenda on "Schumpeterian Dynamics". What is that? A multi-layer, multi-period field of change: corporate innovation processes (micro), industrial rivalry (meso), sectoral transformation (macro). Such changes are horizontally and vertically interrelated; some proceed slowly (long-term), some swiftly (short-term). Our "Schumpeter Clock" model integrates these into a "post-schumpeterian" approach |1| which is viewed in the profession as "neo-neo-classical".

Investment thory is employed as the framework for MMM, in which we combine three new scientific technologies: A compact model architecture for "deap" causation, a mathematical master equation for micro-to-macro effects, and new multi-variate methodology for calculating econometric hypotheses.

The main benefitiaries of the new intellectual apparatus will probably investment planners in industry and finance. Two results, however, appear particularly relevant for economic policy makers and those in charge of "framework-framing":

a) The model-based analysis of industrial data from both the U.S.A. and the F.R.G. reveal field-patterns of change in real economy that appear to be less confusing than apparently assumed by pure macroeconomists (judged from the literature). This knowledge may help reducing some of the "pervasive uncertainty" thus irritating many policy advisors, inducing fear.

b) The field-pattern of change can be approximated by a bi-equilibrium dynamic which postulates (from Management Science) that investors think in terms of a full-employment equilibrium path and a latent under-employment equilibrium path. According to our empirical analysis, German industry branched into the under-employment path in 1974 and stayed there since. If that were true, it is safe to assume that most policy advisors would view as "normal" a continuation of this pattern of low growth and underemployment of labor and capital. As a conjecture follows that risks and chances for a discontinuous development are being greatly misjudged: b1) the down-side risk of a minor or major recession is probably vastly overrated out of fear and b2) the up-side chance for another phase transition (quantum jump up) to faster growth and higher employment is probably vastly underrated - due to past "framework-fixing".

Schumpeterian Dynamics - A Challenge to Economic Science

This book, entitled "The Schumpeter Clock", presents the main results of a ten-year research agenda on Schumpeterian Dynamics. Economists distinguish several types of dynamics - which one is Schumpeterian, and why is it a challenge?

By Schumpeterian Dynamics one usually denotes the rather complex evolution in the real economy of modern industrial societies. It is complex in at least two dimensions - multiplicity of layers of the economy (disaggregate depth) and multi-periodicity of transformation of various subsystems of the economy (different length of change processes). It has become customary to distinguish at least three layers of the economy, denoted micro, meso and macro. It is the trademark of 'Schumpeterian thought' that change processes on all three levels go on simultaneously yet at different speeds of change:

1. On the micro-level of the economy, where projects and enterprises are the organized units of human interaction, firms buy, make and sell products and services. They invest in plant, equipment, manpower, and in the development of human capital as well as in new products. Firms venture into projects that phase in new products and new production techniques (innovations).

2. On the meso-level (markets, industries, sectors, and regions) groups of firms compete for market share, growth and profit margins, and they cooperate intra-sectorally or inter-sectorally. They are rivaling with and against domestic and foreign vendors. To succeed, they differentiate prices, qualities and other conditions, for example, timeliness. In particular, firms differentiate corporate strategy and product policy while competing (industrial rivalry).

3. On the macro-level of the economy (national economy and lead sectors) industries emerge, grow in scale and in scope, stagnate, or decline. Industries evolve in structure and performance - sometimes slowly, sometimes swiftly and turbulently. At times, turbulence may be contained to separate sectors or to isolated regions; at other times, turbulence may spread nationally or internationally (economic instability).

This multi-layer change is complicated, and, therefore, a challenge for the economic scientist, because corporate innovation processes are vertically integrated with processes of industrial competition (pioneer-follower-competition). Similarly, industrial rivalry is vertically integrated with macroeconomic changes in aggregate demand and supply. Moreover, international competition, and shifts in the international division of labor, tend to accelerate or slow-down the rate of change in affected industries, markets, market segments, and in organizations serving those markets. Therefore, the disaggregate Schumpeterian dynamism calls for micro-macro-modelling (MMM). Chapters 1 and 2 deal with the microeconomics and the macroeconomics of the Schumpeterian evolution as we perceive of it - grounded on the empirical work on innovative investments and industrial changes that has gone into the econometric analyses given here.

The second dimension of complexity mentioned above is the dichotomy of continuity and discontinuity. As vertical linkages between processes of change accelerate or deccelerate changes on other aggregates, these interactions may be characterized as gradual or smooth if the Principle of Continuity privails, i.e. if the whole economy is macro-stable despite possible turbulence on disaggregate levels. In fact, macroeconomic instability can be expected to develeop if the rate of innovation in the microeconomic fabric is deficient, or if competition on the meso-level has lost its driving force. Conversely, if innovation and creative competition has slowed down, as it does in periods which Mensch calls Stalemate situations, the rate of structural change in the economy will be low, yet the structure will turn brittle (structural instability). These, then, are the side conditions for discontinuous developments which may all of a sudden interrupt the calm. A burst of radical innovations may be released by the sudden shift. In hindsight, one would then call these innovations 'disruptive'. One feature of Schumpeterian Dynamics is that change, at some lengthy periods of relative calm, may proceed gradually, while at other times, during relatively brief intervalls, change will speed up and become disruptive. Schumpeter insisted that the industrial evolution cannot be described sufficiently well if discontinuity was ruled out of theorizing and model building.

In order to capture this dimension in our research, we actually took the Principle of Discontinuity as the rule and we perceived of continuity as the exception, albeit an exception observed quite frequently in econometrics. In fact, in time series analysis of change, as we have provided for in Chapters 3 and 4, observations of continuity or discontinuity depend heavily on the time scale of the time series used. In Chapter 3, entitled "The Schumpeter Clock" again, we present our new model of business fluctuations (business cycles of short-term duration). In Chapter 4, we present our formal theory of long-term transformation which, however, is intergrated longitudenaly with short-term changes. In fact, the engine of change, in our theory it is the composition of strategic investments, interlocks changes in the short and the long both in the macro and in the micro. In order to emphasise this interlocking of change as well as the existence of a prime mover, the analogy of a clock has been chosen.

This book's title, "The Schumpeter Clock", thus refers to the characteristic of our contribution, which is an empirically validated desription of the industrial evolution after the 2nd world war in Western countries, a description based on our new methodology. In actual fact, it is a three-pack parcel of new scientific technology and material, denoted here T_1, T_2, T_3:

T_1 - the compact model architecture, which breaks with the usual large-scale model tradition, where scholars employ vast numbers of equations and variables. Instead we employ a "deap" model architecture, one that holds micro and macro with the same variables, namely, with strategic investment expenditures measured at the level of the firm, the industry, and the economy on the whole.

T_2 - the mathematical master equation of disequilibrium motion of the economy. It incorporates non-additive effects of the investment mix. That is, while investment expenditures can be added-up in aggregation, the capacity effects and employment effects, measured in physical terms, cannot be added, because competitive investments sometimes cancel one another (overinvestment), sometimes stimulate one another (underconsumption). The master equation, employed in our dynamic model, thus overcomes inconsistencies which traditionally have haunted nonlinear causal models, such as the accelerator-multiplier models.

T_3 - the multi-variate method for computing latent variables from manifest indicators of economic activity. In this regard, our econometric work has greatly benefited from conjoint data analysis and model building efforts conducted in parallel with Professor H. Wold since 1979. Results from that conjoint research will be published elsewhere. It suffices to mention here that there we utilize different T_2 and T_3, that is, Wold's engenious "soft modeling".

The clock is a metaphor that signifies the methodolical fit of the three parts into a parcel of working components. The metaphor itself has been used as early as 1950 by Professor Richard Goodwin and other deciples of Schumpeter. Then, nonlinear multiplier models had been tried but found logically inconsistent when applied to detrended data. The Schumpeter Clock models of those years did not work satisfactorally, although they remained the only serious,mathematically perfect theories of business cycles. It took a whole new approach, denoted the alternator-coordinator theory, to arrive at a Schumpeterian description that does not suffer from those defects. We keep our fingers crossed that our approach does not suffer from another yet unknown defect.

A Dynamic Policy-Framework for Framework-Setting

What happens when new theory emerges? This has been explained by Thomas Kuhn in his seminal "Structure of Scientific Revolutions". If the theory is insignificant, or rubbish, very little will happen. If, however, the emerging theory adresses some burning issue, or vexing problem, then the scientific estate, according to Kuhn, will give it more attention than the less learned would expect considering the fact that there probably exist already a host of explanations of that problem. It depends, according to Kuhn, upon the paradigmatic scope of the new theory. Of course, Schumpeter's name signifies large scope.

Presently, mainstream economics enjoyes an onslought of new hypotheses adressing parts of the real-economic evolution. Some of these hypotheses are being generated within Schumpeterian quarters, others within rivaling schools of thought, such as New Classicism, Neo-Keynesianism, and other lines of theory-based pursuit of better understanding reality. Our approach is "post-schumpeterian" as differentiated from "neo-schumpeterian" industrial organization, and as distinct from Schumpeter by a rigorous employment of new methodology. While mainstream economis is presently invigorated by a refreshing competition of ideas, opinion leaders engage in what Kuhn predicted: They withhold judgement while first considering the practical implications of noteworthy theories before making-up their minds about validity, reliability, etc. in a second step.

We shall honor that sequence by first indicating some perceived policymaking implications. This judgemental aproach has tradition and makes sence. Over a centure ago, at a time of intellectual turmoil created by breakthroughs in the medical sciences, the great French physiologist, Claude Bernard, stated: "There is no such thing as a right theory or a wrong one, there only exist useful theories and useless". In this century we would not go that far; in particular, we definitely are anxious to beware of wrong theories even if they appear useful, to some, in some special interest situation. Yet, even today only a utopist would insist that new theory that appears useful to some will be used right then and there if and when and where its apparent truth reveals itself. There still may be a long way ahead even if the theory is good and solid. Moreover, at a time of emergence of rivaling theories, socio-economic conditions may induce that one is being favored over others, as it happened to a bunch of European contributions made sometimes around the two world wars, for example, by the Austrian, the Dutch and the Swedish schools. Right or wrong, they were brushed aside by the Keynesian Revolution, wrong or right. For example, Professor Robert Lucas, the prominent New Classicist, is quoted by Newsweek (Feruary 4, 1985, p. 60) as having said that Keynesian economics had gained its popularity because of "political needs, not economic truth". When indicating policy implications, we should, however, keep in mind that not only political needs must be perceived as pluralistic but economic truth as well. In fact, Karl Popper's teaching forces us to conclude that one economic truth is non-existant, whereas depending on the intellectual apparatus employed there exist several "economic truths" - depending on the point of view which predisposes opinion leaders to prefer one apparatus over other apparatuses also available and "free to choose" (M. Friedman).

Over the past four decades, policy making has evolved from Keynesian macroeconomic demand management via a brief interlude of fine-tuning that triggered the Monetarist Counterrevolution, during the seventies, to a policy mix of supply-side economics coupled with micro-economic positive adjustment, on the one hand, and international cooperation, on the other hand. During those years, alternative partial descriptions of the very same general economic process have come into prominence and have declined again when rivaling descriptions emerged from basic research, or were adopted because of better fit with facts or opportunities. On the meta-level it is safe to say that through all those years, what has changed is the specific framework adopted for policy making, but what remained constant is the very process of framework revision, reframing, and renewed framework-setting as needs and opportunities shifted.

Our "post-schumpeterian" approach implies: continuity in this wholistic process of framework-setting but discontinuity at its stages as the real economy evolves over time, necessitating occasional reframing based on perpetual monitoring.

The main proposition advanced in this book is that the occasions that will prompt a revision of frameworks and reaffirmation or reframing will not emerge completely at random, or unexpectedly, but in a partially ordered fashion. This pattern is not completely regular, yet the irregularities in the micro, the meso and the macro can be explained, described in policy-relevant terms, and possibly acted upon in a benefitial manner. In the First Part of this book, we delineate in Chapter 1 the microeconomic foundation of short-term and long-term changes in the context of investment theory. The intellectual setting is largely neoclassical, yet the short-term profit-maximizer is replaced by the strategic investor who a) is an actor in the sence of neo-Austrian theory of action, one who contemplates short- and long-term results, risks, and returns on investment under rivalry. Moreover, b) the investor strategizes largely in a contrarian fashion: since project plans take effect only after some gestation period, and competitive advantages must at least in plan stay-on for several years to come, the Schumpeterian investor acts distinctly non-Keynesian. He does not simply react to present shifts in demand but proacts in accordance to unfolding patterns of change in the target markets - cognitive patterns that we all know are not, and cannot be true pictures of tommorrows reality. Instead, these patterns are mental images of future windows of opportunities which will eventually open up and later close again when the industrial evolution is running its course as expected - based on experience with evolutionary patterns observed over the short run (Chapter 3) and the long haul (Chapter 4). Hence, time and timing matter.

Of course, policy designs and announcements of targets that either indicate framework affirmation or framework resetting will have consideral impact upon expectation formation of both rational expectation draftsmen and of blind followers of these. In Schumpeterian dynamics, the market process is one of both competition among vendors and buyers and cooperation between rivaling investors: as examples for cooperation we may remind you of collective bargaining of wages, collective bargaining of trade conditions, taxation rules. Another splendid example of investors' interaction is cost-sharing of precompetitive research; technological R&D as well as market research, industry analysis, and other information processing meant to assist decision makers in expectation formation and decision making. It is hardly surprising that public policy feeds strongest into those investment decisions which, in their cost-benefit calculations, prove the most sensitive to externalities, namely the hightened risks that follows from rivals' actions when public influences orchestrate these actions, on one end, and the uncertain benefits from serving customer groups which might take sudden turns under the influence of policy change, on the other.

In the aggregate, collective phenomena in microeconomic investment behavior may show up in macroeconomic indicators of production, employment, income, prices, wages, interests etc. Non-Schumpeterian theories usually predict how changes in the rate of interest, wages, profits et.al. induce investors to increase or decrease their level of investment. Some theories may even predict how autonomous changes in the level of investment induce other macroeconomic variables to change: the Keynesian multiplier theory being the most obvious case in point. However, as the saying goes under younger economists, "the multiplier is dead" as a concept, because as a macro-to-macro concept it is shallow in meaning and impossible to measure, and as a micro-to macro- concept, it is flawed by logical inconsistencies emanating from nonlinear causality. The joint effects of corporate investments under cooperation and rivalry are non-additive, and that means, theses synergies or negsynergies create pervasive uncertainties, and fear of disappointment, in rational investors as well as in rational policy framework setters.

The key result of our macroeconomic modeling and econometric analyses of West German and North American industry data since 1949 is presented in Chapter 2. A macroeconometric potential function describing the rate and direction of change in production and prices serves as our tool. It describes these changes as joint effects of changes not in the level but in the composition of investment. This composition is grounded in the microeconomic behavior, which in turn is the result of several determinants such as demand pull, cost pressure, and other variables which impact upon individual investors; however, depending upon situational circumstances, such as the competitive direction of innovation, international development etc. these determinants generate differing joint effects within groups of investors. For example, in most modern (Schumpeterian) industries (where progress is fast and structures are unstable), oligopolistic competition prevails in transnational market segments, investors take part in an innovation race, and this race proceeds at times in fast advancement and market growth, at other times in slow motion and friction caused by temporary overinvestment or underconsumption if buyers are not sure which version to select from an overabundance of choices as in personal computers, pocket calculators, and many other durable goods.

In such Schumpeterian sectors, where innovation, industrial competition, import substitution, structural change and an unfavorable price and wage dynamic determine the collective investment behavior of "strategic groups" of producers, nonlinear causality of an endogenous, bottom-up nature tends to generate surprisingly swift and sharp swings and reversals in aggregate accounts such as the growth rates of production and the employment rate of labor. The following graphs depict such motions in production and prices for U.S. industry and West German industry after 1949.

The noteworthy capability of the underlying model is that it combines the ability to represent relatively steady motions in production and prices with an ability to also represent the larger shifts resulting from structural change and transition to another trend (discontinuities). For example, within the production series of German industry, depicted in Figure 1, we observe a fairly continuous long-term motion from around 1949 to 1965, and thereafter some discontinuous motions. The mathematics in Chapter 2 suggest that during the first intervall, West German industry advanced on a full-employment equilibrium path, where cyclical fluctuations (modeled in Chapter 3) stayed near the production levels attainable with physical capacity and available human skill. In Chapter 3 we prove with our new alternator-coordinato model that during those years side-conditions of investment in industry prevailed which contained the joint disequilibrating effects of aggregate investment within the neighborhood of the full-employment level.

After the mid-sixties, though, German industry apparently entered a treacherous territory. The investment mix became les expansionary and more cost economizing. On balance, industrial production tended to grow slower as cost-cutting investment became the overriding force, whereas cyclical ups and downs tended to become reinforced in whatever direction the entire industrial sector tended to swing. Today, there exists a large number of mathematical models which can describe such fluctuations by pol'nomial regression and other "blind" models - models that calculate parameters from the symptoms rather than by applying a "deap" causal model that suggests these parameters from theory and measures of causes, such as the rate and composition of strategic investment determined by dynamical expectations and political framework-setting at the respective junctures. Our micro-macro-model and analyses presented in the book do exactly that.

For example, there are two switches depicted by the production index in Figure 1 that deserve our particular attention, because they can illustrate the key result of our post-schumpeterian analysis as compared to interpretations attained with other leading theories. Observe that in 1966/67 and in 1974/75, apparently as a result of a distortion, or convulsion of the macroeconomic potential function, the production index dropped down by a multiple of the cyclical variation experienced during preceding years. Standard mainstream economists might argue when looking at the 1966/67 down-turn that the extent of the down-fall was an outlier, a random blip. Indeed, after the 1967 recession reversed, the economy appeared to return to the long-term trend spanning - in Figure 1 - the years from 1955 to 1973. In contrast, our stability analysis underlying the evolution depicted there suggests another interpretation: The 1966/67 recession was a cyclical reinforcement of a structural shift from full-employment equilibrium to underemployment equilibrium; a shift which was temporarily reversed during the groth-push policy adopted by the Schiller-Strauß bi-ministerial concerted action which looked so good as long as it lasted as an artificially induced superboom. The deaper was the drop 1974/75.

According to our bi-equilibrium model of economic motion, the underemployment equilibrium path was actually reached by a major slump in 1975 (Friedman: "by slumpflation") after in 1974 a cyclical variation in the downward direction prepared investors for a down-scaling of expected long-term growth. It can be argued that this discontinuity did not come all of the sudden; one might have expected it if the development in preceeding years had been taken as a warning. The warning message would have read than: In previous years, actually as far back as 1966, the West Germany economy was structurally instable, or split, with one central tendency (trend) striving for full employment and fast growth, and another central tendency striving for slow-down and easing performance. Rational investors would, if they perceive of such undercurrents, respond with a reduction of sales plans, stricter cost controls, and a focus of their innovative efforts at process innovations, which are factor saving, rather than product innovations, which need an expanding demand side if expected to succeed safely.

This brings us to the most interesting point depicted by, for example, Figure 1. If the reader would utilize his or her imagination and draw the two linear trends suggested by the production indicators, he would realize that between 1966 and 1975, two trends ran separately but simultaneously. This may mean utter nonsence to some, because reality can follow one tendency only, right?

The fundamental contribution of post-schumpeterian investment theory is that in the eye of the planner, as he forms expectations of possible future developments, two latent developments can surely coexist. In fact, it is the characteristic of fools and people with tunnel vision that they cannot perceive of contingent developments, future options, and forks in the road. The least we must assume of investment planners to hold in their heads as alternative courses are two, one optimistic and one pessimistic path. The trademark of post-schumpeterian theory, we maintain, is that planners hold the two ways in their head simultaneously; hence, the model must represent the two simultaneously as well, that is, in terms not of manifest variables but in latent variables. Only in latent variables is the conjoint description of two mutually exclusive states not a contradiction but a true representation of cognitive but not yet manifest reality.

For example, those of us who remember the artificial superboom of around 1970 may remember the shared awareness, then popular among investment decision makers, that without the investment boosting public expenditures and tax concessions the aggregate demand would have been much lower, and employment might in fact have stayed in the vicinity of the underemployment equilibrium trend latent there, then. In fact, only some sunshine industries, such as automotive, benefitted superproportionately from the public demand management activities; the majority of industries did not feel the "trickle-down effect". Nor did many small and medium sized U.S. companies, one decade later, received much "trickle-down", or subcontractors' revenues, as had been promised by the Reagan administration around 1981.

The ability of models and methods to handle nonlinearity in latent variable systems permit, for example, analyses of structural instability (brittleness of economic sectors), and may thus give us early warning of treacherous terrain ahead.

But not only that, and at this juncture we return to the issue of reframing. After the stock market crash in 1977, and the wild swings in exchange rates, the frame of mind for reviewing frameworks for policymaking has become one of optimism in saying and pessimism in spending. Investors do not believe policy makers optimistic verberations, because, when putting their own money at stake, they will ask on what such statements can possibly be based. They know that they can be based only on trend extrapolation, and in all models used in public research institutions and decision support institutions in the OECD countries, none has the capability as yet to deal with co-existing latent developments simultaneously and, thus, causaly meaningful. Therefore, the two alternative scenarios that can be espoused from such models is first, business as usual (continuity scenario), and second, a non-defined derailment of the present tendency (discontinuity). The continuity scenario thus reads, in terms of growth rates: maybe zero to two percent deviation, which translates into honestly saying: nearly no forseeable change. Of course, nobody offers any evaluation of the level of significance, or validity and reliability of such a simple trend extrapolation.

In contrast, phantasies about the discontinuous scenario run the gamut from minor business slow-down to severe recession and deap depression, which some even think might last for more than a decade from now on. To the investor, the wide variety of contingent developments not covered by standard models is much more frustrating, and incredulous, than the narrow spread in the continuity scenario. What investors and policy decision makers need is precise analysis of latent developments that in the minds of the public might be waiting for us just around the next bend. What is needed is to recognize forks in the road which lead to alternative roads.

Presently, with no such models and methods in use for policy advice, the only alternative to the incredible continuity scenario are diffuse phantasies about negative developments. A consequence of it is that only few can imagine, and can communicate to those trapped in the beforementioned mind-set, the possibility that events might take a much more desirable turn.

Our analysis suggests: Yes, there appears to loom large a possibility for discontinuous development. The real economy seems structurally unstable, thus ready for a quantum jump. But only those trapped in a framework for framework-setting that attaches a very high probability to a negative development can fail to recognize that a positive development is presently unlikely only because of that mental block. Within the old frame which is blind for any discontinuity, a contingent negative discontinuity is being perceived as the only, albeit incalculable alternative. On a scale of -1 to -10, knowbody can say with any plausibility how deap a fall or how fast a slide might happen. Who in that mind-set asks, instead: On a scale of +1 to +10, how far up a quantum jump might be feasible, and how fast up the scale can such upswing take place?

If our econometric result is correct that the West German economy slumped down on the underemployment equilibrium path as early as 1974/75, possibly already in 1966/67 - as did Great Britain and other european countries - isn't it then fairly safe to assume that the potential for a long-term gradual decline has already been consumated by these countries? What's more, if our econometrics do not err when indicating that the

U.S. economy did not fall down onto an unemployment equilibrium level but stabilized on a path of higher growth and employment, isn't it at least conceivable that a doubling of present rates of growth in european countries can be attained if we discontinued framing Europe as a sclerotic continent?

In a nutshell, these authors deduce from their incomplete yet penetrating analysis that a change in the framework for framework-setting would lift the probability of a positive discontinuity to occur to a level above 50% - as a consequence of present structural instabilities as necessary conditions, and as a consequence of structural instability implying readiness for something new. It appears that the european economy has become structurally ready for another push of basic innovations and another spurt in market integration, offering sufficient conditions for economies of scale and of scope in the years to come.

TRI - EQUILIBRIUM APPROXIMATION OF DISEQUILIBRIUM DYNAMICS

= THE CONFLUENCE OF BUSINESS CYCLE AND STRUCTURAL SHIFTS =

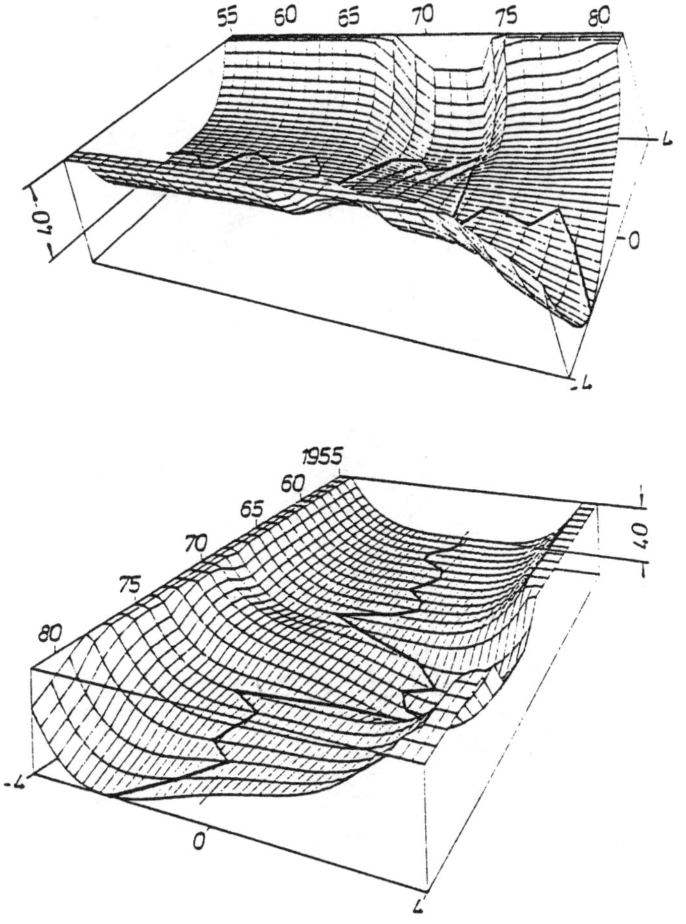

MOTION OF THE INDUSTRIAL PRODUCTION INDEX IN WEST GERMANY, 1955 TO 1981, ACTUAL DETRENDED VALUES AND THEORETICAL ESTIMATES.

FIGURE 1

TRI - EQUILIBRIUM APPROXIMATION OF DISEQUILIBRIUM DYNAMICS

= THE CONFLUENCE OF BUSINESS CYCLE AND STRUCTURAL SHIFTS =

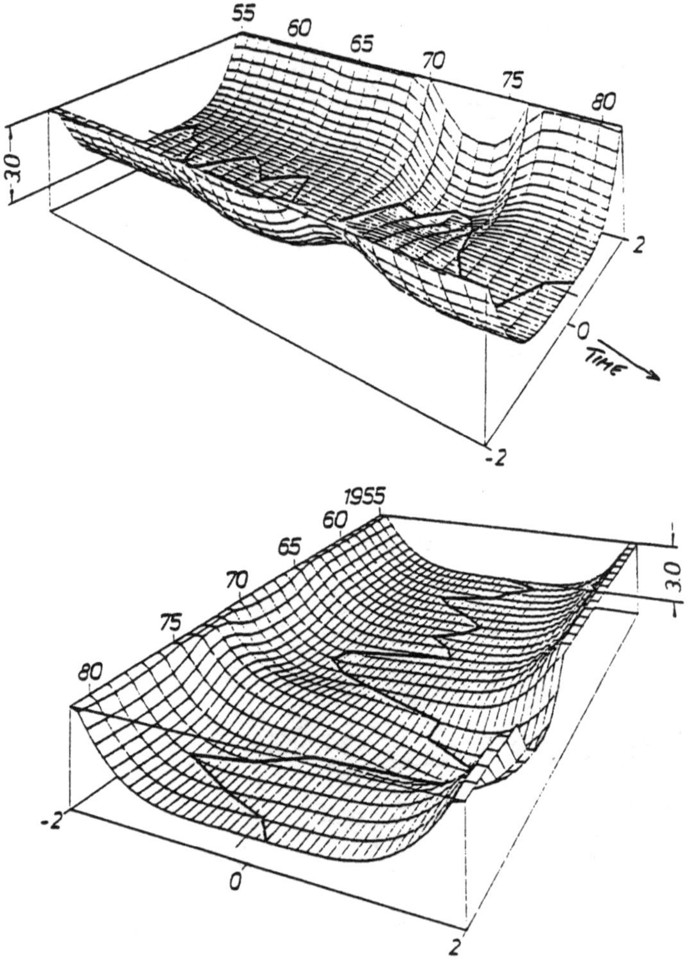

MOTION OF THE PRODUCERS' PRICES IN WEST GERMANY, 1955 TO 1981
(DETRENDED ACTUAL VALUES COMPARED WITH THEORETICAL ESTIMATES)

FIGURE 2

TRI - EQUILIBRIUM APPROXIMATION OF DISEQUILIBRIUM DYNAMICS

+ THE CONFLUENCE OF BUSINESS CYCLE AND STRUCTURAL SHIFTS =

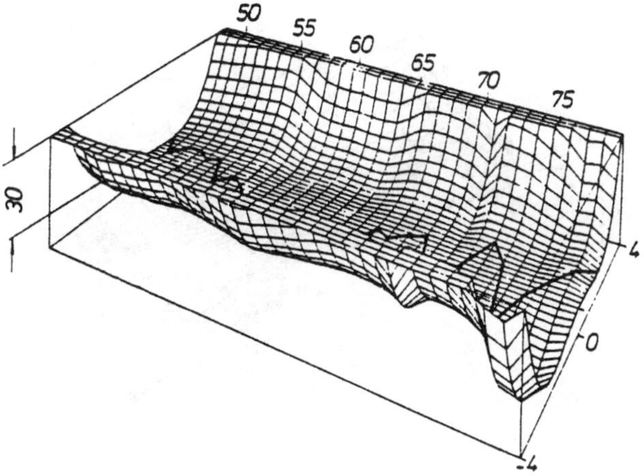

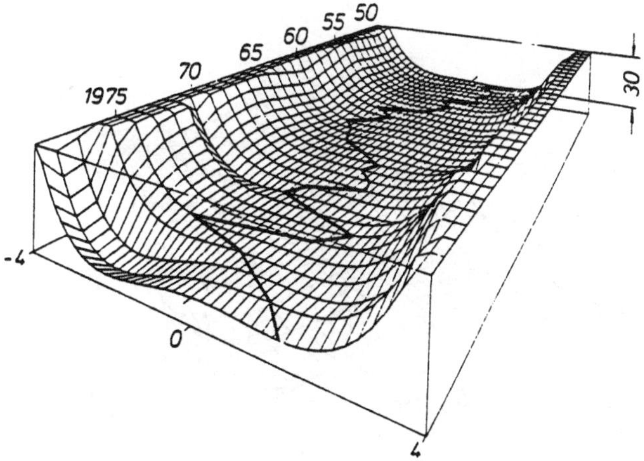

MOTION OF THE INDUSTRIAL PRODUCTION INDEX, UNITED STATES, 1950 TO 1979, ACTUAL DETRENDED VALUES AND THEORETICAL ESTIMATES.

FIGURE 3

TRI - EQUILIBRIUM APPROXIMATION OF DISEQUILIBRIUM DYNAMICS

= THE CONFLUENCE OF BUSINESS CYCLE AND STRUCTURAL SHIFTS =

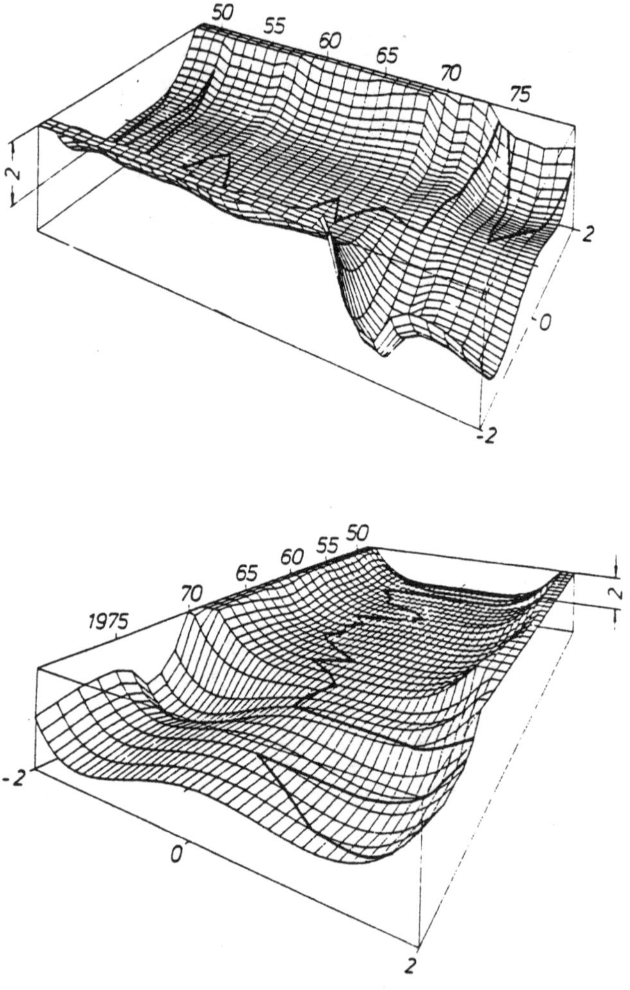

MOTION OF THE PRODUCERS' PRICES IN THE UNITED STATES, 1950 - 79,
(DETRENDED ACTUAL VALUES COMPARED WITH THEORETICAL ESTIMATES)

FIGURE 4

C. Investment, Asset Markets, and Economic Policy

DYNAMICS OF A KEYNESIAN ECONOMY UNDER DIFFERENT MONETARY REGIMES

PETER SKOTT, INSTITUTE OF ECONOMICS, UNIVERSITY OF AARHUS.

INTRODUCTION

Keynesian theory has always regarded investment as an independent variable in the short run. Firms decide the level of investment and saving then adapts passively to bring about the *ex post* identity between saving and investment. But *can* investment be independent of saving? Firms need *finance* in order to execute their investment plans and the availability of finance has been linked to saving by many economists.

In pre-Keynesian economics the level of saving was clearly seen as a constraint on investment. In fact investment was not always distinguished as a separate category and saving was often identified with the growth in productive capital. Following the monetarist revival, similar views are widespread today. Recently the independence of investment from saving has also come under attack from within the Keynesian camp [Asimakopulos (1983)].

The Keynesian/Kaleckian tradition accepts that finance and liquidity - as opposed to saving - may constrain investment. Several writers, however, have suggested that the constraint is unlikely to become operational. Money, it is argued, takes the form of credit money and the availability of credit adapts to financial needs. Thus, the money supply is not, as suggested by the monetarists, an exogenous variable, it is endogenous [Kaldor (1982), Moore (1984)].

In this paper I wish to examine some of these issues in the context of a formal model. I shall consider three types of agents, firms, banks and households. Firms decide their own levels of employment, production and investment and obtain finance from three different sources: retained earnings, new issues and bank loans.

Households supply labour, consume output and hold financial assets. There may be involuntary unemployment but households are not quantity constrained in any other market.

Banks extend loans to firms and accept deposits from households. Neither firms nor households hold cash and the amount of bank loans equals the amount of deposits. For simplicity, it is assumed that there are no costs involved in banking and that lending and borrowing rates coincide. Banks thus have neither costs nor profits: interest payments by firms to banks are exactly equal to the interest received by households. It is assumed that firms can borrow as much as they wish at the prevailing interest rate. The rate of interest, however, may vary over time, and this assumption makes an elastic supply of bank credit compatible with a range of different monetary regimes.

The paper is in five sections. Section 1 describes the basic model and the following three sections analyse the dynamic behaviour of the economy under different monetary regimes. In section 2 the real rate of interest is constant over time. Section 3 adopts the monetarist assumption that the interest rate is manipulated so as to ensure a constant rate of growth of the demand for money, and in section 4 it is assumed that the nominal rate of interest is kept constant. The final section summarises the main conclusions.

1. THE MODEL

Households

Households receive wage income as well as a return on their financial wealth. They own no physical capital goods and their (non-human) wealth is held in the form of money (deposits with banks) and securities. Interest is being earned on bank deposits, and the return on securities comprises both dividend payments and capital gains. Household incomes are either spent on consumption or used to augment the financial assets.

The desired stocks of financial assets are related to current income flows and the saving/consumption decision aims to achieve these desired stock flow ratios. Algebraically, this aspect of household behaviour is described as follows,

$$p\,C + vN(\hat{v}+\hat{N}) + M\,\hat{M} = W + (1-s_p)\,P + vN\,\hat{v} + iM \qquad (1)$$

$$\alpha\,(P-\delta pK-rM) = v\,N \qquad (2)$$

$$\beta\,pY = M \qquad (3)$$

where Y, C and K represent output, consumption and the capital stock in real terms, N and M are the number of securities and the amount of bank deposits held by households, W and P are aggregate nominal wages and gross profits, s_p is the corporate retention rate, δ the rate of depreciation, p the price of output, v the price of securities, i the nominal interest rate, and r the real rate of interest ($r=i-\hat{p}$). Carets are used to denote logarithmic derivatives.

Equation (1) is households' budget constraint, and equations (2)-(3) describe the behavioural assumptions. It is assumed (eq. (3)) that the demand for money is proportional to national income. Some limitations of this simple quantity theory specification are discussed in section 3.

The demand for financial securities (eq. (2)) is proportional to the level of profits net of depreciation and (real) interest payments. In other words, it is assumed that the economy-wide p/e ratio is constant over time. This specification has the virtue of simplicity. It may also have a somewhat neoclassical flavour: if the level of profits is given then the share valuation is independent of firms' pay-out decisions. But in fact no important conclusions of the model depend on the specification: it would be easy to relate vN to, say, total distributed income or total consumption, and this change would not affect the qualitative results of the analysis. More questionable is the assumption that α and β are constant. Both these parameters are likely to depend on the relative rates of return on financial assets and these returns will not in general be constant. The simplifying assumption of fixed parameters is thus not entirely satisfactory. As a first approximation, however, it may not by too misleading.

Equations (1)-(3) can be used to derive saving and consumption functions. Substituting (2) and (3) into (1), we get

$$pC = pY - s_p P - vN \hat{N} - M(\hat{Y}-r)$$

$$= pY \left[1 + \alpha\hat{N}(\beta r+\delta/\sigma) - \beta(\hat{Y}-r) - \pi(s_p+\alpha\hat{N})\right] \quad (4)$$

where π and σ denote the share of profits in income and the output capital ratio, respectively. From (4) it follows that

$$S/Y = A + B\,\pi \quad (5)$$

where

$$A = -\alpha\hat{N}(\beta r+\delta/\sigma) + \beta(\hat{Y}-r)) \quad (6a)$$

$$B = s_p + \alpha\hat{N} \quad (6b)$$

The composite parameters A and B depend on $\hat{Y}$ and σ as well as on parameters describing the behaviour of households, firms and banks. Apart from this modification, the saving/consumption system is similar to a simple Kaldorian specification, and indeed, equations (1)-(3) represent a generalised version of Kaldor's neo-Pasinetti theorem, Kaldor (1966). Note, in particular, that firms' financial decisions - the values of s_p and $\hat{N}$, influence the average saving propensity of the economy.

Pricing

As far as pricing and distribution are concerned, I shall follow a simple mark-up specification. Furthermore, I shall (i) ignore material inputs and (ii) assume that excess capital capacity is the normal state of affairs and that unit labour costs are constant below full capacity utilisation. With these assumptions[1], the pricing equation is given by

$$p = mw/\lambda \tag{7}$$

where m is the mark-up factor, λ represents labour productivity, and w is the nominal wage rate. From equation (7) it follows that the share of profits is given by

$$\pi = 1-\frac{1}{m} \tag{9}$$

<u>Production</u>

Production is not instantaneous and at any moment the rate of output, Y, is predetermined by past production decisions. Discrete time lags are difficult to handle analytically but the effects of production lags can be approximated within a continuous time framework by assuming that the rate of growth of Y at time t, $\hat{Y}_t$, is the decision variable at time t. The approximation is close if the production lag is short and the time path of Y is smooth (differentiable). Smoothness, in turn, may be justified by an appeal to the existence of adjustment costs. Numerically high rates of growth of employment will be associated with high adjustment costs and the standard convexity assumption seems reasonable. The position of the adjustment cost function will, however, depend on the state of the labour market.

The rate of employment influences the position of the adjustment cost function through its effects on the social relations of production (high rates of employment strengthen workers vis-a-vis management) as well as on the availability of labour with the desired qualifications. These effects combine to make it less attractive and more costly to expand production the higher is the rate of employment. High employment and a high turnover of the labour force, on the other hand, allow firms to contract production and employment more rapidly without incurring the costs associated with compulsory redundancies. For positive rates of expansion the cost of adjustment will therefore be an increasing function of the rate of employment while for negative rates of expansion it will be a decreasing function of the rate of employment.

These considerations lead to the following specification of the 'output expansion function'

$$\hat{Y} = h(\frac{I-S}{Y}, e); \; h_1 > 0, \; h_2 < 0. \tag{9}$$

The growth rate of output depends on the state of demand - the difference between ex ante investment and saving - and on the rate of employment in the economy, e.

Investment

With respect to investment, I shall adopt a simple stock adjustment formulation.[2] It is assumed that the desired capital stock, K*, is proportional to the level of output

$$K^* = \mu Y \tag{10}$$

but that the proportionality factor may depend on the real rate of interest,

$$\mu = \mu(r); \qquad \mu' \leq 0 \tag{11}$$

There is a gradual adjustment toward the desired capital stock, K*, and the share of investment in income depends on the ratio K*/K.[3]

$$\frac{I}{Y} = \tilde{f}\left(\frac{K^*}{K}\right) = \tilde{f}\left(\mu\frac{Y}{K}\right) = f(\sigma, r); \; f_\sigma > 0, \; f_r \leq 0 \tag{12}$$

It is assumed that investment plans are realised. Equation (12) thus describes actual investment as well as desired investment.

Investment needs to be financed and in analogy with households' budget constraint, firms face a financial constraint,

$$pI = s_p P + vN\hat{N} + M(\hat{M}-i) \tag{13}$$

Equation (7) states that investment must be financed through retained earnings ($s_p P$), new issues ($v\hat{N}N$), or an increase in bank loans which exceeds current interest payments ($M(\hat{M}-i)$). The parameters s_p and $\hat{N}$ reflect the financial decisions of firm and to simplify the exposition it will be assumed that $\hat{N}=0$.

The model

In order to complete the model we need to add the following accounting equations describing changes in the capital stock and the employment rate,

$$\hat{K} = \frac{I}{K} - \delta \tag{14}$$

$$\hat{e} = \hat{Y} - n \tag{15}$$

In equation (15) it is assumed that the rate of growth of the labour force is constant (or reinterpretting the parameters that there is a constant rate of Harrod neutral technical progress) and that employment is proportional to total production: the production function has fixed coefficients and there is no labour hoarding.

Using equations (5), (9) and (12) as well as the simplifying assumption that $\hat{N} = 0$ we get

$$\hat{Y} = h(f(\sigma,r) - \beta\hat{Y} + \beta r - s_p(1-\frac{1}{m}), e) \qquad (16)$$

or

$$\hat{Y} = \varphi(\sigma, e, r); \ \varphi_\sigma > 0, \ \varphi_e < 0, \ \varphi_r \gtrless 0 \qquad (17)$$

Substituting (17) into (15) yields

$$\hat{e} = \varphi(\sigma, e, r) - n \qquad (18)$$

and from the definition of σ — we get

$$\hat{\sigma} = \hat{Y} - \hat{K} = \varphi(\sigma, e, r) - \sigma f(\sigma, r) + \delta \qquad (19)$$

2. A CONSTANT REAL RATE OF INTEREST

When the real rate of interest is kept constant, equations (18) -(19) form a two dimensional system of differential equations. Setting $\sigma = e = 0$ and substituting (18) into (19) yields the steady state value of σ and, using this value of σ, equation (18) can be solved for e*.

The dynamic behaviour of the system in the neighbourhood of equilibrium is determined by the value of the Jacobian (at the equilibrium)

$$J(\sigma, e) = \begin{bmatrix} \sigma(\varphi_\sigma - f - \sigma f_\sigma) & \sigma\varphi_e \\ e\varphi_\sigma & e\varphi_e \end{bmatrix} \qquad (20)$$

and

$$\text{Det}(J) = -\sigma e(f + \sigma f_\sigma)\varphi_e > 0 \qquad (21)$$

$$\text{Tr}(J) = \sigma(\varphi_\sigma - f - \sigma f_\sigma) + e\varphi_e$$

$$= \sigma\left(\frac{\partial\hat{Y}}{\partial\sigma} - \frac{\partial\hat{K}}{\partial\sigma}\right) + e\frac{\partial\hat{Y}}{\partial e} \qquad (22)$$

The equilibrium will be locally asymptotically stable (unstable) when Tr is negative (positive). The sign of Tr is apparently ambiguous, but in fact Tr will almost certainly be positive and the equilibrium unstable.

A positive value of Tr is ensured if an implicit assumption underlying standard short run macroeconomics is satisfied. Short run macroeconomics examines variations in output which are rapid relative to any movements in the capital stock: the level of output is related to the level of investment and hence to the growth of the capital stock. This assumption in static short run theory has a dynamic equivalent. If a short run equilibrium is established so quickly that the capital stock can be taken as fixed in short run analysis then the rise in $\hat{Y}$ following an increase in $\hat{K}$ must be many times greater than the original increase in $\hat{K}$. This implies that $\partial\hat{Y}/\partial\sigma \gg \partial\hat{K}/\partial\sigma$ (see Skott (1988b) for details).

From the expression for TR(J) it therefore follows that the equilibrium is unstable unless the sensitivity of desired output growth to changes in employment is very high. Although stability cannot be ruled out, instability thus seems the more likely outcome. Assuming that a compact and positively invariant set exists, instability implies - by the Poincaré-Bendixson theorem - that a limit cycle exists.

In conclusion, with a fixed real rate of interest the steady growth path is likely to be locally asymptotically unstable and the economy will either diverge or exhibit bounded fluctuations around the steady growth path.

One aspect of the model deserves special comment. A Phillips curve is usually one of the main building blocks of macroeconomic theory and recent debates in macroeconomics have focused strongly on the correct specification of this particular relation. It may therefore be surprising that the analysis has made no reference at all to Phillips curves, to absolute prices and wages, or to the rate of inflation. Neglect of these aspects has been possible because the real variables (employment, investment, production, distributive shares) depend upon neither absolute prices nor inflation. If we want to know the rate of inflation then a Phillips curve may be needed, but the model is recursive and the shape of the Phillips curve does not matter for the determination of the real side of the economy.

The model thus has 'neutrality' properties which superficially resemble those of many monetarist models: real outcomes are homogeneous of degree zero with respect to proportional variations in wages, prices and the stock of money. Unlike monetarist theory, however, the present model assumes that the stock of money is an endogenous variable and this difference is crucial. Monetarists argue that 'if the money supply is doubled then prices and wages will also double and the new equilibrium will leave real quantities unchanged'. In this model on the other hand, the corresponding argument is that 'if money wages double then the prices of output and securities will also double leaving real quantities unchanged'.

One important implication of the reversal should be noted. The neutrality of the real economy with respect to changes in money wage rates implies that unemployed workers are involuntarily unemployed in Keynes's sense. The problem of involuntary unemployment arises exactly because "[t]here may exist no expedient by which labour as a whole can reduce its real wage to a given figure by making revised money bargains with the entrepreneurs" [Keynes (1936), p. 13]. There may be unemployment, workers may be willing to accept a cut in real wages, firms may want to expand employment even at the prevailing real wages if the demand for output were to increase, but neither real wages nor aggregate demand in real terms respond to changes in money wages.

Keynes discussed changes in money wage rates in chapter 19 of the General Theory. He analysed a number of different effects of such changes, and Leijonhufvud has subsequently expanded the list and provided a catalogue of the sundry effects which have been identified in the literature [Leijonhufvud (1968), pp. 319-331)]. Some of the effects only apply to open economies and many of the closed economy effects are by general consent either insignificant or their direction cannot be unambiguously determined.

The two most important economic arguments against money wage neutrality focus on the Keynes effect and the Pigou effect (or real balance effect[4]). The Keynes effect is the effect of falling interest rates on aggregate demand, where the fall in interest rates is itself caused by falling money wages. The Pigou effect describes the effect on real consumption of an increase in real balances, where again the rise in real balances is brought about by a fall in money wages.

Neither the Keynes effect nor the Pigou effect has so far appeared in the model. The Pigou effect is absent because there are no outside assets whose value is fixed in nominal terms. The value of securities is tied to nominal income, and we have assumed that the real rate of interest on bank deposits (money) is constant. The latter assumption implies that changes in nominal prices are completely neutralised in their effect on the real value of total bank deposits. The assumption that real rates of interest are constant also cuts out the Keynes effect. If we wish to remove money wage neutrality we must thus abandon the assumption of fixed real interest rates.

3. EXOGENOUS MONEY

Monetarists view the stock of money as exogenous (or directly under the control of the central bank). As a literal description of actual monetary institutions this view is clearly anachronistic.[5] The endogenous character of the money supply - the infinitely elastic supply of bank loans at the prevailing rate of interest - does not, however, per se rule out the possibility that banks may control the growth of the money stock.

So far it has been assumed that the supply of bank loans to firms is perfectly elastic at a given time-invariant real rate of interest. This assumption is now reversed. Instead it is assumed that banks vary the interest rate to whatever extent is required in order to induce firms to expand their total bank loans at some given growth rate, m. Individual firms still have unused overdraft facilities with banks but the interest rate on these overdrafts is adjusted so as to make firms want to increase the balance on these accounts at the given rate. Since firms hold no money balances themselves, this implies that the money stock of the household sector must grow at the rate m.

Using the demand for money, equation (3), and the pricing assumption, equation (7), we have

$$m = \hat{Y} + \hat{p} = \hat{Y} + \hat{w} \tag{23}$$

What determines $\hat{w}$? For present purposes we may assume that a simple Phillips curve applies

$$\hat{w} = \eta(e); \quad \eta' > 0, \ \eta \rightarrow \infty \text{ for } e \rightarrow 1;$$

$$\eta \rightarrow a < m-n \text{ for } e \rightarrow 0 \tag{24}$$

Equations (23)-(24) imply that

$$\hat{e} = \hat{Y}-n = (m-n)-\eta(e) \tag{25}$$

The steady growth solution is determined by setting $\hat{e} = 0$ and solving equation (25) for e*. The steady growth path depends on the Phillips curve. With an exogenous growth rate of the nominal stock of money the system has, not surprisingly, lost its money wage neutrality. The real rate of growth is given by the growth rate of the labour force, and the long run rate of inflation is also independent of the shape and position of the Phillips curve: it is equal to the difference between the growth in the money supply and the real rate of growth. A shift in the Phillips curve will, however, affect the rate of employment as well as the output capital ratio along the steady growth path.

The militancy of workers now influences the outcome directly via the determination of money wages as well as through the output expansion function. The reserve army of labour is a restraint on militancy and following an upward shift in the Phillips curve the reserve army needs to be expanded in order to maintain the given rate of wage inflation. The change in the rate of employment in turn affects the output capital ratio along the steady growth path: if the position and shape of the output expansion function is unchanged then the reduction in employment leads to a decline in the output capital ratio. Assuming more realistically that the increase in militancy affects both the Phillips curve and the output expansion function, the effect on employment of an increase in militancy remains unambigous but the net effect on the capital ratio becomes indeterminate.

The dynamics of the monetarist system are extremely simple. From (25) it is readily seen that the employment rate converges monotonically to its equilibrium value: the equilibrium is globally stable.

Before jumping to monetarist conclusions, it should be noted, however, that the success of 'constant m policies' in generating stable growth in nominal output depends on several questionable assumptions. If we introduce a simple modification and let the parameter β vary with the rate of interest then fluctuations in the rate of interest will entail fluctuations in the growth of nominal income when m is constant. 'Goodhart's law' also complicates matters. Outside simple models there is a range of monies and near-monies and the particular money stock which is being controlled may cease to bear any relation to nominal output as agents switch to assets outside the chosen definition of 'money'.

More importantly, it was assumed that banks could and would adjust interest rates to whatever extent is needed in order to induce firms to increase their loans at the given rate. Recent experience in a number of countries, however, suggests that extremely volatile interest rates may be required in order to generate stable growth in the money stock and a high degree of volatility in interest rates may endanger the stability of the financial system [Kaldor (1982)]. Let us therefore look more closely at the way interest rates affect the demand for money.

The demand for money is proportional to nominal income and in order to control money growth the monetary authorities need to control firms' production and employment decisions. Production decisions in turn are related to the demand for output as determined by the difference between ex ante investment and saving. From equations (10) and (23) we get

$$h(\frac{I-S}{Y}, e) = m-\eta(e) \tag{26}$$

A rise in interest rates has direct effects on both saving and investment. It raises households' interest income which tends to reduce the saving rate through a real balance effect: the rise in interest rates implies that real balances will expand unless there is an increase in consumption and a fall in saving out of current non-interest income.[6)]

The desired capital stock may depend on the cost of finance [cf. equation (11)] and investment will therefore also be affected by changes in interest rates. This effect is closely related to the Keynes effect. The only difference is that in this case the change in the rate of interest is directly under the control of the monetary authorities and not the indirect result of a change in the level of money wages. The direction of the effect is unambigous - a rise in interest rates depresses investment - but the strength of the effect and the speed with which it comes into operation have been the subject of considerable debate [see e.g. Desai and Weber (1986)].

The saving and investment effects are contradictory in their influence on aggregate demand and hence on the demand for money: a rise in interest rates stimulates consumption and depresses investment. It is therefore not surprising that large fluctuations in interest rates may be needed in order to control the growth in the stock of money. Furthermore, successful control of the money stock depends on the relative strength of the saving and investment effects. If the Keynes effect dominates the real balance effects then interest rates should be raised in order to curb excessive growth in the money stock (and nominal incomes); if on the other hand the real balance effect is dominant then reductions in interest rates are required. Monetarist policies have prescribed a rise in interest rates in order to lower the growth rate of the money stock and have thus been based on the implicit assumption that the Keynes effect on investment exceeds the real balance effect on saving.

This conclusion may need to be modified if one allows for induced changes in financial behaviour. It is not, however, obvious that all such changes facilitate monetarist control. The financial behaviour of households is described by the parameters α and β representing the ratio of the stock of securities to profit income and the ratio of the money stock to total current income. A rise in interest rates raises households' return on money holdings and causes a capital loss on securities. One might therefore expect rising interest rates to be associated with a tendency for α to fall and β to rise. The net effect on household saving of these changes is uncertain: a rising value of β will raise the saving propensity whereas a fall in α reduces saving if $\hat{N}>0$.

Changes in interest rates may affect firms' financial decisions too. A rising interest rate makes bank loans more expensive but it also reduces the stock market valuation of firms. The latter effect is hardly conducive to higher levels of new issues and it also militates against a higher retention rate and reductions in dividends. The net effect of a rise in interest rates on the retention rate and the amount of new issues is therefore unclear.

4. CONSTANT NOMINAL INTEREST RATES

The final case to be considered is that of constant nominal interest rates. Adopting again the simple Phillips curve, equation (24), equations (18)-(19) can be rewritten

$$\hat{\sigma} = \varphi(\sigma, e, i - \eta(e)) - \sigma f(\sigma, i - \eta(e)) + \delta \tag{27}$$

$$\hat{e} = \varphi(\sigma, e, i - \eta(e)) - n \tag{28}$$

where $\varphi_1 > 0$, $\varphi_2 < 0$, $\varphi_3 \lesseqgtr 0$, $\eta' \gtrless 0$, $f_1 > 0$, $f_2 < 0$. Calculated at a steady growth path where $\hat{\sigma} = \hat{e} = 0$ the Jacobian for the system (27)-(28) is given by[7)]

$$\tilde{J}(\sigma, e) = \begin{bmatrix} \sigma[\varphi_\sigma - f - \sigma f_\sigma] & \sigma[\varphi_e - \varphi_r \eta' + \sigma f_r \eta'] \\ e\varphi_\sigma & e[\varphi_e - \varphi_r \eta'] \end{bmatrix} \tag{29}$$

and

$$\begin{aligned} Det(\tilde{J}) &= -\sigma e(f + \sigma f_\sigma)(\varphi_e - \varphi_r \eta') - \sigma e \varphi_\sigma \sigma f_r \eta' \\ &= Det\ (J) + \sigma e[(f + \sigma f_\sigma)\ \varphi_r - \varphi_\sigma \sigma f_r]\eta' \end{aligned} \tag{30}$$

$$\begin{aligned} Tr\ (\tilde{J}) &= \sigma[\varphi_\sigma - f - \sigma f_\sigma] + e\varphi_e - e\varphi_r \eta' \\ &= TR(J) - e\varphi_r \eta' \end{aligned} \tag{31}$$

where J is the Jacobian for a system with the same steady growth path as (27)-(28) but where the real rate of interest is kept constant at its steady growth value.

Comparing the fixed nominal interest rate system with the case where real rates are fixed, a negative value of φ_r thus increases the trace and reduces the determinant. In other words, negative (positive) values of φ_r tend to destabilise (stabilise) the steady growth path. Referring back to the definition of φ, equations (16)-(17), it is readily seen that $\varphi_r < 0$ (>0) corresponds to the case where the Keynes effect on investment dominates the real balance effect on saving (the saving effect dominates the investment effect). There is a straightforward intuitive explanation for this result. In order to stabilise the system one needs to reduce the growth rate of output and employment when the economy is booming and above average employment leads to above average inflation rates and hence to low real interest rates. A stabilising influence therefore requires that excess demand in the product market respond positively to changes in real interest rates: the saving effect must dominate the investment effect.

In terms of stability, a constant real interest rate thus represents an intermediate case. If investment effects are stronger than saving effects then a necessary condition for monetarist stabilisation is satisfied but the stability condition for a regime of fixed nominal interest rates will be more stringent than those for the constant real interest regime. Conversely, if saving effects are dominant then standard attempts to control the money stock via variations in interest rates will be unsuccessful and destabilising while constant nominal interest rates will have a better chance of stabilising the economy than a fixed real rate of interest.

5. CONCLUSIONS

This paper has analysed a simple model in the Keynesian tradition. Three features of the model should be noted. The labour market plays a more prominent role than is usually the case in Keynesian theory, but the treatment of the labour market is inspired by Marxian theory rather than by the traditional neoclassical approach. The analysis, secondly, allows for short run disequilibrium in the product market: the production process takes time and employment and output adjust only gradually to excess demand in the product market. Finally, and perhaps most importantly given the aim of the paper, the model pays particular attention to monetary and financial aspects.

The dynamic properties of the model were analysed for three different monetary regimes. It was assumed, first, that banks maintain a constant real rate of interst. This regime implies the existence of a unique (non-trivial) steady growth path. Stability of the growth path cannot be ruled out but (local asymptotic) instability is the more likely outcome.

We then considered a 'monetarist' regime where the rate of growth of the money stock is kept constant. This change in the monetary system led to fundamental changes in the qualitative properties of the economy: if banks succeed in adjusting interest rates so as to induce firms to expand their bank liabilities by a constant rate then the steady growth path becomes stable. This result is not surprising given the simple quantity theory specification of the demand for money. Stability of the steady growth path, however, depends critically on both the precise specification ot the demand function as well as on the successful manipulation of interest rates to attain the constant growth in the money stock: a rise in interest rates will not reduce the demand for money unless the Keynes effect dominates the real balance effect.

As a third possibility we considered a regime with fixed nominal interest rates. It was shown that a switch from a regime with constant real interest rates to one where nominal rates are fixed will be destabilising (stabilising) if the Keynes effect on investment dominates (is dominated by) the real balance effect on saving.

In terms of stability the constant real interest regime is thus an intermediate case. This regime is also the one which best fits the stylised facts: nominal interest rates as well as the growth rate of (the different measures of) the money supply have historically shown larger fluctuations than real interest rates. Analysis of the two alternative specifications does, however, indicate the potential importance of changes in the overall monetary and financial structure.

FOOTNOTES

1. See Skott (1988a, 1988b) for analysis of an alternative specification with flexible prices and profit margins. Auerbach and Skott (1988) discuss some problems and limitations of the simple monopoly capital approach to the determination of the mark-up.

2. See Skott (1988b) for further discussion of the investment function.

3. This specification generalizes the simple linear case where $I = \nu(K^*-K)$. Dividing through by Y we get $I/Y = \nu(\mu(r)-1/\sigma)$.

4. The argument that unanticipated changes in prices cause a redistribution of wealth between creditors and debtors (which could cause bankruptcies) is also powerful. It reinforces the Keynesian argument against money wage changes as a means for securing full employment.

5. See Chick (1986) for a discussion of the evolution of the banking structure and its implications for economic theory.

6. An increase in the rate of interest (given the rate of inflation) is equivalent in its effects on saving and consumption to a decline in the rate of inflation (given the nominal rate of interest). If households do not earn interest on their money holding (or if the rate of interest on money holding is invariant with respect to changes in the loan rate) then variations in the loan rate have no real balance effects.

7. Without additional assumptions, the steady growth path may not be unique.

REFERENCES

Asimakopulos, A. (1983): "Kalecki and Keynes on Finance, Investment and Saving". Cambridge Journal of Economics, 7, pp.221-233.

Auerbach, P. and Skott, P. (1988): "Concentration, Competition and Distribution". International Review of Applied Economics, vol. 2, no. 1.

Chick, V. (1986): "The Evolution of the Banking System and the Theory of Saving, Investment and Interest". Economies et Societies, Cahiers de l'I.S.M.E.A. Serie Monnaie et Production, no. 3.

Desai, M. and Weber, G. (1986) "Money, Inflation and Unemployment - an Econometric Model of the Keynes Effect". Discussion Paper A59, LSE.

Kaldor, N. (1966): "Marginal Productivity and the Macroeconomic Theories of Distribution". Review of Economic Studies, 33, pp. 309-319.

Kaldor, N. (1982): The Scourge of Monetarism. Oxford: Oxford University Press.

Keynes, J.M. (1936): The General Theory of Employment, Interest and Money. London and Basingstoke: Macmillan.

Leijonhuvud, A. (1968): On Keynesian Economics and the Economics of Keynes. Oxford University Press.

Moore, B.J. (1984): "Unpacking the Post Keynesian Black Box: Wages, Bank Lending and the Money Supply". Thames Papers in Political Economy, Spring.

Skott, P. (1988a): "Effective Demand, Class Struggle and Cyclical Growth". International Economic Review, forthcoming.

Skott, P. (1988b): Class Struggle and Effective Demand - a Theory of Capital Accumulation. Cambridge: Cambridge University Press (forthcoming).

TAX REFORM AND INVESTMENT DECISIONS IN A NEOKEYNESIAN INVESTMENT MODEL

by

Michael Funke*

Science Center Berlin
Labour Market and Employment

I. Introduction

Taxation has always been an important item on the policy agenda in the OECD countries, but it is now receiving unusual attention.[1] This growing concern partly reflects the view that the tax systems in the Western world are inefficient and distort the allocation of resources but the reform plans in the United States also had a great influence on the thinking about the need and the practicality of tax reforms.[2]

In Germany the economy has picked up since spring 1987, but by less than expected. The OECD anticipates real GNP to grow by no more than 2.0% in 1987 as a whole, but some of the major economic research institutes reckon it will be lucky to reach 1.5%. Unemployment, which fell in 1986, has started to increase again. In this situation, the agreed tax-reform package, nonetheless, provides an opportunity to give a boost to demand at a time when it is clearly needed to reach a higher growth path of the domestic economy and to reduce the world's current-account imbalances. As a first stage of the long-term tax reform a 14 billion DM package has been planned to go into effect on January 1, 1988. The announced big tax cuts will not happen until 1990, when individuals and companies will be cut by an annual 39 billion DM - equivalent to an impressive 1.75% of GNP. However, the finance minister plans to finance most of the package by reducing subsidies, rising some indirect taxes and holding back other public spending. The stimulus to the economy is therefore less certain.

* The paper is part of the project "Tobin's Q and Industry Investment in West Germany and the UK" which was carried out under DFG grant, number Fu-178/1-1. I wish to acknowledge considerable research assistance from my co-workers Dr. A. Ryll, S. Wadewitz, and D. Willenbockel. Mistakes are, of course, my own.

This paper examines the influences of changes in the tax system on the *investment policy* of companies in the West German *manufacturing industry*. In the context presented, the effects of these changes in the tax system will thereby not be considered in isolation but in relation to other factors. The basis of investigation is a Neokeynesian investment model which captures the complex and changing structure of tax rules.

Section II of the paper outlines the Q-model of investment that provides the basis for the simulations reported in the paper. In section III we analyze the relationship between investment and Q, and in section IV the impact of various changes in the tax system on Q is examined. This requires calculating the impact of these policies on the market value of capital and the tax parameters.

II. The Analytical Set-Up

In this section of the paper we develop a theoretical model of the determinants of private investment based on the so-called Q-theory of investment. The basis for this Q-I-relationship has been established by James Tobin[3], who examined in some well-known articles the channels through which changes in prices in the financial markets influence underlying real variables. In this context, "Q" is the central variable. Tobin's Q measures the value set by the financial markets for equity and loan capital, relative to the replacement value of the underlying real capital. According to Tobin's theory of investment, changes in Q are crucially important for the incentive to invest. Against this general background we provide a rigorous foundation of the Q-theory of investment in the following. The model assumes that investment is set by the firm at a profit-maximizing level in the presence of taxation. Following Hayashi, Abel, Summers, and Yoshikawa[4] we assume that firms seek to maximize the expected present value of the firm's equity. At t = 0, the value of the firm's shares V(o) is given by

$$(1) \qquad V(o) = \int_{o}^{\infty} \theta \, Div(t) \, \mu(t) \, dt$$

where

$$(2) \qquad \mu(t) = \exp \left(\int_{o}^{t} - r(s) ds \right)$$

and Div(t) = nominal dividends on shares, r = exogenous equity return (discount rate), θ = King's tax discrimination variable. In the context of our model θ is introduced to capture the different tax treatment of retentions vis-à-vis distributions in West Germany in a consistent manner (both under the split-rate system in force before the year 1977 and under the imputation system afterwards), while preserving a unique profit tax rate in the model equations.[5] Additionally we impose the transversality condition

$$(3)\ \lim_{t \to \infty} V(t) \exp \int_{o}^{t} - r(s)ds = 0$$

which leads the system to follow a unique path to the steady state. In the presence of taxation and exogenous sources of funds dividends may be derived as after-tax profits less investment expenses plus new net debt. That is,

$$(4)\ Div(t) = (1-\tau)(pF(K,N)-wN-iB)+\tau D(t)+B^{n}(t)-\varepsilon B(t)-(1-k)pI(t)$$

where p is the price level, $F(\cdot)$ is the production function which relates the firm's usage of labour (N) and capital (K) to output, τ is the corporate tax rate on retained earnings, k captures investment grants, D represents the amount of depreciation allowances, i is the interest rate on debt B, ε is the amortization rate on debt, $B^{n}(t)$ is new debt and I represents real investment. The capital accumulation constraint is given by

$$(5)\ \dot{K}(t) = I - \delta K$$

where the dot over K indicates the time derivative, I and K give gross investment and the capital stock respectively, and δ is the depreciation rate. A crucial feature of the model is that there exist some positive installation costs. The installation costs rise with the rate of capital accumulation, thereby preventuring jumps in the demand for capital. In the following we are assuming that the installation cost function takes the form of wasted capital and can be written as $\Psi(I,K,t)$[6]

$$(6)\ \dot{K} = \Psi(I,K,t) - \delta K(t).\ \Psi_{I} > 0;\ \Psi_{II} < 0;\ \Psi_{K} > 0$$

If we assume that total debt is exogenous and therefore independent of investment decisions by the firm,[7] the optimizing problem reduces to[8]

$$(7)\ V(0) = \int_{o}^{\infty} \theta[(1-\tau)(pF(\cdot)-wN) - (1-k-z)pI]\ \mu(t)dt + \theta A(o) - \theta B(o)$$

where $$-B(o) = [B(t)\ \mu(t)]_{o}^{\infty}$$

$$A(o) = \int_{o}^{\infty} \tau(t)\ [\int_{-\infty}^{o} d(t-v,v)pI(v)dv]\ \mu(t)dt$$

$$z(t) = \int_{0}^{\infty} \tau(t+s)\, d(s,t)\, \mu(s) ds$$

When maximizing the firm's equity the term A(o) can be ignored because it is independent of any current or future investment decisions. The term z(t) represents the present value of the tax bill saving due to depreciation allowances on a mark of investment installed in period t. The optimization problem consists in maximizing equation (7) subject to the capital accumulation constraint (6) and can be solved using the Potrjagin maximum principle. The optimality conditions from the Hamiltonian are:

(8) $pF_N = w$

(9) $\theta(1 - k - z)p = \Psi_I(I,K,t) \cdot \lambda$

(10) $\dot{\lambda} = \lambda(r + \delta - \Psi_K) - \theta(1 - \tau)\, pF_K$

(11) $\lim_{t \to \infty} \lambda(t)\, K(t)\, \mu(t) = 0$

where λ is the shadow price of capital. Equation (8) gives the familiar marginal productivity condition for the use of labor. Equation (9) is the key equation because it uniquely determines the I-Q relationship. It demonstrates that the rate of gross investment is chosen so as to equate the marginal cost of investment with the shadow price of installed capital. Equation (10) determines the dynamic behavior of the shadow price λ. Under the assumption of linear-homogeneity of $\Psi(\cdot)$ equation (9) can be inverted to give the rate of investment.

(12) $I/K = f\left(\dfrac{\lambda/p}{\theta(1-k-z)}\right) +$

(13) $I/K = f\left(\dfrac{q_m}{\theta(1-k-z)}\right) +$

(14) $I/K = F\ (Q_m)$

where $q_m = \lambda/p$ = marginal Q

$Q_m = q_m/[(1-z-k)\theta]$ = tax-adjusted marginal Q.

In other words: the optimizing model leads to an investment equation linking total gross investment to marginal Q and the tax parameters.[9] Equation (14), however, has no *empirical* relevance on a theory of investment unless an observable counterpart to Q_m can be obtained. By making use of some simple manipulations an exact relationship between average Q[10] and marginal Q and therefore a relationship between (I/K) and the observable average Q value can be determined. Specifically, the following statement can be made:

Proposition: If the production function and the adjustment cost function are linear homogeneous and if the firm is a price taker, then:

$$(15)\quad Q_m = \frac{V(o)}{\theta(1-k-z)pK} + \frac{B(o)-A(o)}{(1-k-z)pK}$$

and

$$(16)\quad \frac{I}{K} = f\left(\frac{V(o)}{\theta(1-k-z)pK} + \frac{B(o)-A(o)}{(1-k-z)pK}\right).$$

Proof: See Appendix I

Equation (16) represents the observable counterpart to (14) and shows that (I/K) is an upward sloping curve in the (I/K) and Q_a space, too. A major advantage of the above outlined Q-model is that it can easily be used to evaluate various sets of tax policy proposals.[11] In order to do that it is convenient to analyze the dynamics of the model in the K-Q-space. Equations (6) and (14) together with the linear-homogeneity assumption about $\Psi(\cdot)$ imply

$$(17)\quad \dot{K} = \phi\,(Q_m)\cdot K - \delta K.$$

For $\dot{K} = 0$ we obtain

$$(18)\quad Q_m^o = \phi^{-1}(\delta)$$

The second system's equation of motion can be derived from equation (10)

$$(19)\quad \dot{q}_m = (r + \delta - \Psi_K)\, q_m - (1 - \tau)\, F_K$$

$$(20)\quad \dot{Q}_m = (r+\delta-\Psi_K+(\dot{k}+\dot{z})/(1-k-z))Q_m-(1-\tau)/(1-k-z)p$$

The condition $\dot{Q}_m = 0$ implies

$$(21)\quad \dot{Q}_m = [(1-\tau)/(1-k-z)p]\;[(F_K p)/(r+\delta-\Psi_K)]$$

By making use of the fact that both $F_{KK} < 0$ and $\Psi_K < 0$ it follows that the $\dot{Q}_m = 0$ curve has a negative slope. Both equations (18) and (21) and the corresponding curves in figure I and II uniquely determine the equilibrium of the system in terms of the endogeneous variables. From the phase diagrams in figure I and figure II we can furthermore deduce that the system exhibits *saddle-point stability*, i.e., there is only one unique path along which the system converges to equilibrium.[12]

Figure I: Introduction of an Investment Grant (or an Increase in the Investment Grant respectively)

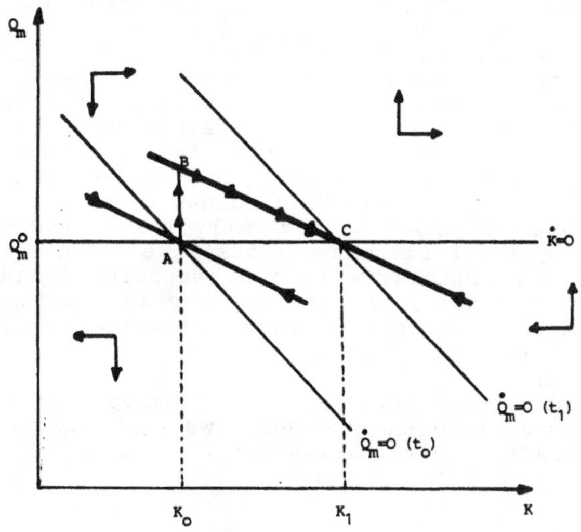

Figure II: Decrease in the Corporate Income Tax Rate

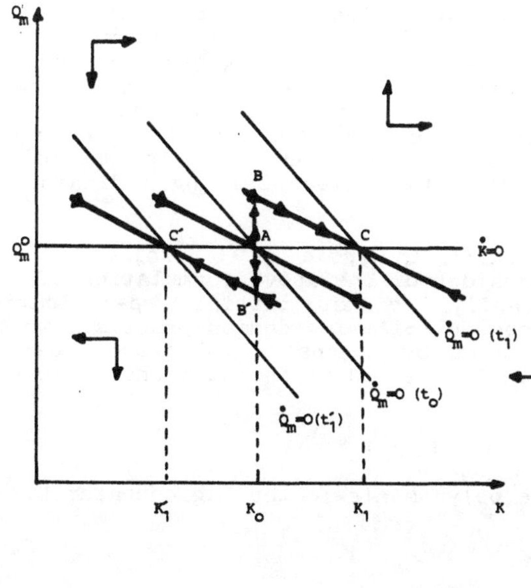

We now use the phase diagram to investigate the response of firms to changes in taxation. First, we consider the introduction of an investment grant (or an increase in k respectively). The Q_m=0-schedule will shift to the right and because the $\dot{K}$=0 locus is unaffected the capital stock K unambiguously increases. Investment initially jumps to point B, thus overshooting the long rund steady state level and then gradually falls to point C. Contrary to the first case, a decrease in the corporate income tax rate has no unambiguous effect on K. Again the $\dot{K}$=0 locus is unaffected while the Q_m=0 locus shifts either to the right or to the left. The phase diagram therefore looks like in figure II. Both points C and C' in figure II are possible solutions. Economically this surprising result is caused by the fact that the positive and expansionary effect of higher profits (due to the reduction in taxation) *may* be totally compensated by the reduced tax-saving due to depreciation (z). This decline in z automatically increase the net price of investment goods.

Having shown the qualitative results of exogenous shocks in the above presented investment model, we will now assess the quantitative impact of alternative tax reforms Q.

III. The Relation Between Investment and Q

Before we will use the theoretical model outlined above to examine the responses of Q and therefore investment to changes in the tax rules, we will first estimate the Q-I-relationship for *total industry and total manufacturing* in West Germany. Figure III and IV graph the gross investment rates[13] and the marginal tax-adjusted Q values[14] from 1960 to 1982. A visual inspection of the graphs show the pronounced drop in all four time series over the sample period. When comparing the series the prevailing impression is one of overall similarity although Q is a clear leading indicator compared with I/K.

In order to evaluate the role of Q explicit econometric models must be considered. The above formulation of the model lends itself easily to an initial specification for econometric purposes. Following common practise, we captured various possible lags by unconstrained distributed lags on the Q values. We then fitted different econometric equations of the form

$$(22) \qquad A(L)\,(I/K)_t = B(L)\,Q_t$$

where A and B are polynomials in the lag operator L.

Figure III: Gross Investment Rate (in %) and Tax Adjusted Marginal Q for Total Industry in West Germany, 1960-1982 (Note: * I/K; - Q)

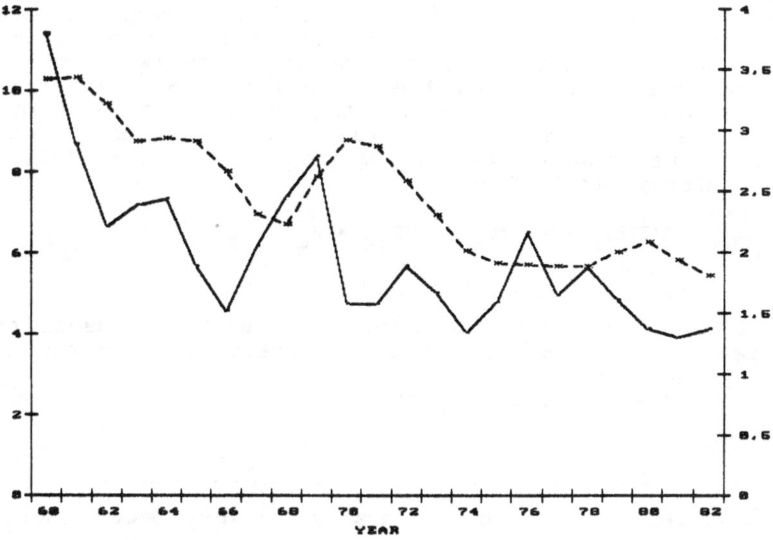

Figure IV: Gross Investment Rate (in %) and Tax-Adjusted Marginal Q for Total Manufacturing in West Germany, 1960-1982 (Note: * I/K; - Q)

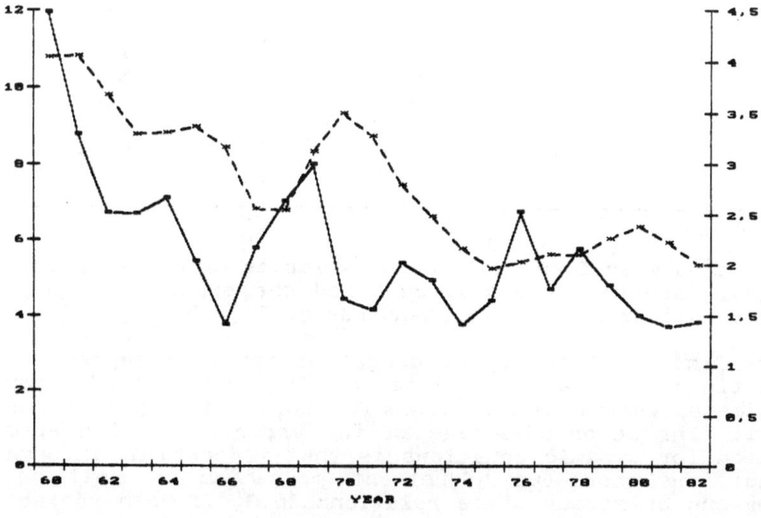

Experiments with various lags have shown that the current value of Q was generally insignificant and that two lags were sufficient and did not oblige us to reduce the degrees of freedom too much. As with most other econometric investigations of investment behaviour, we found strong evidence of autocorrelation in the unadjusted estimates of the model in equation (22). Because of the lag length of two years it is not surprising that adjustment for first-order and second-order autoregressive time series pattern is significant in the fitted equations. The parsimonious model estimated for total industry and manufacturing industry therefore takes the particularly simple form

$$(I/K)_t = a + bQ_{t-1} + cQ_{t-2} + u_t \qquad (23)$$

$$u_t = r_1 u_{t-1} + r_2 u_{t-2} + \varepsilon_t$$

where ε_t is a stochastic error term with the usual properties. The estimated Q-investment relations for the 1965-1982 period were:[15]

Table I:

Variables	**Total Industry**	**Total Manufacturing**
Constant	4,85	3,98
	(6,1)	(4,5)
Q_{-1}	0,45	0,69
	(2,0)	(3,4)
Q_{-2}	0,50	0,61
	(2,4)	(3,0)
r_1	1,33	1,30
	(6,2)	(6,2)
r_2	-0,56	-0,51
	(-2,6)	(-2,3)
R^2	0,91	0,91
D-W	1,35	1,52

As can be seen from (24) the Q variables have the anticipated signs, are highly significant and the explanatory power of the equations, judged by standards of R^2, is very high.[16]

Given the warnings of the danger of spurious repressives in equations which autocorrelated errors, we have additionally estimated *error correction models* for both industry subgroups as well. The major advantage of the error correction specification for dynamic equations is that it contains information about the short-run adjustment mechanism as well as the long-run or steady state relationship.[17] If both variables Q and I/K follow a random-walk and are co-integrated[18] the error correction model can be written as

$$(24)\ \Delta(I/K)_t = a+b\Delta Q_t-c\ [(I/K)_{t-1}-d\ Q_{t-1}]\ +e\Delta(I/K)_{t-1}+u_t$$

where Δ is the first difference operator. The short-run (marginal) adjustment coefficient is given by b while d gives the steady state relationship between Q and I/K.

Table II

Variables	**Total Industry**	**Total Manufacturing**
Constant	-0,27	-0,82
	(-0,7)	(-1,7)
ΔQ_t	0,53	0,55
	(2,6)	(2,8)
$(I/K)_{t-1}$	-0,29	-0,29
	(-5,6)	(-5,5)
Q_{t-1}	1,22	1,41
	(6,5)	(7,3)
$\Delta(I/K)_{t-1}$	0,54	0,48
	(4,3)	(4,1)
R^2	0,83	0,85
Q(10)	11,43	9,23
Parameter b	0,53	0,55
Parameter d	4,21	4,86

Sample period: 1963-1982; t-values is parantheses

Table II contains the results of the estimated error-correction model (24). The most important feature of the results is that again the data are quite sympathetic to the functional form presented in (24). The Q theory therefore seems to be quite robust.

IV. Tax Reform and Corporate Valuation

Given the fact that the model "explains" the recent course of investment rather well we are now using the presented framework to examine in an *ex-post analysis* how various alternative tax measures might have altered investment of *total manufacturing* spending via their influence on Q. The principal difficulty in these calculations comes in estimating the effect of each tax modification on the market value of equity V.[19]

Table III: Effects of Tax Changes on Q for Total Manufacturing

Year	Actual Q Value	Modification No.			
		(1)	(2)	(3)	(4)
1960	4.484	4.484	4.484	4.484	4.848
1961	3.292	3.292	3.292	3.292	3.292
1962	2.519	2.519	2.519	2.519	2.519
1963	2.509	2.509	2.509	2.509	2.509
1964	2.663	2.663	2.663	2.663	2.663
1965	2.032	2.032	2.032	2.032	2.032
1966	1.407	1.407	1.407	1.407	1.407
1967	2.165	2.166	2.166	2.166	2.166
1968	2.636	2.636	2.636	2.636	2.636
1969	3.001	3.001	3.001	3.001	3.001
1970	1.660	1.660	1.660	1.660	1.660
1971	1.558	1.558	1.558	1.558	1.558
1972	2.020	2.020	2.020	2.020	2.020
1973	1.846	1.846	1.846	1.846	1.846
1974	1.406	1.406	1.406	1.406	1.406
1975	1.645	1.645	1.645	1.645	1.645
1976	2.529	2.528	2.528	2.528	2.528
1977	1.756	1.756	1.756	1.756	1.756
1978	2.162	(3.093)	(2.153)	(2.177)	(3.412)
1979	1.793	(2.437)	(1.786)	(1.809)	(2.805)
1980	1.495	(1.980)	(1.491)	(1.504)	(2.299)
1981	1.384	(1.865)	(1.381)	(1.392)	(2.246)
1982	1.427	(1.998)	(1.418)	(1.422)	(2.152)

Note: All modifications are explained in the text. The brackets indicate the years where changes have taken place.

In general, we have always estimated the effect of tax changes on after-tax profits and have then capitalized these marginal profit streams to get the marginal market value. Together with the recalculated A and z-terms this procedure leads to the modified Q values. Table III summarizes the simulated effects of the tax changes and/or fiscal measures considered. The first column gives the actual Q values to show the differences between the reference economy and the simulations in column (2) to (5). In column (2) of Table III (modification (1)) the effect of the introduction of an investment grant of 7,5% on equipment investment[20] from 1978-1982 is examined. Because the introduction of investment grants lowers the effective price of capital goods this tax measure directly influences the profitability of prospective investment projects. When comparing column (1) and (2) it can be seen that the Q value is increasing significantly, i.e., this possible fiscal measure will encourage investment significantly. The most surprising feature of the simulations is the result of a hypothetical reduction in the corporate tax rate on *retained earnings* from 56% to 53% from 1978-1982. The comparison of column (1) and (3) illustrates that Q is rather insensitive to slight changes in the corporate tax rate. This result may at first seem counter-intuitive, but it reflects the working of the countercurrent effects discussed in the last section.[21] A similar weak investment stimulus results from the hypothetical elimination of local profit taxation ("Gewerbeertragsteuer") from 1978-1982 as can be seen in column (4) (modification (3)). The last modification simulates the effect of the introduction of a 100% first year depreciation allowance for equipment investment from 1978-1982 on Q. As one would expect, the results in column (5) suggest that this measure is predicted to have the greatest impact on Q and therefore the desire for investment.

Taken together, the simulations indicate that the various tax measures have very different effects on Q. When ranking the four tax changes and/ or fiscal stimuli the important morale for policy from the analysis is that the most desirable investment incentives are probably those that operate by reducing the effective purchase price of investment goods,[22] such as grants or rapid depreciation allowances. On the other hand changes in the taxation of corporate income will lead, if at all, to a less pronounced effect on Q and therefore investment.[23]

V. Conclusions

Using the Q theory approach this article examines the impact of four alternative tax reforms and how they might influence the rate of capital formation. The simulation results confirm that tax policies can have large effects on the valuation ratios. They also indicate that the quantitative incentives for business investment spending are likely to differ substantially across the measures considered. While these results are encouraging, further research has to be done to make more accurate forecasts. One important area for further investigations is the question how the tax reforms may alter households' attitudes to save or consume their disposible income and the economy wide feedback effects resulting from these decisions.

Notes

1 Compare Boss, A. (1987), Kurz, R. (1987), Leibfritz, W. (1987), Weichert, R. (1986).

2 The planned radical tax reform in the U.S. is widely discussed in Germany, but there is considerable scepticism

about the political feasibility of adopting a similar approach.

3 See Brainard, W., Tobin, J. (1968), Tobin, J. (1969), and Tobin, J. (1982).

4 See Hayashi, F. (1982), Abel, A.B. (1983), Abel, A.B. (1984), Abel, A.B. (1985), Yoskikawa, H. (1980), and Summers, L.H. (1981).

5 For the rationale of θ see King, M. (1977), pp. 48-53, and King, M./ Fullerton, D. (eds.) (1984).

6 We assume that $\Psi(\cdot)$ is linear-homogeneous in K and I.

7 We are requiring that the transversality condition lim $\mu(t)\ B(t) = o$ for $t \rightarrow \infty$ holds. For an alternative model which treats the choice of financial policy in more detail, see Chirinko, R.S. (1987).

8 Time subscrips are omitted whenever possible in order to simplify the exposition. It is assumed that $i(1-\tau) = r$.

9 It must be mentioned that a crucial assumption in the preceding deviations is that capital is both malleable and homogenous.

10 Average Q is defined as $Q_a = (V(o) + B(o))/pK$.

11 For the following compare Summers, L.H. (1981).

12 In the figures this downward sloping path is indicated by a dark line.

13 We define the gross investment rate as gross investment over the end year's gross capital stock (both in 1980 prices).

14 The calculation of Q is described in detailed way in Funke, M., A. Ryll and D. Willenbockel.

15 The t-statistic is given in parantheses.

16 Given the relatively small number of observations it is necessary to be cautious in interpreting the coefficients.

17 For an overview about error correction models and specially the differences in the steady state properties of partial adjustment models and the error correction mechanism compare Nickell, S. (1985), Patterson, K.D. (1987) and Salmon, M. (1982).

18 The assumption of co-integration has been tested using the test statistics presented in Engle, R.F. and C.W.J. Granger (1987), pp. 266-270.

19 The procedure followed in estimating these *marginal* market values of equity dV's is similar to Salinger, M.A./ Summers, L.H. (1983), pp. 261-268. Another noticeable study is Summers, L.H. (1981), and Summers, L.H. (1985).

20 According to the "Stabilitäts- und Wachstumsgesetz" from 1967.

21 Larger effects of corporate tax reductions on Q can be generated if a parallel reduction in the personal income tax rates is assumed.

22 This finding is consistent with Summers, L.H. (1981), and Summers, L.H. (1985).

23 It must be mentioned, however, that the analysis is based on a partial equilibrium model and neglects economy wide feedback-effects.

Appendix I: The Relationship between Average and Marginal Q

We consider the expression

(A1) $(d/dt)(\lambda(t)\ K(t)\ \mu(t)) = \mu(t)\ (\dot{\lambda}K + \lambda\dot{K} - r\lambda K).$

Equation (10) tells us that $\dot{\lambda}K$ must obey the following equation

(A2) $\dot{\lambda}K = \lambda(r + \delta - \Psi_K)\ K - \theta\ (1 - \tau)\ pF_K K$

$= \lambda(r + \delta)\ K - \Psi\lambda + pI\theta\ (1 - k - z) - \theta\ (1 - \tau)\ (pF - wN)$

Now we consider the term $\lambda\dot{K}$:

(A3) $\lambda\dot{K} = \Psi\lambda - \delta K\lambda$

It follows that equation (A1) can be transformed as follows:

(A4) $(d/dt)(\lambda(t)\ K(t)\mu(t))=\mu(t)(-\theta(1-\tau)(pF-wN)+pI\theta(1-k-z)).$

Next we are integrating equation (A4) from 0 to ∞:

(A5) $$\int_o^\infty (d/dt)(\lambda(t)K(t)\mu(t))dt = [\lambda(t)K(t)\mu(t)]_o^\infty$$

$$= -\int_o^\infty [\theta(1-\tau)(pF-wN) - \theta(1-k-z)pI]\ \mu(t)dt$$

With the assumption $\lim_{t \to \infty} \lambda(t)K(t)1(t) = 0$ for $t \to \infty$ and using equation (7) we derive

(A6) $\lambda(o)K(o) = V(o) - \theta A(o) + \theta\ B(o) \to$

(A7) $$\frac{\lambda}{p} = \frac{V(o) - \theta\ A(o) + \theta\ B(o)}{pK} \to$$

(A8) $$\left(\frac{I}{K}\right) = f\left(\frac{V(o)}{\theta(1-k-z)pK} + \frac{B(o) - A(o)}{(1-k-z)pk}\right)$$

We have thus shown the above presented proposition. It is interesting to note that the proposition is essentially an extension of the results derived by Hayashi. This can easily be demonstrated because it follows directly that in the special case of $\theta = 1$ equation (16) and (A8) reduce to Hayashi's equation (13") (See Hayashi, F. (1982), p. 220).

Appendix II: Variables and Data Sources

I: Gross fixed investment in 1980 prices. Source: Statistisches Bundesamt (ed.): Fachserie 18, Reihe S. 8, Revidierte Ergebnisse 1960 bis 1984 (VGR).

K: Gross capital stock in 1980 prices. Source: VGR.

Q: The construction of the time series for Q is based on stockmarket corporations' balance-sheet and profit and loss account data compiled and published annually by the Statistical Office (Statistisches Bundesamt) in "Fachserie 2, Reihe 2, Abschlüsse der Aktiengesellschaften". Detailed information about the calculation of the Q series is contained in Funke, M., A. Ryll and D. Willenbockel (1988).

References

ABEL, A.B. (1983): "Optimal Investment under Uncertainty", *American Economic Review, 73*, 228-233.

ABEL, A.B. (1984): "The Effects of Uncertainty on Investment and the Expected Long-Run Capital Stock", *Journal of Economic Dynamics and Control, 7*, 39-53.

ABEL, A.B. (1985): "A Stochastic Model of Investment, Marginal q and the Market Value of the Firm", *International Economic Review, 26*, 305-322.

ALAM, K.F., AND STAFFORD, L.W.T. (1985): "Tax Incentives and Investment Policy: A Survey Report on the United Kingdom Manufacturing Industry", *Managerial and Decision Economics*, 6, 27-32.

BOSS, A. (1987): "Die Vorschläge zur Steuerreform in den USA", *Konjunkturpolitik*, Beiheft 33, 215-232.

BRAINARD, W., AND TOBIN, J. (1984): "Pitfalls in Financial Model Building", *American Economic Review, Papers and Proceedings, 58*, 99-122.

CHIRINKO, R.S. (1987): "Tobin's Q and Financial Policy", *Journal of Monetary Economics, 19*, 69-87.

ENGLE, R.F., AND C.W.J. GRANGER (1987): "Co-Integration and Error Correction: Representation, Estimation, and Testing", *Econometrica, 55*, 251-276.

FUNKE, M., A. RYLL, AND D. WILLENBOCKEL (1988): "Taxation and Capital Formation in West German Industries: A Q-Theory Approach", in: *"Factors in Business Investment"*, ed. by M. Funke. Forthcoming, Berlin/Heidelberg/New York (Springer).

HAYASHI, F. (1982): "Tobin's Marginal q and Average q: A Neoclassical Interpretation", *Econometrica, 50*, 213-224.

KING, M.A. (1977): *"Public Policy and the Corporation"*, London.

KING, M.A., AND FULLERTON, D. (eds.) (1984): *"The Taxation of Income from Capital. A Comparative Study of the United States, the United Kingdom, Sweden, and West Germany"*, Chicago/London .

KURZ, R. (1987): "Einkommensteuersenkungen in den USA und ihre Auswirkungen auf die wirtschaftliche Entwicklung", in: *Konjunkturpolitik*, Beiheft 33, 185-207.

LEIBFRITZ, W. (1987): "Die Unternehmensbesteuerung im internationalen Vergleich", *Konjunkturpolitik*, Beiheft 33, 101-131.

NICKELL, S. (1985): "Error Correction, Partial Adjustment and All Mat: An Expository Note", *Oxford Bulletin of Economics and Statistics, 47*, 119-129.

PATTERSON, K.D. (1987): "Growth Coefficients in Dynamic Time Series Models", *Oxford Economic Papers, 39*, 282-292.

SALINGER, M.A., AND SUMMERS, L.H. (1983): "Tax Reform and Corporate Investment: A Microeconometric Simulation Study", in: Feldstein, M. (1983), *"Behavioural Simulation Methods in Tax Policy Analysis"*, Chicago, 247-281.

SALMON, M. (1982): "Error Correction Mechanism", *The Economic Journal, 92*, 615-629.

SUMMERS, L.H. (1981): "Taxation and Corporate Investment: A Q-Theory Approach", *Brooking Papers on Economic Activity, No. 1/1981*, 67-140.

TOBIN, J. (1969): "A General Equilibrium Approach to Monetary Theory", *Journal of Money, Credit, and Banking 1*, 15-29.

TOBIN, J. (1982): "Money and Finance in the Macroeconomic Process", *Journal of Money, Credit, and Banking 14*, 171-204.

WEICHERT, R. (1986): "Zur Besteuerung von Risikokapital in der Bundesrepublik Deutschland", *Die Weltwirtschaft*, 1986, 89-105.

WITTMANN, F. (1986): "Der Einfluß der Steuern auf die Investitionsentscheidungen der Unternehmen", *WSI-Mitteilungen, No. 12/86*, 782-790.

YOSHIKAWA, H. (1980): "On the 'q'-Theory of Investment", *American Economic Review, 70*, 739-743.

A Consistent Analysis of Government Financing in a Continuous-Time IS-LM Model

By H.-Dieter Wenzel,
Kora Kristof and Rolf Tschernig
Ludwig-Maximilians-Universität München

1. *Introduction*

In macroeconomic modelling one can choose between a continuous or a discrete formulation of variables. At first glance, discrete models seem to correspond naturally with economic activities because the latter have to take place at specific moments of time. Economic activities however, are not carried out at the same moments of time but are, instead, overlapping. A popular approximation of overlapping activities are continuous-time models. This is not too surprising since for them ample mathematical tools are available.

For choosing an appropriate model structure, the criterion of analytical tractability can only hold as long as the essence of the economic problem as well as its qualitative properties of the solution stay invariant in the face of simplifying methodical variations.

Thus, according to Foley (1975), our choice of the discrete approach as a starting point requires that the length of the time period must not have any impact on the economic solutions.

If this consistency requirement is also accepted for the construction of continuous-time models, these models should be derived by taking the time limit of discrete models which are independent of variations in their time periods. This approach has been developed by Hayakawa (1982a), (1982b), (1983), (1984). By evaluating the limit within the IS-LM framework he obtains flow equilibrium conditions for the commodity market and the asset markets. Having accomplished a uniform flow dimension for all markets, one is able to remove an inherent defect of the traditional stock-flow approach. This results from the differing dimensions of the equilibrium conditions where the flow equilibrium condition for the commodity market contrasts with the stock equilibrium conditions for both asset markets.

Assuming that income is also a variable of the stock demand functions for both assets, it follows, from the adding-up condition that the income elasticities of the asset demand functions must be of the opposite sign. However, this contradicts the results of positive income elasticities for both assets in the theory of transaction demand[1].

Since this contradiction does not evolve in a discrete period model with end-of-period planning and since the limit taking process conserves - among other things - this essential property, this contradiction does not show up in a continuous-time flow-flow model. This suggests that we should define continuous-time models as consistent if they can be interpreted as the limit of discrete-time models.

Formulating a consistent continuous-time version of the IS-LM model however, leads to unfamiliar efficiency and stability results for various public financing regimes. Their analysis is the purpose of the present study.

The paper proceeds as follows. Starting with a discrete-time model with end-of-period planning in section 2 we will derive the properties of the continuous-time limit. By obtaining the Walras Identity in flow form we are able to establish a flow-flow version of an IS-LM model. In section 3 we will analyse the efficiency and stability properties of government expenditure policies financed either by issuing bonds or printing money and compare these results with those of a traditional but corresponding stock-flow model. A summary will be given in section 4.

2. *The Model*[2]

2.1. *A discrete-time macro model with end-of-period planning*

In a closed economy all three sectors, the household sector (H), the firm sector (F), and the government sector (G) decide at the beginning t of a time period (t+h,t), h>0, on their planned supply and demand for the end t+h of this period. This end-of-period planning is consistent since the budget constraint of each sector is taken into account. The planning period (t+h,t) also represents the market period for all sectors. There are two assets, outside money (M) and bonds (B) which are represented by consols[3]. Both assets are only demanded by the household sector. Private bonds (B_p) and public bonds (B_g) are perfect substitutes.

Firms completely finance their investments (I) by issuing private bonds (B_p). The government, however, finances its budget deficit by issuing government bonds (B_g) or by printing money (M). Households consume (C) or save (S) their income flow (Y) earned over period (t+h,t) and, thus, decide on the accumulation and allocation of wealth. The income flow is obtained by adding up the flows of the factor income and interest payments paid by the firm and government interest payments ($B_g h$)and by subtracting the flow of tax payments (U). Furthermore, expected changes in the value of the existing wealth stock are taken into account.

There is no depreciation on the capital stock. The price level is constant. Each sector j = H,F,G forms expectations about the market interest rate (r).

To avoid complications from different ways of forming

expectations, we assume the axiom of weak consistency[4] with

Assumption 1: $\hat{r}_j(t,t) = r_j(t) > 0$ for $j = H,F,G,$

Δ

i.e the expected interest rate $\hat{r}(t,t)$ at time t for time t equals the actual interest rate $r(t)$. We also assume conformity among the expectations of each sector with

Assumption 2: $\hat{r}_H = \hat{r}_F = \hat{r}_G = \hat{r} > 0.$

Δ

They allow us to derive the budget constraints for each sector. The household budget constraint can be written as

$$\text{(HBC)} \quad M(t) + \frac{B(t)}{r(t)} + B(t)\left[1/\hat{r}(t+h,t) - 1/r(t)\right] + B_g(t)h + Y(t+h,t) = M^d(t+h,t) + \frac{B^d(t+h,t)}{\hat{r}(t+h,t)} + C(t+h,t) + U(t+h,t).$$

For the firms' budget constraint one obtains

$$\text{(FBC)} \quad \frac{B_p^s(t+h,t) - B_p(t)}{\hat{r}(t+h,t)} = I(t+h,t).$$

The government budget constraint can be expressed as

$$\text{(GBC)} \quad M^s(t+h,t) - M(t) + \left(B_g^s(t+h,t) - B_g(t)\right)/\hat{r}(t+h,t) = G(t+h,t) + B_g(t)h - U(t+h,t).$$

The first conclusions can be drawn [5].

Conclusion 1: Adding up all three budget constraints yields the Walras Identity in discrete form

$$(1) \quad \left\{Y(t+h,t) - C(t+h,t) - I(t+h,t) - G(t+h,t)\right\} + \left\{M^s(t+h,t) - M^d(t+h,t)\right\} + \left\{\left(B_p^s(t+h,t) + B_g^s(t+h,t) - B^d(t+h,t)\right) / \hat{r}(t+h,t)\right\} = 0.$$

Δ

Thus, for consistent end-of-period planning the excess supply over all three markets equals zero. If one defines household wealth (W) as

$$W(t) := M(t) + B(t) / r(t)$$

and saving (S) out of the disposable income flow (Y^V) as

$$S(t+h,t) := Y^V(t+h,t) - C(t+h,t)$$

and considers $W^d(t+h,t) = M^d(t+h,t)+B^d(t+h,t)/\hat{r}(t+h,t)$, one can come up with

Conclusion 2:

$$(2) \qquad S(t+h,t) = W^d(t+h,t) - W(t)$$

holds if and only if

$$(3) \qquad Y^V(t+h,t) = Y(t+h,t) + B_g(t)h - U(t+h,t) + B(t)\left[1/\hat{r}(t+h,t) - 1/r(t)\right].$$

Δ

This means that savings and wealth accumulation are only consistent if in calculating the disposable income flow the value change of bonds held at time t is taken into account. This important property will be maintained if the limit is evaluated as we will see below.

2.2. *The continuous-time limit*

In the following we will derive a continuous-time model from the outlined discrete-time model by taking the limit ($h \to 0$). This requires a precise definition of the concept of the flow over the period (t+h,t). We define for the income flow[6]

$$(4) \qquad Y(t+h,t) := \int_t^{t+h} y(\upsilon,t)d\upsilon$$

where $y(\upsilon,t)$ denotes the (finite) rate of the income flow $Y(t+h,t)$ over the period considered. It is planned at time t for time υ. Using also the following definition

$$(5) \qquad y(t) := \lim_{h\to 0} Y(t+h,t)/h,$$

we obtain[7]

Conclusion 3: The income flow at time t is

$$(6) \qquad Y(t,t) = 0$$

and the rate of the income flow at time t is given by

$$(7) \qquad y(t) = y(t,t) < \infty.$$

Proof: It follows from (4) and (5).

Δ

We are now prepared to derive the Balance Sheet Identity (BSI) which makes sure that wealth held at time t corresponds with the quantities planned for time t.

Conclusion 4: Under assumptions 1 and 2 one obtains

$$\text{(BSI)} \qquad M(t) + \frac{B(t)}{r(t)} = M^d(t,t) + B^d(t,t)/r(t)$$

or equivalently

$$\text{(BSI')} \qquad W(t) = W^d(t,t).$$

Proof: Taking $h \to 0$ in (HBC) and using (6) yields the (BSI)[8] which is identical to (BSI') per definition.

Δ

The following will also be helpful.

Conclusion 5: For $h \to 0$ one obtains

$$\text{(8.1)} \qquad M^s(t,t) = M(t)$$

$$\text{(8.2)} \qquad B_g^s(t,t) = B_g(t)$$

$$\text{(8.3)} \qquad B_p^s(t,t) = B_p(t).$$

Proof: (8.1) and (8.2) follow directly from the (GBC)[9] and (8.3) follows from the (FBC).

Δ

Because of the budget constraints, the assets supplied at time t are also demanded at the beginning of the period.

Finally, we need the partial derivatives of the stock demands and the expected interest rate with respect to their first argument, i.e. the time for which plans are made and actions are carried out. The partial derivative of the stock demand for money is given by

$$\text{(9)} \qquad M_1^d(t,t) := \lim_{h \to 0}\left(M^d(t+h,t) - M^d(t,t)\right)/h$$

This notation is applied according to the demand of bonds and the expected interest rate. For simplification we now assume that the firms and the government are always able to realize their planned supply of money and/or bonds.

Assumption 3:

$$\text{(10.1)} \qquad M^s(t+h,t) = M(t+h)$$

$$\text{(10.2)} \qquad B_g^s(t+h,t) = B_g(t+h)$$

(10.3) $$B_p^s(t+h,t) = B_p(t+h).$$

Δ

As a first result, we are able to state the budget constraints in flow form.

Conclusion 6: The continuous-time limit of the budget constraints of each sector is given by [10]

(11) $$y(t) - c(t) - u(t) + B_g(t) - B(t)\hat{r}_1(t,t)/r(t)^2$$
$$= M_1^d(t,t) + \frac{B_1^d(t,t)}{r(t)} - B^d(t,t)\hat{r}_1(t,t)/r(t)^2$$

(12) $$\dot{B}_p/r(t) = i(t)$$

and

(13) $$\dot{M}(t) + \dot{B}_g/r(t) = g(t) + B_g(t) - u(t).$$

Proof: The limit $h \to 0$ is taken for each budget constraint after its division by h. Then, (11) is obtained from the (HBC) if conclusions 3 to 5 are considered. If assumption 3 is also used, (12) and (13) follow accordingly from the (FBC) and (GBC).

Δ

This enables us to write the Walras Identity in its continuous-time version.

Proposition 1: Under the assumptions made so far one obtains the Walras Identity in its continuous time form as

(14) $$\{y(t) - c(t) - i(t) - g(t)\}$$
$$+ \{\dot{M}(t) - M_1^d(t,t)\}$$
$$+ \{[(\dot{B}_p(t) + \dot{B}_g(t))/r(t) - B(t)\hat{r}_1(t,t)/r(t)^2]$$
$$- [B_1^d(t,t)/r(t) - B^d(t,t)\hat{r}_1(t,t)/r(t)^2]\} = 0.$$

Proof: By adding up the flow budget constraints of conclusion 6.

Δ

This result was derived by Hayakawa (1982a). If one compares (14) with its traditional continuous-time version which is well known in the literature of IS-LM models and interprets the expressions of (14) in brackets as the excess supply on the commodity, money, and bond markets, one notices that the excess supply of all three markets are defined in a pure flow form (in the dimension [money unit/time unit]). Furthermore, the expected change in the value of the actual and planned stock of bonds is taken into account in determining the flow demand and supply on the bond market. This avoids a separation of the flow equilibrium condition for the commodity market and the stock equilibrium conditions for the asset markets. As a result, the inconsistence of contrary income elasticities of the demand functions for bonds and money which has been a consequence of the stock equilibrium conditions for the asset market[11] disappears.

The Walras Identity (14) allows us (in the tradition of Hicks) to drop one equilibrium condition. We now can write our continuous-time version of a three sector market model as the limit of the discrete-time model. We choose, without the loss of generality, to eliminate the bond market so that the model includes the equilibrium conditions for the commodity and the money markets

$$\text{(M1)} \qquad y(t) = c(t) + i(t) + g(t)$$

$$\text{(M2)} \qquad \dot{M}(t) = M_1^d(t,t)$$

and the budget constraints of the firm and government sector

$$\text{(M3)} \qquad \frac{\dot{B}_p(t)}{r(t)} = i(t)$$

$$\text{(M4)} \qquad \dot{M}(t) + \frac{\dot{B}_g(t)}{r(t)} = g(t) + B_g(t) - u(t).$$

Our first goal, to derive a continuous-time model from a discrete-time model by taking limits, has been reached. Observe that this model structure resembles the traditional IS-LM structure extended by budget constraints. The main difference occurs in the equilibrium condition for the money market (M2) which is defined in a flow form instead of a stock form. This is a consequence of our approach.

Before we can analyse our model however, we need to assume behavioral hypotheses in order to specify the rates of the flow demands of money M_1^d and consumption c, the flow of taxes u and government expenditure g. The rate of the flow demand of money will be derived from an assumption made for the discrete-time version.

<u>Assumption 4:</u> The stock demand for money for time t+h is given by

$$\text{(15)} \quad M^d(t+h,t) := m_y\, Y(t+h,t) + m_r\, \hat{r}(t+h,t) + m_w(h)W(t)$$

where $m_y > 0$, $m_r < 0$, $m_w(h) > 0$ and $\frac{d}{dh} m_w(h) < 0$ for $h \geq 0$.

Δ

The coefficient $m_w(h)$ of the initial wealth $W(t)$ is positive but the weight of the initial wealth is declining with an increasing period length[12]. A higher income flow or decreasing expectations of the interest rate cause a higher end-of-period demand for the money stock.

Further note that in the specification of (15) the period income $Y(t+h,t)$ instead of the income rate is included which avoids the possible inconsistency of the (BSI) since one obtains for $Y(t+h,t)$ as the argument of $M^d(t+h,t)$ and $B^d(t+h,t)/\hat{r}(t+h,t)$ in

$$\begin{aligned} W(t) &= \lim_{h\to 0} \left(M^d(Y(t+h,t),\dots) + \frac{B^d(Y(t+h,t),\dots)}{\hat{r}(t+h,t)}\right) \\ &= M^d(Y(t,t),\dots) + B^d(Y(t,t),\dots) \;/\; \hat{r}(t,t) \\ &= M^d(0,\dots) + B^d(0,\dots) \;/\hat{r}(t,t). \end{aligned}$$

Thus, for the stock restriction (BSI) there does not exist a meaningful partial derivative with respect to the income flow over the period. If one includes an income rate however, as for example $Y(t+h,t)/h$, both expressions, $M^d(t,t)$ and $B^d(t,t)/\hat{r}(t,t)$ remain functions of the income variable after the limit $h\to 0$ is taken. This leads to the initial inconsistency problem.

Thus it makes sense to start with the discrete-time demand for money (15). To facilitate the analysis we assume for the expectated interest rate[13]

<u>Assumption 5:</u> There is perfect foresight

$$(16) \qquad \hat{r}_1(t,t) = \dot{r}(t).$$

Δ

Now, the flow demand for money M_1^d can be derived.

<u>Conclusion 7:</u> Under assumptions 4 and 5 the flow demand is given by

$$(17) \qquad M_1^d(t,t) = m_y\, y(t) + m_r\, \dot{r}(t) + m_w W(t)$$

where $$m_w := \frac{dm_w}{dh}(0) < 0.$$

<u>Proof:</u> See appendix 1.

Δ

This completes the consistent specification of the flow equilibrium condition for the money market (M2).

It remains to be pointed out that our approach for the commodity market condition (M1) which includes the disposable income rate $y^V(t)$ implicates

<u>Conclusion 8:</u> If the disposable income flow is consistent according to (3) and if assumption 5 holds, then the disposal income rate is

$$(18) \qquad y^V(t) = y(t) + B_g(t) - u(t) - B(t)\dot{r}(t)/r(t)^2.$$

<u>Proof</u>: See appendix 2.

Δ

In order to complete the equation of the commodity market equilibrium, we proceed as we did with the flow demand of consumption with

<u>Assumption 6:</u> The flow demand of consumption over the period (t+h,t) is postulated as

$$C(t+h,h): = c_y\, Y^V(t+h,t) + c_w(h)\, W(t)$$

where $$\frac{dc_w(h)}{dh} > 0 \text{ for } h \geq 0 \text{ and } c_w(0) = 0.$$

Δ

This yields

<u>Conclusion 9:</u> If assumption 6 holds, then the rate of consumption is given by

$$(19) \qquad c(t) = c_y\, y^V(t) + c_w W(t), \quad c_w: = \frac{dc_w}{dh}(0) > 0.$$

<u>Proof:</u> See appendix 3.

Δ

We assume a traditional investment function which is linear in Y and r. To complete the specification of the government budget constraint (M4) we assume a linear tax function u(t) consisting of a flat tax rate τ, $0 \leq \tau \leq 1$, on income from factor supply and interest which is paid by firms and the government and a flat tax rate θ, $0 \leq \theta \leq 1$, on the stock of bonds. The government expenditure g is the main policy variable of the model. With these behavioral assumptions we are able to analyse how our modified approach towards a continuous-time IS-LM model changes the traditional results for fiscal policies which are financed either by bonds or money. This will be done in the following section.

3. *The Analysis of the Model*

Inserting the behavorial assumptions and derivations into the model structure (M1) - (M4) gives us the following system of equations

(20.1) $$y = c_y\{(1-\tau)(y+B_g) - \theta B/r - B\dot{r}/r^2\} + c_w\{M+B/r\} + \{i_y y + i_r r\} + g$$
$$c_w, i_y > 0, i_r < 0, 0 < c_y < 1$$

(20.2) $$\dot{M} = m_y y + m_r \dot{r} + m_w(M+B/r)$$
$$m_y > 0, m_r, m_w < 0$$

(20.3) $$\dot{B}_p/r = i_y y + i_r r$$

(20.4) $$\dot{M} + \dot{B}_g/r = g + B_g - \tau(Y+B_g) - \theta B/r, 0 \leq \tau, \theta \leq 1$$

To distinguish (20) from the traditional stock-flow approach of the IS-LM model we call the model (20) a flow-flow version.

It differs from the common continuous-time model in several ways. First, the budget constraint of firms and the government are taken into account which should give us interesting insight into the intrinsic dynamics of these financing constraints. We know this from the literature which has until now only been dealing with the (GBC). Second, the equilibrium condition for the money market (M2) includes flow but not stock variables. The flow demand for money is dependent on changes in the interest rate instead of the interest level and is negatively instead of positively correlated with wealth. As has been shown in the previous section, this specification follows directly from our assertion, to provide a "discrete time foundation" for continuous-time models by using end-of-period planning. Third, the disposable income rate y^v which also determines consumption is dependent on the change in the value of financial assets so that changes in the interest rate have to be considered. This is another consequence of our previous assertion.

As a result, this flow-flow version of an IS-LM model can be expected to have richer dynamics than its traditional stock-flow versions. In fact, stability properties will change which will shed new light on the two financing regimes considered here. Deficits generated by additional government expenditures are either purely financed by issuing bonds (BF) while the money supply remains constant ($M=\overline{M}$) or by printing money (MF) in which case the number of bonds is held constant ($B_g=\overline{B}_g$). To analyse the multiplier and stability properties we define an equilibrium state as one in which there is no change in any variable of the model through time, i.e.

(21) $$\dot{M} = \dot{B}_g = \dot{B}_p = \dot{r} = 0.$$

For the non-linearity in B/r we choose an appropriate linear approximation so that we obtain the linear system

(22) $$Ax = b$$

where $x = (y,r,M,B_g,B_p)^T$ denotes the vector of variables, $A = (a_{ij})_{1 \leq j \leq 5}^{1 \leq i \leq 4}$ the matrix of the system and $b = (g,o,g,o)^T$ the policy vector[14]. By eliminating the third or the fourth column in A, respectively, one obtains the relevant matrix of the system A_{BF} for bond financing and A_{MF} for money financing.

To facilitate the notation of the following proposition about the existence of equilibria we define

(23) $$\mu := (1-c_y)m_w + m_y c_w.$$

This strategy parameter of the commodity and money market is an important one.

<u>Proposition 2:</u> There exists a unique equilibrium of (2o) if and only if

a) $\mu \neq 0$ and $\tau < 1$ for the BF-regime

and

b) $\mu \neq 0$ and $\theta > 0$ for the MF-regime.

<u>Proof:</u> See appendix 5 [15].

A comparative-static analysis is only well defined if $\mu \neq 0$ is satisfied. This is not trivial due to the opposite signs of m_w and c_w that have been derived in the previous section. At the same time, these wealth effects in the demand for consumption and money are necessary so that a comparative-static analysis of the flow-flow model makes sense. The proposition also stresses the importance of the tax parameters. In particular, the existence of a money financed equilibrium requires the taxation of bond holdings.

3.1. *Comparative-static results*

Our interest focuses on the multiplier effects of additional government expenditure. It is well known that in the traditional Solow-Blinder model the efficiency is highly dependent on the choice of the financing regime. However, this is not true for our version of a continuous-time model.

Proposition 3: If there exists a unique equilibrium for the money and bond financing regime, the expenditure multipliers are given by

$$\text{(24)} \qquad \left.\frac{dy}{dg}\right|_{BF} = (1-c_y)m_w/\mu$$

and

$$\text{(25)} \qquad \left.\frac{dy}{dg}\right|_{MF} = \left.\frac{dy}{dg}\right|_{BF}.$$

Proof: See appendix 6.

Δ

The choice of the financing regime has no influence on the comparative-static efficiency of additional public demand. In our model a discussion of both regimes regarding their efficiency becomes irrelevant. What is the reason for this surprising result ?

The following proposition shows that additional government expenditure has the same effect on the interest rate in both regimes.

Proposition 4: The effect on the interest rate of additional government expenditure is given by

$$\text{(26)} \qquad \left.\frac{dr}{dg}\right|_{BF} = -\left(\frac{i_y}{i_r}\right)\left.\frac{dy}{dg}\right|_{BF}$$

and independent of its financing regime

$$\text{(27)} \qquad \left.\frac{dr}{dg}\right|_{BF} = \left.\frac{dr}{dg}\right|_{MF}.$$

Proof: See appendix 7.

Δ

Equation (26) follows directly from the firms' budget constraint (20.3) and holds analogously for the money financing regime. There is no impact on the interaction between the income rate and the interest rate from the equilibrium condition of the money market (20.2) since the latter is written in flow form. The flow demand for money only changes with interest rate variations if one neglects the indirect effect of the interest rate through wealth.

Furthermore, in the equilibrium state with $\dot{M}=0$ the money supply M cannot be a determining factor but instead operates symmetrically through the wealth effect for both regimes. This, we assume, causes the equivalent interest rate effect in

(27). If the equivalent income effect (26) is added, one directly obtains the indifference in efficiency (25) which contrasts with traditional results.

More information about the multipliers provides

Conclusion 10: For bond and money financing the size of the multipliers is given by

(28)
$$\left.\frac{dy}{dg}\right|_{BF(MF)} \geq 1 \Leftrightarrow \mu < 0$$

and

$$\left.\frac{dy}{dg}\right|_{BF(MF)} < 0 \Leftrightarrow \mu > 0 \ .$$

Proof: It follows from (24) with (23).

Δ

Thus the sign of μ determines the sign of the multiplier. If the multiplier is positive, it increases with increasing c_w, m_y, c_y and m_w. A special case is derived in

Conclusion 11: (Haavelmo-case) For $c_w = 0$ the multiplier is

(29)
$$\left.\frac{dy}{dg}\right|_{BF} = \left.\frac{dy}{dg}\right|_{MF} = 1.$$

Proof: It follows from proposition 3.

Δ

In contrast to the literature on the (GBC), the magnitude of the multiplier (24) is not dependent on the income tax rate. It is solely determined by the parameters of the demand functions for money and goods. However, the tax parameters influence the size of the deficit generated by an increase in government expenditure. For the BF-regime this is shown in the following

Proposition 5: For $\Theta > 0$ there is

(30)
$$0 < \left.\frac{dy}{dg}\right|_{BF} \gtrless \frac{1}{\tau - \Theta m_y/m_w} \Leftrightarrow \left.\frac{dB^g}{dg}\right|_{BF} \gtrless 0.$$

Proof: See appendix 8.

Δ

In order to find out more about the consequences of our approach we will compare these results with a traditional stock-flow model denoted by (20'). We eliminate from (20) all properties which are generated by the evaluation of the limits. The stock-flow model (20') differs from (20) in two respects. First, the term of the disposable income in (20') looses its dependence on changes in the interest rate

$$(31) \qquad y^{v'} = (y+B_g)(1-\tau) - \theta B/r.$$

Second, the flow equilibrium condition for the money market (20.2) is replaced by the stock condition (20.2')

$$(20.2') \qquad M = L(y,r,W)$$

$$L_y > 0,\ L_r < 0,\ 0 < L_W < 1.$$

The change from y^v to $y^{v'}$ only effects the dynamics of the stock-flow model. As assumed before, the second difference will also effect the comparative-static properties. Since the stock-flow version (20') diverges from the traditional models discussed in the literature, it explicitely takes into account the firms' budget constraint (FBC) and introduces the tax θ on wealth in the (GBC), we cannot invoke the literature so that we present our results[16].

As with the model (20) the sign of the parameter

$$(32) \qquad \lambda := L_y - L_r i_y / i_r$$

is important[17] in calculating the multipliers.

<u>Proposition 6:</u> In the stock-flow model the expenditure multiplier is either given by

$$(33) \qquad \left.\frac{dy}{dg}\right|_{BF}^{SF} = \frac{(1-c_y)L_W}{c_W\lambda + (1-c_y)L_W} \qquad \text{for } \tau < 1$$

or by

$$(34) \qquad \left.\frac{dy}{dg}\right|_{MF}^{SF} = \frac{c_W + \theta\,(1-L_W)(1-c_y)}{c_W\tau + \theta[(1-L_W)(1-c_y) - c_W\lambda]} \qquad 0 \leq \tau,\ \theta \leq 1 \text{ and not } 0 = \tau = \theta\ .$$

<u>Proof:</u> See appendix 9.

Δ

It is remarkable that in both versions the BF-multipliers have a similar structure since λ can be replaced by m_y. Both parameter express the income effect on the money demand. However, λ is, in addition to the direct effect, determined by an indirect effect through interest rates and investment.

Nevertheless, there remain differences. The parameters describing the wealth effect on the demand for money have opposite signs. In the flow-flow model a high income effect amplifies a positive multiplier (24) through m_y but reduces it in the stock-flow model (33). A similar result holds for the wealth effect on consumption c_w. In the flow-flow version the wealth effect reinforces the multiplier whereas the stock-flow multiplier is weakened (for $\lambda > 0$).

Note also that the MF-multiplier of the stock-flow model (34) includes both tax parameters and exists for arbitrary values, in particular for $\theta = 0$.

Conclusion 12: In (20') there is

(35) $$\left.\frac{dy}{dg}\right|_{MF}^{SF} = 1/\tau \qquad \text{for } \theta = 0 \text{ and } c_w > 0$$

(36) $$\left.\frac{dy}{dg}\right|_{MF}^{SF} = \left.\frac{dy}{dg}\right|_{BF}^{SF} = 1 \qquad \text{for } c_w = 0,\ \theta > 0 \text{ and } 0 < \tau < 1$$

Proof: It follows from (33) and (34).

Δ

The extended stock-flow model (20') provides the traditional result of the Inverse Income Tax Rule if there is no wealth taxation (θ=0). It also reproduces the Haavelmo case in (36). In this model we also find the well-known difference in the efficiency of financing policies.

Proposition 7: For $\theta \geq 0$ in model (20') the following holds

(a) Is $\lambda \geq 0$, then there is

(37) $$0 < \left.\frac{dy}{dg}\right|_{BF}^{SF} \leq 1 < \left.\frac{dy}{dg}\right|_{MF}^{SF},$$

if and only if

$$\lambda < \frac{\tau}{\theta} + \frac{(1-L_w)(1-c_y)}{c_w}.$$

(b) Is $\lambda < 0$, then there is

(38) $$0 < \left.\frac{dy}{dg}\right|_{MF}^{SF} \leq \left.\frac{dy}{dg}\right|_{BF}^{SF}$$

and $$\left.\frac{dy}{dg}\right|_{BF}^{SF} > 1$$

if and only if the inequality

$$-\frac{(1-c_y)L_w}{c_w} < \lambda \leq -\frac{(1-c_y)(1-\tau)L_w}{c_w + \theta(1-c_y)}$$

is fulfilled.

Proof: See appendix 10.

Δ

The efficiency of a financing regime is exclusively determined by the sign of λ. The sign of λ can be derived from elasticities which can be empirically measured.

Conclusion 13:

$$(39) \qquad \lambda \lesseqgtr 0 \Leftrightarrow \frac{E(L,y)}{E(L,r)} \gtreqless \frac{E(i,y)}{E(i,r)}$$

Proof: Transforming (32) into the notation of elasticities.

Δ

The condition (39) contains information about the slopes of the IS- and LM-curve. It can be said that a flat (steep) LM-curve and a steep (flat) IS-curve work towards $\lambda<0$ ($\lambda>0$) so that bond (money) financing is more efficient.

So far, comparative-static properties of two policy measures have been analysed in two different frameworks of an IS-LM model. An efficient policy becomes irrelevant however, if its corresponding equilibrium is dynamically unstable. We will therefore, proceed with analysing the stability of the flow-flow equilibria.

3.2. *Stability*

The flow-flow model (20) is a non-linear dynamic system which complicates the analysis of its stability properties. Nevertheless, it is possible to obtain information about its asymptotic stability behavior if one linearises the dynamic system[18]. First, consider the bond financing regime for which $\dot{M}=0$. Linearisation of (20) yields the differential equation system

$$(40) \qquad \dot{x} = B_{BF}x$$

where B_{BF} denotes the Jacobian and $x := (r, B_g, B_p)^T$ [19]. According to Fuller (1968), the stability behavior of a three-dimensional linear differential equation system is described by the following auxiliary proposition.

Auxiliary proposition: The linear diferential equation system is asymptotically stable if and only if

(Stab 1) $\qquad$ Spur $B_{BF} < 0$

(Stab 2) $\qquad$ Det $B_{BF} < 0$

(Stab 3) $\text{Det } F(B_{BF}) < 0.$

∆

$F(B_{BF})$ denotes the Fuller-Matrix of B_{BF}[20]. Usually, an analytical analysis of these stability conditions does not produce usable results so that simulations seem to be more promising.

Nonetheless, we are able to obtain a decisive clue about the stability behavior from

Proposition 8: In the BF-regime if $i_r < 0$, there is

$$\text{Det } B_{BF} \lessgtr 0 \Leftrightarrow \mu \lessgtr 0 . \tag{41}$$

∆

The inequality $\mu < 0$ is one condition necessary for stabiliy. At the same time it guarantees positive values of the expenditure multipliers. It also justifies the previous comparative-static analysis of our flow-flow model. More importantly, the flow-flow approach obtains economic relevance. Furthermore, proposition 8 gives with, $\mu > 0$, a sufficient condition for local instabilitiy of the initial system (20) since at least one positive real eigenvalue of B_{BF} is generated so that the instability of the linear system (40) carries over to the original system (20).

The following proposition shows that for particular parameter combinations, which satisfy the parameter restrictions given by (20), the necessary stability condition $\mu < 0$ can contradict the stability requirements resulting from (Stab 1).

Proposition 9: For bond financed budget deficits and "small" values of θ ($\theta < r^*$), r, i_y, and $|i_r|$, the condition

$$\mu > 0$$

is necessary to fulfill (Stab 1).

∆

In these cases the equilibrium state is unstable even for positive multipliers ($\mu<0$). This existence of pathological stability behavior in the bond financing regime corresponds to the results of the traditional stock-flow approach[21]. However, there is no loss of efficiency in the case of an unstable BF-equilibrium in the flow-flow version since a MF-equilibrium is equally efficient. Thus, one cannot derive the famous Blinder-Solow result where for $i_y = 0$ bond financing is more efficient if and only if it has a stable equilibrium.

A necessary stability condition for the money financing regime is included in

<u>Proposition 10:</u> For money financed budget deficits and $i_r < 0$ there is

$$(42) \qquad \text{Det } B_{MF} \lesseqgtr 0 \Leftrightarrow (\mu \lesseqgtr 0 \text{ und } \theta > 0) \; .$$

Δ

B_{MF} denotes the matrix of the linearised system (20) under the assumption of a constant stock of public bonds.

This result coincides with proposition 8 for a bond financing regime with respect to the sign of μ. However, stability requires tax on bond wealth. This seems to be compatible with the comparative-static conclusions.

At this point we terminate our dynamic analysis since our goal has been reached. We wanted to show that a flow-flow approach is well defined and yields reasonable economic results. Furthermore, these results modify traditional conclusions for stock-flow IS-LM models. However, one should not be tempted to deduce more than that. Suggestions for economic policy should be based on a more realistic model structure.

4. *Conclusion*

Our initial task was to provide a "discrete time foundation" for a continuous-time model by using a period model, i.e. to derive a continuous-time version by taking the time limit of a discrete-time model. This model has been based on consistent end-of-period planning of all three sectors (households, firms, government) and for all three markets (goods, money, bonds). For simplification we assumed the same time horizon for planning, deciding, and forming expectations. The limits of the resulting sectoral budget constraints have been evaluated so that the length of the period approaches zero. Errors in expectations or planning have been ruled out so that it was not necessary to distinguish between ex-ante and ex-post restrictions. This invoked a Walras Identity in flow form. Its main characteristic has been that the aggregate excess supply of all markets - including the money market - could be derived without stock variables which has been the prerequisite for the aggregation.

By constructing a macro model in the tradition of Hicks and Hansen which includes market equilibrium conditions, one equilibrium condition can be eliminated according to Walras Law. Eliminating the bond market, the continuous-time model consists of the equilibrium conditions for the commoditiy and money markets and the budget constraints of the government and firms.

This model has been called a flow-flow version of the IS-LM model because the equations of the model are exclusively dimensioned in flows which revokes the usual separation of a stock equilibrium for the money market and a flow equilibrium for the commodity market.

This model is in line with the tradition of Hicks and Hansen in that the commodity and money markets are constitutive elements. However, it deviates substantially in its money market equilibrium condition that is determined by changes of supply and demand in time as well as in the disposable household income which must take into account changes in the interest rate. These dynamic elements add to the intrinsic dynamics due to the budget constraints.

The richer dynamic structure suggests that we reexamine the results on efficiency and stability properties of fiscal policies derived in the common stock-flow model (with a GBC). For both comparative-static and dynamic properties, important differences are deduced. Without repeating them in detail we conclude that a flow-flow analysis of the effects of public financing regimes opens up new methodes and provides new insight in order to represent an interesting alternative to traditional continuous-time macro models.

5. *Appendix*

Appendix 1: Proof of conclusion 7:

$$M_1^d(t,t) = \lim_{h\to 0} [M^d(t+h,t) - M^d(t,t)]/h$$

$$= m_y \lim_{h\to 0}[Y(t+h,t) - Y(t,t)]/h + m_r \lim_{h\to 0}[\hat{r}(t+h,t) - \hat{r}(t,t)]/h$$

$$+ W(t) \lim_{h\to 0}[m_w(h) - m_w(0)]/h$$

$$= m_y\, y(t) + m_r\, \dot{\hat{r}} + m_w W(t).$$

Δ

Appendix 2: Proof of conlusion 8:

If assumption 5 holds and (3) is taken into account, one obtains

$$y^v = \lim_{h\to 0} [Y^v(t+h,t)/h] =$$

$$= \lim_{h\to 0} \frac{1}{h} \Bigg[Y(t+h,t) + B_g(t)h - U(t+h,t)$$

$$+ B(t) \left(\frac{1}{\hat{r}(t+h,t)} - \frac{1}{r(t)} \right) \Bigg]$$

$$= y(t) + B_g(t) - u(t) - B(t)\dot{r}(t)/r(t)^2.$$

Δ

<u>Appendix 3:</u> Proof of conclusion 9:

Using the Rule of L'Hospital one obtains

$$c(t) = \lim \frac{1}{h} C(t+h,t)$$

$$= c_y \lim_{h\to 0} Y^v(t+h,t)/h + W(t) \lim_{h\to 0} \left(\frac{c_w(h)}{h} \right)$$

$$= c_y\, y^v(t) + c_w\, W(t).$$

Δ

<u>Appendix 4:</u> The matrix of the system $A = (a_{ij})_{1 \leq j \leq 5}^{1 \leq i \leq 4}$ is given by

$$A = \begin{bmatrix} [1-c_y(1-\tau)-i_y] & [(c_w-c_y\theta)\frac{B^*}{r^{*2}} - i_r] & -c_w & -[(c_w-c_y\theta)\frac{1}{r^*} + c_y(1-\tau)] & -(c_w-cy\theta)/r^* \\ m_y & -m_w\ B^*/r^{*2} & m_w & m_w/r^* & m_w/r^* \\ \tau & -\theta\ B^*/r^{*2} & 0 & \theta/r^* - (1-\tau) & \theta/r^* \\ i_y & i_r & 0 & 0 & 0 \end{bmatrix}.$$

where * denotes equilibrium values.

Δ

Appendix 5: Proof of proposition 2:

A solution for the linearisation of model (20) exists if and only if

a) Det $A_{BF} \neq 0$ for the BF-regime

and b) Det $A_{MF} \neq 0$ for the MF-regime.

Regarding a): If in matrix A the third column is eliminated, one obtains A_{BF}. After some linear transformations of A_{BF} one yields

$$\text{(A.1)} \qquad \text{Det } A_{BF} = (1-\tau)\ i_r \left(c_w m_y + (1-c_y) m_w\right)/r^* \neq 0$$
$$\Leftrightarrow \mu \neq 0 \text{ and } 1-\tau \neq 0 \quad .$$

The conditions $i_r < 0$ and $r^* > 0$ are fulfilled by assumption.

Regarding b): If in matrix A the fourth column is eliminated, one obtains A_{MF}. After some linear transformations of A_{MF} one gets

$$\text{(A.2)} \qquad \text{Det } A_{MF} = \theta\ i_r \left(c_w m_y + (1-c_y) m_w\right)/r^* \neq 0$$
$$\Leftrightarrow \mu \neq 0 \text{ and } \theta > 0 \ .$$

The conditions $i_r < 0$ and $r^* > 0$ hold by assumption.

Δ

Appendix 6: Proof of proposition 3:

(24) is obtained from the matrix of the system by applying Cramer's Rule if one considers that the policy vector for variations in government expenditure is given by $b = (dg, 0, dg, 0)^T$ and that Det A_{BF} is given by (A.1).
Similarly one produces $dy/dg|MF$ if one uses the matrix of the system A_{MF} and its determinant Det A_{MF} from (A.2).

Δ

Appendix 7: Proof of proposition 4:

The left-hand side of (27) is obtained from the matrix of the system A_{BF} by applying Cramer's Rule and if (A.1) is taken into account. The right-hand side is obtained from the matrix of the system A_{MF} if (A.2) is considered.

Δ

Appendix 8: Proof of proposition 5:

First, one has to determine $\left.\frac{dB_g}{dg}\right|_{BF}$ from A_{BF} by using (A.1). This yields

$$\text{(A.3)} \qquad \left.\frac{dB_g}{dg}\right|_{BF} = - \frac{\{(1-c_y)(1-\tau)m_w + m_y\, c_w - \theta m_y(1-c_y)\}}{(1-\tau)[m_y\, c_w + (1-c_y)\, m_w]}$$

$$= - \left.\frac{dy}{dg}\right|_{BF} \left\{\frac{(1-\tau)(1-c_y)m_w + m_y\, c_w + m_y\theta(1-c_y)}{(1-\tau)\, m_w\, (1-c_y)}\right\} \text{ with (24)}$$

Then the hypothesis follows.

Δ

Appendix 9: Proof of proposition 6:

In the stock-flow version, the Jacobian A' of the system (20') differs from the Jacobian A of the flow-flow version (20) only in its elements of the second row which are

$a'_{21} = L_y > 0,$ $\qquad a'_{22} = L_r - L_w B^*/r^{*2} < 0,$
$a'_{23} = -(1-L_w) < 0,$ $\qquad a'_{24} = a'_{25} = L_w/r^* > 0.$

For the BF-regime one yields after elimination of the third column of A'

$$\text{(A.4)} \qquad \text{Det } A'_{BF} = (1-\tau)\ i_r\ \left(c_w\lambda + (1-c_g)L_w\right)/r^*$$

and for the MF-regime one obtains after elimination of the fourth column of A'

$$\text{(A.5) Det } A'_{MF} = -\ i_r\left\{c_w\tau + \theta[(1-L_w)(1-c_y) - c_w\lambda]\right\}/r^*.$$

Using Cramer's Rule and (A.4) or (A.5) one obtains the multiplier (33) or (35), respectively, since in (20.1) is $i_r < 0$ assumed in general.

Δ

Appendix 10: Proof of proposition 7:

(a) follows directly from (33) and (34); (b) follows analogously after some calculations.

Δ

Appendix 11: Definition of the Fuller-matrix:

If $\qquad A = (a_{ij}),\ 1 \leq i,\ j \leq 3,$

then

$$F(A): = \begin{pmatrix} a_{11} + a_{22} & a_{23} & -a_{13} \\ a_{32} & a_{11} + a_{33} & a_{12} \\ -a_{31} & a_{21} & a_{22} + a_{33} \end{pmatrix}.$$

Δ

6. *Footnotes*

[1] Cf. Baumol (1952) and Hellwig (1975).

[2] The proceeding is based on Hayakawa (1982a, 1982b). The approach was developed by Turnovsky (1977a, 1977b). However, he assumed share financing instead of bond financing.

[3] Thus, the number of bonds held by the households correspond with the interest payments per time unit. In a period of length $h > 0$ the interest payments amount to Bh.

[4] Cf. Turnovsky/Burmeister (1977).

[5] The budget constraints are given in ex-ante form. They are identical to their ex-post forms if there are no errors in expectations. This will be assumed in the following. In particular, the flow of income (Y) and taxes (U) is known.

[6] This definition holds analogously for all flows over the period C, S, U, I, and G whereby the flow of interest payments is expressed by

$$B(t)h = \int_t^{t+h} B(t)d\upsilon.$$

It is always assumed that the integrands are finite and that the integrals exist. There exist two types of flows. By using lower case notation for the rates of flows with the dimension [money unit/time unit] they can be distinguished from flows over the period with the dimension [money unit] which are denoted by capital letters.

[7] We always assume that the limits exist and that all functions are sufficiently smooth so that they are differentiable at least from the right hand side if the limits $h \to 0$ are taken for $h > 0$.

[8] The (BSI) is often characterised as an implication of the Walras Identity. It is, however, a direct implication of the (HBC).

[9] In the following we have assumed either pure money or pure bond financing. This guarantees the independence of (8.1) and (8.2). For mixed financing one can conclude from the (HBC) for $h \to 0$ that the sum of the planned quantities equals the sum of the actual quantities. This does not exclude individual errors in expectations or planning. For further conclusions this distinction is irrelevant.

[10] A dot on a variable denotes its derivative with respect to time.

[11]Cf. Hayakawa (1985).

[12]This formulation originates from Hayakawa (1982a).

[13]Pursuing various hypotheses about the formation of expectations would distract us from presenting the basic idea of our approach.

[14]Cf. appendix 4.

[15]Since linearising the system only makes sense for $r^* \neq 0$, we assume in the following without the loss of generality $r^*>0$. and because of (20.4) $y^*>0$. This guarantees usable economic solutions.

[16]In the following stock-flow results and flow-flow results will be indexed by SF and FF, respectively.

[17]The following results for (20') are derived from the linearised version of (20'). The equilibria considered exist uniquely if the denominator of (33) for the BF-regime and the denominator of (34) for the MF-regime is not zero.

[18]The asymptotic stability properties of the linearised system hold locally for the initial non-linear system if and only if the stability behavior of the linerarised version is distinct, i.e every eigenvalue of the Jacobian has negative real parts or at least one eigenvalue has a positive real part. Cf. Wenzel (1976, p. 155 ff.) and Willems (1973, p. 155 ff.).

[19]The calculations are elementary but tedious. They are available from the authors, as is a more detailed version of this paper which includes the detailed proofs of the following propositions 8 to 10.

[20]Cf. appendix 11 for the definition.

[21]We dispense with the analysis of stability of the stock-flow version (20') since its dynamics fall back into a two-dimensional system. Its stability properties (for a comparable model) have been discussed in Wenzel (1983).

7. *References*

BAUMOL, W.J. (1952): "The Transactions Demand for Cash: An Inventory Theoretic Approach", *QJE*, 66, 545-556.

BLINDER, A.S., SOLOW, R.M. (1973): "Does Fiscal Policy matter?", *JPubE*, 2, 319-337.

FOLEY, D. (1975): "On Two Specifications of Asset Equilibrium in Macroeconomic Models", *JPolE*, 83, 303-324.

FULLER, A.T. (1968): "Conditions for a Matrix to have only Characteristic Roots with Negative Real Parts", *Journal of Mathematical Analysis and Applications*, 23, 71-98.

HAYAKAWA, H. (1984): "Balance Sheet Identity and Walras' Law", *JET* ,34, 187-202.

--- (1983): "Consistency of Flow Equilibrium Specification of Asset Market with Transactions Theory of Demand for Money", *Economic Letters*, 11, 49-55.

--- (1982a): "The Efficacy of Debt Financed Fiscal Policy Under a Flow Equilibrium Model", *Mimeo*. University of Georgia.

--- (1982b): "Conservation Principles and an Alternative Formulation of a Continuous Time Macro Model", *The Economic Studies Quarterly*, 33, 111-125.

HELLWIG, M.F. (1975): "The Demand for Money and Bonds in Continuous-Time Models", *JET*, 11, 462-464.

TURNOVSKY, ST.J. (1977a): "On the Formulation of Continuous Time Macroeconomic Models with Asset Accumulation", *IER*, 18, 1-28.

--- (1977b): "Macroeconomic analysis and stabilization policy", Cambridge.

TURNOVSKY, St.J., BURMEISTER, E. (1977): "Perfect Foresight, Expectational Consistency, and Macroeconomic Equilibrium", *JPolE*, 85, 379-393.

WENZEL, H.D. (1983): "Defizitfinanzierung als Instrument einer zielorientierten Finanzpolitik. Eine dynamische Analyse", Baden-Baden.

--- (1976): "Stabilität und Optimale Stabilisierung Makroökonomischer Systeme", *Mathematical Systems in Economics*, 22, Meisenheim am Glan.

WILLEMS, J.L. (1973): "Stabilität ökonomischer Systeme", München.

IS-LM dynamics and the hypothesis of combined adaptive-forwardlooking expectations.

by

Christian Groth[1)]

University of Copenhagen

I. Introduction

This paper studies the dynamic behaviour of an IS-LM model extended by the expectations augmented Phillips curve and what we call "combined adaptive-forwardlooking expectations".

The theory of IS-LM dynamics arose from the monetarist critique of Keynesian economics for being one-sidedly concerned with flows rather than stocks of assets and for attaching too much importance to short run effects rather than ultimate, cumulative effects (M. Friedman 1968, 1972). The "IS-LM cross" began to be seen as a momentary stage of a dynamic process. One source of the dynamics is the changes in stocks of money and bonds brought about by the financing of government budget deficits (Christ 1968, Blinder & Solow 1973, Tobin & Buiter 1976). Other sources of the dynamics are the change in prices, as derived from an expectations augmented Phillips curve, as well as the change in expectations, usually described by the adaptive expectations formula (Tobin 1975, Buiter & Lorie 1977, Scarth 1977, Taylor 1977, Yarrow 1977). The large body of studies dealing with these problems is surveyed in Currie 1978, Christ 1979, and Rau 1985.

One of the conclusions in this literature is that quite apart form the possible instability arising from the cumulative effects of financing budget deficits by bond issue, instability could arise from the dynamic interaction of output demand and inflation expectations. Given non-fulfilment of some generalized form of the Cagan condition (Cagan 1956) an incipient recession will be self-enforcing; thereby an old Keynesian tenet is confirmed. This outcome rests on the possibility of the nominal and real rates of interest moving adversely, when actual and thereby expected inflation change.

However, one of the well known drawbacks of the adaptive expectations hypothesis used in these studies is that it assumes agents base their inflation forecast on a very limited set of information, i.e. the past development of the inflation rate. This implies that agents can be systematically wrong in their expectations, e.g. when the variable in question has a trend, and that they nevertheless continue to use the same method of forming expectations.

The radical alternative to adaptive expectations is the hypothesis of "rational expectations" (Muth 1961). Incorporating this hypothesis into IS-LM models, whether deterministic or stochastic, together with the expectations-augmented Phillips curve (or, what amounts to the same, the Lucas supply curve), represents, of course, a special case of the "new classical" theory. Output becomes independent of anticipated monetary change and, apart from erratic shoks, output tends towards the "natural level" (Lucas 1972, Sargent & Wallace 1975).[2)]

Feeling that the rational-expectations hypothesis is too strong, Peel & Metcalfe (1979) developed an analysis of a simple "monetarist" model[3)] in which expectations are formed in different ways by different groups of agents (e.g., adaptively by workers and rationally by firms and financial agents). Peel & Metcalfe showed that this being the case the condition for stability of production and inflation tends to become more stringent than in the case of uniformly held adaptive expectations.

Yet another hybrid assumption "the two-part expectations hypothesis", was studied within a similar "monetarist" framework by Frenkel (1975) and Peel (1979). According to this hypothesis changes in expected short-term inflation are based on two elements which are first, the deviation of expected short-term inflation from current inflation and second, the deviation of current inflation from expected long-term average ("normal") inflation. In its turn this expected long-term average inflation is formed according to the standard adaptive formula (Frenkel

1975) or to some kind of long-term "monetarist expectations" (Peel 1979).

The hypothesis introduced below, which we call "combined adaptive-forwardlooking expectations", is akin to this. Both current inflation and the tendentially perceived long-term inflation (equal to monetary expansion) enter. While on the one hand being more crude (using only one exponentially distributed lag), on the other hand our hypothesis allows for a larger interval of parameter values than the Frenkel approach thus giving rise to cases with qualitatively different dynamic behaviour. We study the role of this hypothesis in an otherwise standard IS-LM (Keynesian income-expenditure) model, respecting the government budget constraint. The purpose is partly expository and partly to break new ground.

II. The model

The symbols are

y = real output

y_d = perceived real disposable income

P = price of output

M = nominal money supply (equal to monetary base)

m = real money supply, M/P

$\dot{P}/P$ = actual rate of inflation

x = expected rate of inflation

i = nominal rate of interest (after tax)

T = nominal taxes net of transfer payments

g = real government purchases

a = rate of monetary expansion, $\dot{M}/M$

A dot denotes a time derivative, e.g., $\dot{x} = dx/dt$, time, t, being treated as continuous.

The hypothesis of "combined adaptive-forwardlooking expectations" is:

$$\dot{x} = b((c\frac{\dot{M}}{M} + (1 - c)\frac{\dot{P}}{P}) - x), \qquad b > 0, \ c \geq 0 \qquad (1)$$

This says that expected inflation is lagged on an indicator, which is a weighted average of the tendentially perceived long term inflation rate (equal to monetary expansion) and the current inflation rate. The case $c = 0$ is the simple, one-sidedly backwardlooking, adaptive expectations formula. The case $c = 1$ is considered in Stein (1982) and may be called "monetarist expectations"[4] - corresponding to the monetarist belief in stability of the market economy and convergence towards a steady state. The intermediate case $0 < c < 1$ is a mixture of these two extremes and contains both a forwardlooking and an adaptive element. The case $c > 1$ is included, although we do not find it plausible. It corresponds to a variant (introduced by Peel 1979) of Frenkel's above-mentioned "two-part expectations hypothesis"; the presumption is that people strongly believe in a certain long-term rate of inflation - not only as an ultimate level at which inflation would tend to settle, but also as an average rate of inflation over shorter time intervals.

The remaining seven equations of our short-to-medium run IS-LM model are given below. The derivative of a function, say F, with one argument is denoted F'. The partial derivative with respect to the ith argument of a function, say C, with more than one argument is denoted C_i.

$$y = C(y_d, m) + I(y, i-x) + g, \qquad (2)$$
$$0 < C_1 < C_1 + I_1 < 1, \ C_2 \geq 0, \ I_2 < 0, \ g \geq 0$$

$$y_d = y - T/P - xm \qquad (3)$$

$$T = Pg - \dot{M} \qquad (4)$$

$$m = L(y, i, m), \quad L_1 > 0, \ L_2 < 0, \ 0 \leq L_3 < 1 \qquad (5)$$

$$\dot{M}/M = a, \quad a \geq 0 \qquad (6)$$

$$\dot{P}/P = F(y) + x, \quad F' > 0, \text{ and for some } y^* > 0, \quad F(y^*) = 0 \tag{7}$$

$$m \equiv M/P \tag{8}$$

a, b, c, and g are given constants. From the point of view of the dynamic analysis of the model there are eight endogenous variables, y, i, y_d, T, P, M, m, and x, to be determined from equations (1) - (8) and given initial values of x, P, and M.

y* is the "natural" level of output defined as that level of output which is consistent with actual and expected inflation being equal. It is understood that both the real wage, the capital stock, and the labour supply allow profitable production to take place above the natural level, provided there is sufficient demand. (3) gives perceived disposable income, i.e., real gross income corrected for real net taxes and the perceiced capital loss on the real value of the public's net claims on the government, the socalled inflation tax. (4) is the government budget constraint. It states that government purchases not financed by money creation are financed by taxes, that is by adjusting the tax and transfer instruments. There are no government bonds in the model. Thus, the problem whether government bonds are part of private net wealth or not (see Barro 1974, McCallum 1978, Tobin 1980) does not arise. The remaining equations are self-explanatory.

<u>Short run equilibrium</u>

Inserting (3) in (2), and using (4) and (6) gives

$$y = C(y + (a - x)m - g, m) + I(y, i - x) + g. \tag{9}$$

In the short run the money supply, output price and expectations are historically given. Thus m and x are given for fixed t, and the "IS equation", (9), together with the "LM equation", (5), determine the output level and the rate of interest. Thus short run y can be written as a function

$$y = f(x, m; a, g) \tag{10}$$

with continuous partial derivatives (impact multipliers)

$$f_1 = \frac{-I_2-C_1m}{A}, \quad f_2 = \frac{(1-L_3)I_2/L_2 + C_1(a-x) + C_2}{A}, \tag{11}$$

where

$$A \equiv 1-C_1-I_1 + L_1 I_2/L_2 > 0, \tag{12}$$

cf. (2) and (5). The signs of f_1 and f_2 are in general ambiguous. However, for $x \leqq a$, $f_2 > 0$, a fact which will be useful in what follows.[5)]

Dynamics

In order to study the evolution of short run equilibria through time we derive he differential equations implied by the model. From (1), (6), (7), and (10) we obtain

$$\dot{x} = b[c(a-x) + (1-c) F (f(x,m; a,g))]. \tag{13}$$

Combining (6), (7), (8) and (10) gives

$$\dot{m} = [a - x -F(f(x,m; a,g))]m. \tag{14}$$

A steady state is a time path along which x and m (and therefore y and i) are constant. In a steady state the nominal variables, M, P, and T, are changing at the same constant rate, a. Since $m > 0$, $\dot{x} = \dot{m} = 0$ implies, by (13) and (14), that the steady state value of x equals a, the rate of mometary expansion. Hence, in a steady state, $f(x,m; a,g) = y*$ by (14) and (7), i.e. the steady state value of output is y*, the natural rate. The steady state value of m, m*, is the solution in m to the equation $f(a,m; a,g) = y*$.[6)] Observe that none of the parameters a, b, c, and g influences y*, that only a influences x* and only a and g influence m*.

The phase portrait of the system (13), (14) is depicted in Figure 1. The sloping dotted curves in the diagram represent iso-production-loci obtained from (10). Moving North in the diagram is associated with rising output.

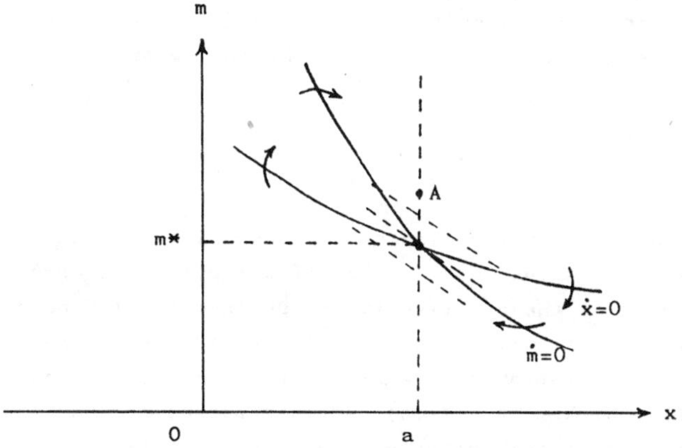

Figure 1. Phase portrait of (13), (14); the case $f_1^* > c/((1-c)F'^*)$.

III. Expectations and stability

The Jacobian matrix of (13), (14) evaluated at the steady state point (a,m*) is

$$\begin{bmatrix} \frac{\delta \dot{x}}{\delta x} & \frac{\delta \dot{x}}{\delta m} \\ \frac{\delta \dot{m}}{\delta x} & \frac{\delta \dot{m}}{\delta m} \end{bmatrix}_{(x,m)=(a,m^*)} = \begin{bmatrix} b[-c+(1-c)F'^* f_1^*] & b(1-c)F'^* f_2^* \\ (-1-F'^* f_1^*)m^* & -F'^* f_2^* m^* \end{bmatrix} \quad (15)$$

The determinant is $bF'^* f_2^* m^*$, which is positive because x = a implies $f_2^* > 0$, by (11). The trace is $b[-c+(1-c)\ F'^* f_1^*]-F'^* f_2^* m^*$, the sign of which is ambiguous. It follows that the steady state is locally asymptotically stable if and only if[7)]

$$b[(1-c)F'^* f_1^* -c] < F'^* f_2 m^*. \quad (16)$$

The possibly destabilizing force comes from the endogenuous inflation expectations, which may, through the feedback in the economy, be self-fortifying such that $\delta\dot{x}/\delta x > 0$.

Before commenting further on (16) we make the reasonable assumption

A1. $C_1^* m^* < - I_2^*$, i.e., $f_1^* > 0$.

Thus we assume that the negative impact on consumption demand of increased expected inflation, due to the inflation tax, is more than counterbalanced by the positive impact on investment demand. Given this assumption the price mechanism is easily seen to have an ambiguous role in relation to stability. While higher prices (or more generally, higher inflation) decrease output demand, the expectation of rising prices (or more generally, of rising inflation) increases demand.

To interpret (16), suppose the economy has for some time been in a steady state. Then some disturbance takes place. One might prefer to look at a contractionary disturbance creating excess capacity - feeling that this is the case for which the IS-LM model is most suitable. It is also the case with which Keynesian theory in general is primarily concerned. However, from the point of view of interpretation of the mathematical logic of (16), it is most straightforward to speak of the effects of <u>increases</u> in actual and expected inflation, that is to start with the case of an expansionary disturbance. When this case has been explained, it is easy to translate to the case of a contractionary disturbance by reversing all signs.

Now, suppose the steady state at time zero is disturbed by, say, an upward shift in the investment function. The immediate effect of this is an upward jumb in short run output. Refering to Figure 1, the initial situation is that x and m are at the historically given values (the old steady state), while the new steady state value of m is below the old (to compensate for the higher propensity to invest). Thus, immediately after the disturbance the economy is at some point, say A, above the new steady state point, cf. Figure 1. The Phillips curve now tells us that the rate of inflation immediately jumbs to a higher level.

i.e. higher than a, the rate of monetary expansion. Real cash balances begin to decrease while expected inflation begins to increase. The question then is whether the further interaction between these dynamic forces is deviation-amplifying or deviation-counteracting. It is convenient to distinguish between four cases according to the value of the parameter c.

<u>Case 1: c = 0</u>. This is the standard case of simple adaptive expectations. The stability condition (16) reduces to

$$bf_1^* < f_2^* m^* \qquad (16.i)$$

which is essentially the condition Tobin (1975) found in a related model. The term on the righthand side of (16.i) measures the <u>price level effect</u> on output which is stabilizing. The term on the lefthand side measures the <u>expectations effect</u> which is potentially destabilizing. Still, one gets more insight, if condition (16.i) is expressed in terms of the basic behavioural functions of the model. Inserting (11), with x = a, in (16.i) and rearranging gives

$$(|I_2^*| - C_1 m^*)\, b \quad < \quad - I_2^*(1-L_3^*)\,\frac{m^*}{L_2^*} + C_2^* m^*. \qquad (16.ii)$$

↑	↑	↑	↑
The potentially destabilizing expectations effect on investment	Inflation tax effect	Keynes effect	Pigou effect

When written in this way we see that the stability condition displays the famous "effects" from the old "Keynes versus the Classics" debate of the late 1930s and the 1940s. Furthermore, if $L_3^* = 0$,

$$b < -\frac{m^*}{L_2^*} \qquad (16.iii)$$

is sufficient, though not necessary for stability. (16.iii) may also be expressed as $b\alpha < 1$, where $\alpha \equiv -L_2^*/m^*$, the "sensitivity" of money demand with respect to the rate of interest. In this form we recognize the condition Cagan found to be necessary and sufficient for stability in his pioneering "monetary dynamics" model developed for the study of historical cases of rapid inflation (Cagan 1956).[8)]

Case 2: $0 < c < 1$. In this case expectations are formed as a convex combination of adaptive and "monetarist" expectations. The parameter c may be called the "degree of monetarist faith". Dividing through by F'^{*} in (16) we see that the stability condition becomes less stingent, the higher is c. Thus, stability is more likely, the more people tend to believe in stability. This is an example of a socalled bootstrap effect of expectations. On the other hand, as long as $c < 1$, i.e., in their expectations formation people put some weight to current inflation, instability cannot be excluded.

The explanation of the stabilizing role of monetarist faith is that increasing c dampens the self-stimulation of x by dampening $\delta\dot{x}/\delta x$. Indeed, the higher c is, the smaller is 1-c, i.e., the weight given to the positive feedback on $\dot{x}$, via actual inflation, $\dot{P}/P = F(f(x,m;a,g)) + x$, from a higher expected inflation. Furthermore, when $1-c < 1$, the negative feedback on $\dot{x}$, via the term $-x$ in (1), is no longer neutralized by the term $+x$ in (7), the net impact being $b[(1-c)x-x] = -bcx$. This explains why, when $c > 0$, the slope, F'^{*}, of the short run Phillips curve matters for stability; the likelyhood of stability being less, the higher is this slope. When $c = 0$, the size of F'^{*} does not matter for stability because in this case F'^{*} affects the two sides of (16) equally much, proportionately. When $c > 0$, this is no longer so. Indeed, increasing F'^{*} now affects the self-stimulation of x, $\delta\dot{x}/\delta x$, by more, proportionately, than it affects $\delta\dot{m}/\delta m$, due to the above-mentioned term $-bcx$, which is independent of F'^{*}.

Case 3: $c = 1$. In this case (used in Stein 1982) we have $\dot{x} = b(\dot{M}/M-x)$, that is pure "monetarist" expectations - corresponding to the monetarist belief in the convergence towards the steady state where the rate of inflation equals the rate of monetary expansion. (16) shows that when $c = 1$ stability is certain. Thus, monetarist expectations are, at least ultimately, self-validating.

Case 4: $c > 1$. This case is a variant of Frenkel's "two-part expectations hypothesis" (Frenkel 1975, Peel 1979). The presumption is that people strongly believe in a certain long-term rate of inflation - not only as an ultimate level at which inflation would tend to settle, but also as an average rate of inflation over shorter time intervals. The implied negative weight to current inflation - which we do not find plausible and which was rejected by the empirical test by Jacobs & Jones (1980) - was wanted by Frenkel in order to revise the Cagan model (Cagan 1956) to make it consistent with the empirical evidence. According to this evidence a shift to a higher monetary growth rate initially raises and eventually lowers real holdings of cash balances.[9)]

Now, whatever the plausibility of the case $c > 1$, if people have such expectations, stability is certain, cf. (16). On the other hand, all these results presuppose asumption A1. If the opposite of A1 holds, then a "reversal of stability" takes place. Stability becomes certain when $0 \leq c \leq 1$ and uncertain when $c>1$.

Finally, we mention a contrast with respect to stabilization policy between the limit case $c = 0$ (adaptive expectations) and the case $c > 0$. It can be shown that, e.g., a fiscal policy of the form $g = \varphi(y)$, $\varphi' < 0$, when $c = 0$, is neutral as to stability, while being favourable to stability, when $c > 0$.[10)]

IV. Rational expectations

For comparison a few comments on the case of rational expectations are added. In our deterministic framework rational expectations means perfect foresight, i.e.,

$$x = \frac{\dot{P}}{P} \qquad \text{for all } t, \tag{17}$$

which now replaces (1) in our model. Then, by (7) $y = y^*$ for all t. This is the wellknown fact that rational expectations in this type of models stabilize output. On the other hand, a complete solution to the model (including assumption A1) shows that rational expectations, by "speeding up" the adjustment of expected inflation to actual inflation, makes the inflationary process definitely instable, if either the initial price level or the initial rate of inflation is regarded as historically given.[11)] There is, however, also another approach to the question of stability under perfect foresight or rational expectations. This is to add to the model the postulate that under perfect foresight the economy must always be on a convergent path. This approach, suggested by Sargent & Wallace (1973), and often used in the "new classial macroeconomics" literature, corresponds, in our set up, to postulating $m = m^*$ immediately after any disturbance. This requires that the output price behaves as an asset price and instantaneously adjusts (by arbitrage) to keep real balances at m^*. In our opinion this begs the stability question.

V. Conclusion

The main part of this paper has dealt with the hypothesis that inflation expectations are formed according to a convex combination of an adaptive and a forwardlooking ("monetarist") formula. The role of this hypothesis as to the question of stability of production and inflation in a Keynesian IS-LM model has been investigated. The existence of bootstrap effects of expectations was observed. The more people believe in the "monetarist" formula, the more likely it is that this formula will be corroborated by the course of events. On the other hand, as long as people, in their expectation formation, put some weight to current inflation, instability of production and inflation cannot be excluded.

It was also shown that just as sluggishness in the adjustment of expectations (small b and small 1-c) promotes stability, sluggishness in price adjustment (small slope of the short run Phillips curve) is in the regular case ($0 < c < 1$), favourable to stability. In the pure adaptive case ($c=0$), however, the size of the slope of the short run Phillips curve is neutral as to stability.

We may add finally, that in practice an important "automatic stabilizer" is the tendency to downward rigidty of prices which seems to be a typical feature of the real world. Thus, if the restriction $\dot{p} \geqq 0$ is added to the model and the money growth rate is positive, then any depression will tend to be only temporary. The explanation is that the contractionary fall of expected inflation ceases sooner or later, while the expansionary increase in real balances continues. Once again we see that sluggishness promotes stability.

Notes:

1) I am grateful to Jorgen H. Gelting, Hans Jorgen Jacobsen, Soren Bo Nielsen, Christian Schultz, Peter Skott, Peter Birch Sorensen, and Troels O. Sorensen for discussions and comments on an earlier draft of this paper.

2) Nevertheless, <u>nominal</u> instability in no way disappears. Indeed, the whole price path is instable if the initial price is given, and is otherwise indeterminate (see section IV).

3) "Monetarist" - because the real rate of interest is assumed fixed.

4) Or "asymptotically rational expectations" (Stein 1982, p. 52).

5) Thus, respecting the government budget constraint implies that in a neighbourhood of the steady state (where $x \simeq a$), $f_2 > 0$ unambiguously, and therefore the problems with the "Pigou argument", stressed by Sargent (1979, pp. 64-65), disappear.

6) Sufficient conditions for the <u>existence</u> of a steady state and for the existence of economically meaningful short run equilibria in a neighbourhood of the steady state are given in Groth (1986). These conditions are not too demanding and need not occupy us here.

7) The "only if" part of this statement is only true, of cource, when we ignore the possibility of the singular case: trace = 0. This is so because a two-dimensional <u>non-linear</u> dynamic system with trace, evaluated at the steady state point, equal to zero <u>may</u> be stable. For simplicity, we abstract from this formal detail throughout the text.

8) Cagan (1956) tested the model on monthly data from seven European countries during hyperinflation using the fact that the adaptive expectations assumption implies that the expected rate of inflation is an exponentially-weighted average of all past actual rates of inflation. Cagan obtained a resonable fit for his model and found that the stability condition was met in five or six out of the seven examined cases of hyperinflation. His procedure and results are, however, rejected in Jacobs (1975) and B. Friedman (1978).

9) Interestingly, it can easily be shown that the Keynesian IS-LM model does not need c > 1 to be consistent with the empirical evidence (Groth 1986 p. 27).

10) This corresponds to the absence of I_1 in (16.ii) versus the presence of I_1, with a positive weight, on the left hand side of (16), when inserting (11) and multiplying through by A, cf. (12).

11) This is familiar from the money and growth literature (Tobin 1965, Sidrausky 1967) and other literature on multiple assets (Hahn 1966). If on the other hand neither the initial price or the initial rate of inflation is historically given, then an infinity of solutions to the model exists, an example of the general indeterminacy problem in self-fulfilling expectations models (see, e.g., Shiller 1978, Geanakoplos & Polemarchakis 1986).

REFERENCES

Barro, Robert J: Are government bonds net wealth? Journal of Political Economy 82, 1095-1117, 1974.

Blinder, Alan S., & Robert M. Solow: Does fiscal policy matter? Journal of Public Econimics 2, 319-337, 1973.

Buiter, Willem H., & Henri R. Lorie: Some unfamiliar properties of a familiar macroeconomic model. Economic Journal 87, 743-754, 1977.

Cagan, Phillip: The monetary dynamics of hyperinflation. In Milton Friedman (ed.): Studies in the quantity theory of money. Chicago, 1977.

Christ, Carl F.: A simple macroeconomic model with a government budget constraint. Journal of Political Economy 76, 53-67, 1968.

Christ, Carl F.: On fiscal and monetary policies and the government budget restraint. American Economic Review 69, 526-38, 1979.

Currie, David A.: Macroeconomic policy and government financing. In M.J. Artis & A.R. Nobay (eds.): Contemporary econmic analysis. London, 1978, 65-99.

Frenkel, Jacob A.: Inflation and the formation of expectations, Journal of Monetary Economics 1, 403-421, 1975.

Friedman, Benjamin M.: Stability and rationality in models of hyperinflation. International Economic Review 19, 45-64, 1978.

Friedman, Milton: The role of monetary policy. American Economic Review 58, 1.17, 1968.

Friedman, Milton: Comments on the critics. Journal of Political Economy 80, 906-950, 1972.

Geanakoplos, J.D., & H.M. Polemarchakis: Walrasian indeterminacy and Keynesian macroeconomics. R.E.S. 53, 1986.

Groth, C.: Centripetal and centrifugal forces in IS-LM dynamics. A reconsideration. Memo nr. 149, University of Copenhagen, Institute of Economics, 1986.

Hahn, Frank H,: Equilibrium dynamics with heterogenous capital goods. Quarterly Journal of Economics 80, 633-646, 1966.

Jacobs, Rodney L.: A difficulty with monetarist models of hyperinflation. Economic Inqüiry 13, 1975.

Jacobs, Rodney L., & Robert A. Jones: Price expectations in the United States: 1947-75. American Economic Review 70, 269-277, 1980.

Lucas, Robert E. Jr.: Expectations and the neutrality of money. Journal of Economic Theory 4, 1972.

McCallum, Bennet T.: On macroeconomic instability from a monetarist policy rule. Economic Letters 1, 121-124, 1978.

Muth, John F.: Rational expectations and the theory of price movements. Econometrica 29, 1961.

Peel, D.A.: On dynamic stability in monetary models which incorporate short- and long-run expectations of inflation in the demand for money function. Economics Letters 2, 131-136, 1979.

Peel, D.A. & J.S. Metcalfe: Divergent expectations and the dynamic stability of some simple macroeconomic models. Economic Journal 89, 789-798, 1979.

Rau, Nicholas: Simplifying the theory of the government budget restraint. Oxford Economic Papers 37, 210-229, 1985.

Sargent, Thomas J.: Macroeconomic theory. New York, 1979.

Sargent, Thomas J., & Neil Wallace: The stability of models of money and growth with perfect foresight. Econometrica 41, 1043-1048, 1973.

Sargent Thomas J., & Neil Wallace: "Rational" expectations, the optimal monetary instrument, and the optimal money supply rule. Journal of Political Economy 83, 1975.

Scarth, William M.: Expectations and the wage-price spiral in a simple monetary model. The Manchester School of Economic and Social Studies 45, 1977.

Shiller, Robert, J: Rational expectations and the dynamic structure of macroeconomic models. Journal of Monetary Economics 4, 1-44, 1978.

Sidrausky, Miguel: Inflation and economic growth. Journal of Political Economy 75, 796-810, 1967.

Taylor, Dean: A simple model of monetary dynamics. Journal of Money, Credit and Banking 9, 107-111, 1977.

Tobin, James: Money and economic growth. Econometrica 33, 671-684, 1965.

Tobin, James: Keynesian models of recession and depression. American Economic Review 65, Proceedings, 195-202, 1975.

Tobin, James: Asset accumulation and economic activity. Oxford 1980.

Tobin, James, & Willem H. Buiter: Long-run effects of fiscal and monetary policy on aggregate demand. In J.R. Stein (ed.): Monetarism: Studies in monetary economics, vol. 1. Amsterdam, 1976, 274336.

Yarrow, G.K.: The demand for money function and the stability of monetary equilibrium. Economic Journal 87, 114-123, 1977.

DEMAND-SIDE POLICIES AND SUPPLY-SIDE POLICIES IN A SMALL OPEN ECONOMY WITH NOMINAL OR REAL WAGE RIGIDITY*

Theo van de Klundert
Tilburg University, Postbox 90153,
5000 LE Tilburg, The Netherlands.

Frederick van der Ploeg
London School of Economics, Houghton Street,
London, WC2A 2AE.
Tilburg University, Postbox 90153,
5000 LE Tilburg, The Netherlands.

Contents

* We thank P. Peters for his helpful discussions, G. Staarink for performing the computations and J. Pynenburg for preparing the figures.

1. Introduction

European countries suffer from long-lasting unemployment problems. This raises questions like whether unemployment can be cured by demand-side policies and/or supply-side policies. The answer to such questions depends to a large extent on the way labour markets function in different countries.

In the present paper the issues raised will be studied by analysing a model for a small open economy under nominal wage rigidity, which is not too unreasonable for the US, and under real wage rigidity, which may be more relevant for the European economies (e.g., Branson and Rotemberg, 1983; van der Ploeg, 1987). Real wage rigidity can reverse the effectiveness of fiscal and monetary policy (cf., Casas, 1975; Argy and Salop, 1977; Sachs, 1980). In particular under real wage rigidity monetary policy has no real effects and a fiscal expansion leads to an appreciation of the real exchange rate which ultimately increases output and employment. Real wage rigidity also gives a more interesting role to supply-side policies.

The macroeconomic model will be based on sound micro foundations by introducing an infinitely-lived utility-maximising representative household, a profit-maximising firm, and a government. The asset menu consists of domestic currency, non-traded bonds and traded bonds. Foreign and domestic goods are imperfect substitutes. The model allows for current-account and wealth dynamics and imperfect capital mobility along the lines of Turnovsky (1985). In Section 2 the model is derived by showing the implications of the assumptions made with regard to the optimal behaviour of economic agents and market clearing. The steady-state properties of the model are analysed in Section 3. As a full analytical solution of the model is not possible, numerical examples have to be worked out to study impact effects of policy shocks and transient properties of the dynamic system. Numerical simulations for demand-side policies (fiscal expansion) and supply-side policies are therefore presented in Section 4. Section 5 concludes the paper.

2. A macroeconomic model with optimising agents

In this section we derive a macroeconomic model for an open economy with optimising households and firms. As most results are derived from numerical simulation, we adopt specific functional forms for the utility and production functions right away. A more general presentation of the model can be found in van de Klundert and van der Ploeg (1987).

Labour supply is assumed to be exogenous. The household's remaining decision problem is split up in two stages. In the first stage it decides on its aggregate consumption, its savings and its portfolio choice. In the second stage the household divides aggregate consumption into consumption of home goods and consumption of foreign goods.

Households maximise a discounted utility stream:

$$\underset{\substack{\{\tilde{c},\tilde{m}\}\\ a,b}}{\text{Max}}\ U = \int_{o}^{\infty} e^{-\delta t}\left[\gamma_c \log(\tilde{c}) + \gamma_m \frac{\tilde{m}^{1-\xi}}{1-\xi}\right]dt \qquad (2.1)$$

subject to the intertemporal budget constraint:

$$\dot{m} + \dot{a} + \dot{b} = (r-p)a + (\underline{r}^*-p+e)b + w\ell + z - c - \tau - pm - \tfrac{1}{2}\alpha b^2 \qquad (2.2)$$

and the initial conditions $m(0) = \frac{M_0}{P(0)}$, $a(0) = \frac{A_0}{P(0)}$ and $b(0) = \frac{E(0)B_0^*}{P(0)}$. Real values are expressed in domestic goods, unless otherwise stated. The symbols have the following meaning:

C = nominal consumption
c = real consumption $(\frac{C}{P})$
$\tilde{c}$ = consumption deflated by consumers' prices (C/P_c)
M = nominal money balances
m = real money balances $(\frac{M}{P})$
$\tilde{m}$ = money balances deflated by consumers' prices (M/P_c)
τ = real lump-sum taxes
A = nominal stock of non-traded bonds
a = real stock of non-traded bonds (A/P)
B^* = nominal stock of traded bonds denominated in foreign currency
b = real stock of traded bonds (EB^*/P)
ℓ = labour; $1-\ell$ = leisure
z = real profits
r = home nominal interest rate
$\underline{r}^*$ = foreign nominal interest rate
p = home inflation rate $(\dot{P}/P)$; P = home price level
$\underline{p}^*$ = foreign inflation rate $(\dot{P}^*/P^*)$; $\underline{P}^*$ = foreign price level
P_c = consumption price level in domestic country
e = anticipated (and actual) rate of exchange depreciation $(\dot{E}/E)$; E = nominal exchange rate
δ = home rate of time preference
w = real producers' wage (W/P); W = nominal wage rate.

Exogenous variables are underlined. Instantaneous utility is separable in consumption and real money balances, so that the real part of the system is separated from the nominal part. The household's budget constraint says that disposable income, that is profit and labour income plus interest income on traded and non-traded bonds minus lump-sum taxes and inflation taxes, minus consumption equals real savings. The latter are invested in cash, traded bonds and non-traded bonds. Following Turnovsky (1985), the certainty-equivalent version of the household budget constraint also includes a risk premium, $\frac{1}{2}\alpha b^2$, where α depends on the degree of risk aversion and the variance of exchange risk. It is assumed that the problem has an interior solution. The first-order conditions for a maximum of U are derived in the usual way. Denoting the costate variable by μ (marginal utility of wealth), we have:

$$\frac{\gamma_c}{\tilde{c}} = \mu \frac{P_c}{P} \Rightarrow c = \frac{\gamma_c}{\mu} \tag{2.3}$$

$$\delta\mu - \dot{\mu} = \gamma_m \tilde{m}^{-\xi} \frac{P}{P_c} - p\mu = (r-p)\mu = (r^*-p+e-\alpha b) \ . \tag{2.4}$$

Equations (2.4) yield:

$$\gamma_m m^{-\xi} \left(\frac{P}{P_c}\right)^{1-\xi} = r\mu \tag{2.5}$$

$$b = \frac{r^*+e-r}{\alpha} \tag{2.6}$$

$$\dot{\mu} = [\delta - (r-p)]\mu. \tag{2.7}$$

Equation (2.3) gives the aggregate consumption function. Demand for real cash balances follows from equation (2.5), so that the semi-elasticity of money demand with respect to the opportunity cost of holding money balances, r, equals $-1/\xi$. Equation (2.6) shows that holdings of traded bonds increase when the differential between the expected return on foreign assets and the return on home assets increases. Limiting cases are perfect capital mobility or zero risk (α=0) and zero capital mobility or infinite exchange risk ($\alpha \to \infty$). Finally, equation (2.7) gives the "tilt" of the consumption function.

The decision problem of the second stage relates to the choice between domestic goods (c_d) and foreign goods (c_m). It will be assumed that households maximise the instantaneous utility function:

$$\max_{\{c_d, c_m\}} u = \left(\frac{c_d}{\gamma}\right)^{\gamma} \left(\frac{c_m}{1-\gamma}\right)^{1-\gamma} , \quad 0 < \gamma < 1 \tag{2.8}$$

subject to the budget constraint:

$$Pc_d + P^*Ec_m = P_c\tilde{c} = Pc \quad \text{or}$$

$$c_d + \nu c_m = c, \tag{2.9}$$

where $\nu \equiv \frac{P^*E}{P}$ denotes the real exchange rate. The solution of this problem is easily obtained:

$$c_d = \gamma c \tag{2.10}$$

$$c_m = (1-\gamma)\frac{c}{\nu} . \tag{2.11}$$

Substitution of equations (2.10) and (2.11) and taking account of $u = \tilde{c}$ results in an expression for the ideal cost-of-living index:

$$\frac{P_c}{P} = \nu^{(1-\gamma)} . \tag{2.12}$$

Firms maximise short-run profits. The stock of capital is constant. Assuming a Cobb-Douglas production function the decision problem can be stated as:

$$\text{Max } z = \varepsilon\ell^{\beta} - \ell w, \quad \beta < 1 . \tag{2.13}$$

The first-order condition reads

$$\beta\varepsilon\ell^{\beta-1} = w,$$

which gives the demand for labour as

$$\ell = \left(\frac{\varepsilon\beta}{w}\right)^{\frac{1}{1-\beta}} . \tag{2.14}$$

The supply of outside money is exogenous; the rate of growth of the money stock is denoted by $\underline{\theta}$. Money market equilibrium therefore requires

$$\frac{\dot{m}}{m} = \underline{\theta} - p \tag{2.15}$$

Goods market equilibrium can be written as

$$\varepsilon\ell^{\beta} = c_d + \underline{g} + c_m^* = c + \underline{g} + c_m^* - \nu c_m \tag{2.16}$$

where $\underline{g}$ denotes the exogeneous level of government spending. Foreign demand for home goods (exports) is expressed in an analogous way to home demand for foreign goods:

$$c_m^* = (1-\gamma^*)\nu\underline{c}^* , \quad 0 < \gamma^* < 1, \tag{2.17}$$

where $\underline{c}^*$ is exogenously given foreign demand.

Accumulation of traded bonds follows from the balance on the current account:

$$\dot{b} = (\underline{r}^*-p+e)b + \varepsilon\ell^{\beta} - c - \underline{g} \qquad (2.18)$$

It is assumed that the economy cannot play Ponzi games with the rest of the world, which requires

$$\lim_{s\to\infty} b(s) \exp\left(-\int_t^s (\underline{r}^*-p+e)ds'\right) = 0.$$

The stock of non-traded bonds (a) is assumed constant. The government budget constraint is then taken account of by varying the level of lump-sum taxes according to

$$\tau = \underline{g} - (r-p)\underline{a} - \underline{\theta}m \qquad (2.19)$$

Two alternatives are considered with regard to the labour market: nominal wage rigidity (NWR) and real wage rigidity (RWR). The NAIRU is indicated by $\underline{u}_n$, which is assumed exogenous. NWR corresponds to the inflation-augmented Phillips curve:

$$\frac{\dot{W}}{W} + p = \psi \log\left(\frac{\underline{u}_n}{1-\ell}\right) + \underline{\theta}, \qquad (2.20a)$$

where the core or trend inflation is set equal to the growth rate of money ($\underline{\theta}$). Alternatively, one could assume that core inflation is a weighted average of past rates of inflation, but this would only lead to minor differences with respect to dynamic adjustment. Equation (2.20a) implies that households are always rationed when they sell labour and firms are never rationed. In the case of NWR we have $W(0) = W_0$. RWR corresponds to

$$\frac{\dot{w}_c}{w_c} = \psi \log\left(\frac{\underline{u}_n}{1-\ell}\right) - \eta \log\left(\frac{w_c}{\underline{w}_d}\right) \qquad (2.20b)$$

where $w_c \equiv (W/P_c)$ denotes the real consumers' wage and $\underline{w}_d$ denotes the desired long-run target for real wages. Equation (2.20b) corresponds to an error-correction mechanism of the type estimated by Sargan (1964) and Layard and Nickell (1985). Such an equation is termed by Layard and Nickell a "level" wage equation because the real wage is determined in the long-run. This contrasts with equation (2.20a) which is a pure "change" equation, in the sense that wages disappear in the long run, and the equation simply sets the natural rate of unemployment equal to the actual rate of unemployment. In a more elaborate model of real wage determination the long-run target w_d can be related to union preferences, whereas u_n captures push factors like search unemployment, the effects of different taxes, union power, benefits and mis-match. Integration of (2.20b) yields

$$\log\left(\frac{w_c(t)}{\underline{w}_d}\right) = \log\left(\frac{w_c(0)}{\underline{w}_d}\right) \exp(-\eta t) + \int_0^t \psi \log\left(\frac{\underline{u}_n}{1-\ell(s)}\right)$$

$$\exp[\eta(s-t)]ds.$$

This shows that the real wage recovers from past disequilibria in the labour market and gradually adjusts towards its long-run target value. The extreme case of a completely rigid wage, $w_c = \underline{w}_d$, corresponds to $\eta \to \infty$. In this case RWR implies complete indexation to the cost of living. The case $\eta = 0$ is identical to NWR with "rational expectations" or immediate adjustment of price expectations to the cost of living. ($\dot{P}_c/P_c$ is substituted for θ in equation (2.20 a)).

The complete model can now be summarised by:

$$c = \frac{\gamma_c}{\mu} \quad (2.3)$$

$$c_d = \gamma c \quad (2.10)$$

$$c_m^* = (1-\gamma^*)\nu\underline{c}^* \quad (2.17)$$

$$\gamma_m m^{-\xi}\nu^{(\xi-1)(1-\gamma)} = r\mu \quad (2.5)$$

$$b = \frac{1}{\alpha}(\underline{r}^*+e-r) \quad (2.6)$$

$$\ell = \left(\frac{\varepsilon\beta}{w}\right)^{\frac{1}{1-\beta}} \quad (2.14)$$

$$\varepsilon\ell^{\beta} = c_d + \underline{g} + c_m^* \quad (2.16)$$

$$\dot{\mu} = [\delta - (r-p)]\mu; \quad \mu(0) \text{ free} \quad (2.7)$$

$$\dot{m} = m(\theta-p); \quad m(0) \text{ free} \quad (2.15)$$

$$\dot{b} = (\underline{r}^* - p + e)b + \varepsilon\ell^{\beta} - c - \underline{g}; \quad b(0) = B_0^*\nu(0)/P_0^* \quad (2.18)$$

$$\frac{\dot{w}}{w} + p = \psi \log\left(\frac{\underline{u}_n}{1-\ell}\right) + \underline{\theta}; \quad W(0) = W_0 \quad (2.20a) \text{ or}$$

$$\frac{\dot{w}}{w} = (1-\gamma)\,\frac{\dot{\nu}}{\nu} + \psi \log\left(\frac{\underline{u}_n}{1-\ell}\right) - \eta \log\left(\frac{w_c}{\underline{w}_d}\right), \qquad (2.20b)$$
$$w_c(0) = w_{c_0}$$

$$\frac{\dot{\nu}}{\nu} = p^* + e - p. \qquad (2.21)$$

These 12 equations can be solved for the 12 endogenous variables, viz., c, c_d, c^*_m, m, r, b, p, e, ν, ℓ, w and μ. In addition, the level of lump-sum taxes (τ) can be found from equation (2.19).

3. Steady-state properties of the model

The steady-state properties of the monetary side of the model are $p = p_c = \underline{\theta}$, $e = \underline{\theta} - \underline{p}^*$ and $r - p = \delta$. Long-run inflation is entirely a monetary event. The steady-state rate of exchange depreciation equals the difference between home monetary growth and foreign inflation. Finally, the steady-state real interest rate equals the pure rate of time preference. It is assumed, following Turnovsky (1985), that

$$\delta = (\underline{r}^* + \underline{p}^* + \alpha\bar{\delta})/(1+\alpha)$$

where $\bar{\delta}$ is a constant, so that $\underline{\delta} = \underline{r}^* - \underline{p}^*$ under perfect capital mobility (α=0) and $\delta = \bar{\delta}$ under zero capital mobility ($\alpha\rightarrow\infty$). This device ensures that consumption in the steady state is well-defined under perfect capital mobility ($\frac{\dot{\mu}}{\mu} = \delta - (r-p) \rightarrow 0$ as $t \rightarrow \infty$) and that the real interest rate is independent of the rate abroad under zero capital mobility. If $\underline{r}^* - \underline{p}^* = \delta^*$, the economy has a long-run debt (credit) vis-à-vis the rest of the world when its pure rate of time preference is greater (less) than the foreign discount rate: $b = (\delta^*-\delta)/\alpha = (\delta^*-\bar{\delta})/(1+\alpha)$.

Under NWR in the long run demand-side policies have no real effects, because the labour market will be in equilibrium at the natural rate of unemployment, hence $\ell = 1 - \underline{u}_n$ and $w = \beta\varepsilon(1-\underline{u}_n)^{\beta-1}$. However, for the case of RWR, fiscal policy is non-neutral in the long run. This can be seen from the long-run aggregate supply (AS) schedule, which can be derived from equation (2.20b) and the marginal productivity condition for labour:

$$(1-\gamma)\log\nu = -(1-\beta)\log\ell + \frac{\psi}{\eta}\log(1-\ell) + A, \qquad (3.1)$$

where $A \equiv \log(\beta\varepsilon) - \frac{\psi}{\eta}[\log\underline{u}_n + \log\underline{w}_d]$.

The aggregate supply curve (3.1) is upward-sloping in $\frac{P}{P^{*}E} - \ell$ space ($\frac{\partial \nu}{\partial \ell} < 0$) and is drawn in Figure 1b. This can be explained as follows. A real exchange appreciation reduces the relative price of foreign goods, diminishes the wedge between the consumers' and producers' real wage and therefore increases employment and output ($\nu\downarrow$, $\frac{P_c}{P}\downarrow$, $w^c\uparrow$, $w\downarrow$, $\ell\uparrow$). It is easily seen that an increase in the natural rate of unemployment ($\underline{u}_n\uparrow$) or an increase in the long-run target wage ($\underline{w}_d\uparrow$) shifts the AS-curve to the left. An improvement in the flexibility of the labour market ($\psi/\eta\uparrow$) reinforces these effects. A positive productivity shock ($\varepsilon\uparrow$) has an opposite effect and shifts the AS-schedule to the right. Obviously, for NWR the long-run supply schedule,

$$\ell = (1-\underline{u}_n), \tag{3.2}$$

is vertical (see Figure 1a). An increase in $\underline{u}_n$ shifts the AS-schedule to the left.

The long-run aggregate demand schedule can be found by observing that the long-run solution for consumption is $c = \frac{\delta^{*}(\delta^{*}-\bar{\delta})}{1+\alpha} + \varepsilon\ell^{\beta} - \underline{g}$. Substitution of this result along with the equations (2.10) and (2.17) into the equilibrium relation (2.16) results in

$$(1-\gamma^{*})\underline{c}^{*}\nu = (1-\gamma)\varepsilon\ell^{\beta} - B, \tag{3.3}$$

where $B \equiv \frac{\gamma\delta^{*}(\delta^{*}-\bar{\delta})}{1-\alpha} + (1-\gamma)\underline{g}$.

The AD-schedule (3.3) is downward-sloping in $\frac{P}{P^{*}E} - \ell$ space ($\frac{\partial \nu}{\partial \ell} > 0$), given that the (marginal) propensity to consume domestic goods in the home country and abroad are less than unity (γ, $\gamma^{*} < 1$). Under these circumstances a reduction in the relative price of home goods increases exports and aggregate demand. As can be checked easily, an increase in public spending ($g\uparrow$) or foreign consumption ($\underline{c}^{*}\uparrow$) increases aggregate demand and shifts out the AD-schedule. A rise in the foreign discount rate or real interest rate ($\delta^{*}=\underline{r}^{*}-\underline{p}^{*}\uparrow$) means that the economy can sustain a higher level of aggregate consumption without suffering a loss in foreign reserves, hence the AD-schedule shifts out, as is obvious also from equation (3.3).

Intersection of the long-run AD- and AS-schedules gives for NWR the result mentioned above that unemployment is at its natural level. For RWR we get the solution:

$$\ell = f(\overset{+}{\underline{g}}, \overset{+}{\underline{c}}^*, \overset{+}{\delta^*}, \overset{-}{\underline{u}_n}, \overset{-}{\underline{w}_d}, \overset{-}{\psi/\eta}, \overset{+}{\varepsilon}) \tag{3.4}$$

It follows that with RWR there is no unique solution for the equilibrium rate of unemployment, since fiscal policy, foreign events and supply-side factors affect the long-run levels of employment and output. Fiscal expansion is effective in the long-run under RWR, because it requires an appreciation of the real exchange rate which in turn reduces the wedge between the consumers' and producers' real wage $(\frac{w_c}{w} = \nu^{-(1-\gamma)})$ and therefore increases employment (see Figure 1b). Supply-side policies which reduce the natural rate of unemployment $\underline{u}_n$ induce a higher level of employment under NWR and RWR. In the latter case the effect will be smaller the lower the flexibility in the labour market (the lower ψ and the higher η). An increase in productivity $(\varepsilon\uparrow)$ means that firms are more able to pay a higher wage, so that there is less conflict and the equilibrium unemployment rate falls $(\ell\uparrow)$.

Figure 1: Effects of an increase in government spending

(a) NWR (b) RWR

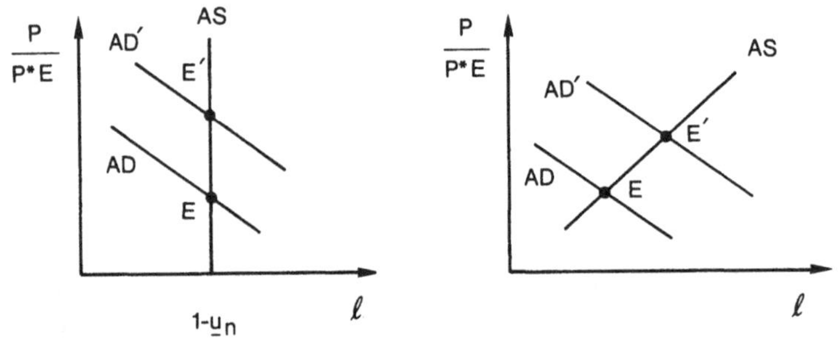

4. Demand-side and supply-side policies: numerical simulations

To study the impact effects and the adjustment over time of policy measures, numerical examples must be worked out as a complete analytical solution of the model is infeasible. The model has two forward-looking state variables, m and $_m u$, and two backward-looking state variables, b and $z \equiv \frac{m}{w}$ (for NWR) or w_c (for RWR). The saddlepoint property therefore requires two unstable and two stable eigenvalues.

The numerical values of the parameters, exogenous variables and initial conditions presented in Table 1 generate the required characteristics for saddlepoint stability. A multiple shooting algorithm (Lipton, et.al., 1982) based on routine DD03 of the Culham Laboratory is used to solve the model (England, 1976).

Table 1: Values of parameters, exogenous variables and initial conditions

Parameters:

$\beta = 0.75$, $\varepsilon = 1$, $\psi = 0.005$, $\eta = 0$, 0.5 or 2, supply-side parameters, $\delta = 0.05$, $\gamma_c = 0.5$, $\gamma = 0.75$, $\gamma^* = 0.75$, $\gamma_m = 0.05$, $\xi = 2$, demand-side parameters, $\alpha = 0.1$, degree of capital mobility

Exogenous variables

$\underline{u}_n = 0.05$, $\underline{w}_d = 0.76$, supply-side policies
$\underline{g} = 0.2$, $\underline{\theta} = 0.1$, demand-side policies
$\underline{r}^* = 0.15$, $\underline{p}^* = 0.1$, $\underline{c}^* = 10$, foreign variables

Initial conditions

$B_0^* = b_0 = 0$, $z_0 = 1.066$ (for NWR), $w_{c_0} = w_d$ (for RWR)

4.1 Fiscal expansion

The effects of an increase in government expenditure by 5% are shown in Figure 2. The figures for consumption (c), the real exchange rate (ν) and the nominal price level (P) are expressed as percentage deviations from their initial steady-state values. The variable $\tilde{u}$ measures changes in unemployment as a percentage of the labour force employed in the initial steady state. Solid lines give results for RWR under different assumptions for η, dotted lines relate to NWR.

A rise in government expenditure leads on impact to a fall in private consumption and a real exchange rate appreciation, which is necessary to balance foreign trade. The price level jumps upwards as real cash balances decline along with private consumption. Unemployment decreases under RWR, because the appreciation of the real exchange rate increases the wedge between the consumers' and the producers' real wage. The consumers' real wage rate is predetermined. Therefore, the producers' real wage falls and employment increases. Under NWR the producers' real wage declines even further as the price level rises and nominal wages are predetermined. Consequently, the impact effect on unemployment is greater under NWR than with RWR.

Figure 2: Fiscal policy; $\Delta \underline{g}/\underline{g}$ = 0.05

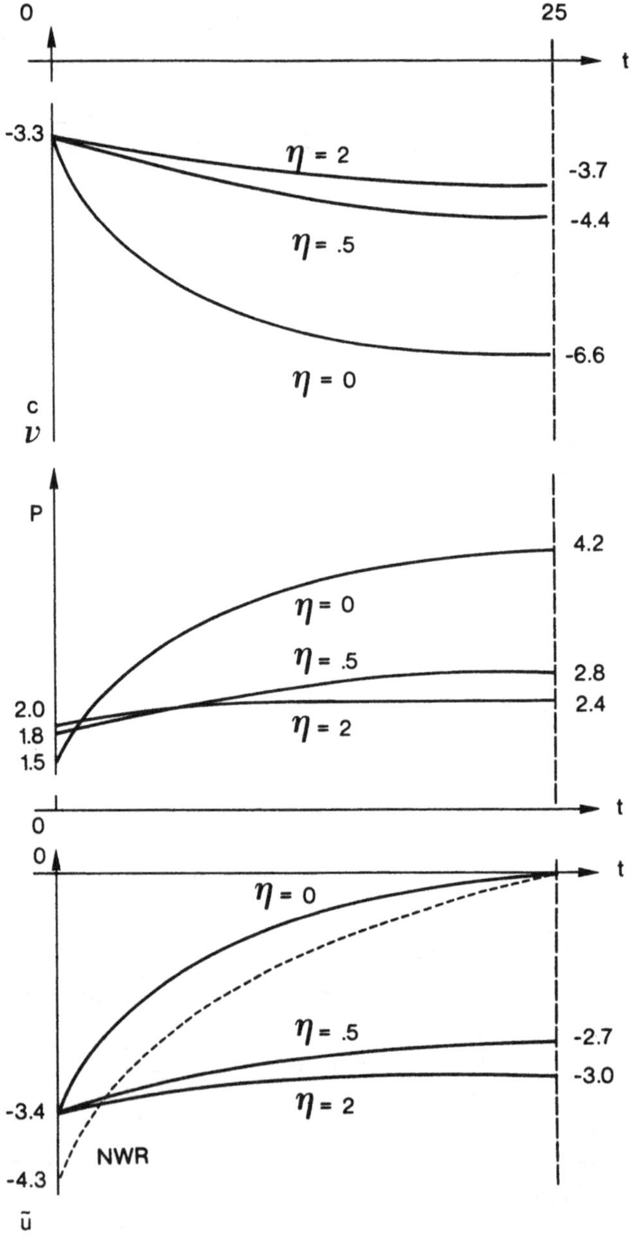

Figure 2 shows that the adjustment path is rather sensitive to different values of η. The real consumers' wage rises under the influence of the excess demand for labour. The effect is greater as the labour market is more flexible (η is lower). An increase in real wages reduces employment and output, which in turn reduces private consumption. The crowding-out phenomena is most pronounced for $\eta = 0$. In this case unemployment returns to the natural rate. For higher values of η the adjustment path is more smooth, implying less inflation and less crowding-out of private spending. As shown in Section 3, fiscal policy has a long-run effect on unemployment for $\eta > 0$. The long-run results which are attained approximately at $t = 25$ can be read off Figure 2. The transition path under NWR resembles the results for RWR with $\eta = 0$, which will not come as a surprise. As argued in Section 2 this variant can be considered as a special case of NWR, that is nominal wage rigidity with "rational expectations" with regard to core inflation.

The results of this section are consistent with Casas (1975), Argy and Salop (1977) and Sachs (1980), who find that fiscal policy under perfect capital mobility and flexible exchange rates is effective (ineffective) in the long-run when there is real (nominal) wage rigidity.

Finally, note that even though a fiscal expansion reduces unemployment, it does not necessarily increase economic welfare as both consumption and holdings of real money balances fall. Of course, if government spending yields direct utility, then it may be possible that welfare increases despite the crowding out of private consumption and money holdings.

4.2 Supply-side improvements

In this section the results of a reduction in the natural rate of unemployment ($\underline{u}_n$) by 1 percentage point are discussed. Such a reduction in the natural rate may be brought about by measures aimed at increasing labour mobility, at reducing mis-match between vacancies and unemployed, at reducing unemployment benefits, or at decreasing the wedge caused by tax tariffs. The effects of such a change in $\underline{u}_n$ are shown in Figure 3.

The nominal exchange rate depreciates on impact, which is reflected in an upward change of the price level for a given real exchange rate. Under RWR the real exchange rate and the other real variables do not change on impact of the supply shock. The nominal exchange rate depreciation illustrates the well-known overshooting phenomenon. In the long-run the real exchange rate depreciates, but the nominal exchange rate appreciates. Such an appreciation is necessary to slow down the fall in prices, which would otherwise decline too much.

Figure 3: Supply-side policy; $\Delta \underline{u}_n = -0.01$

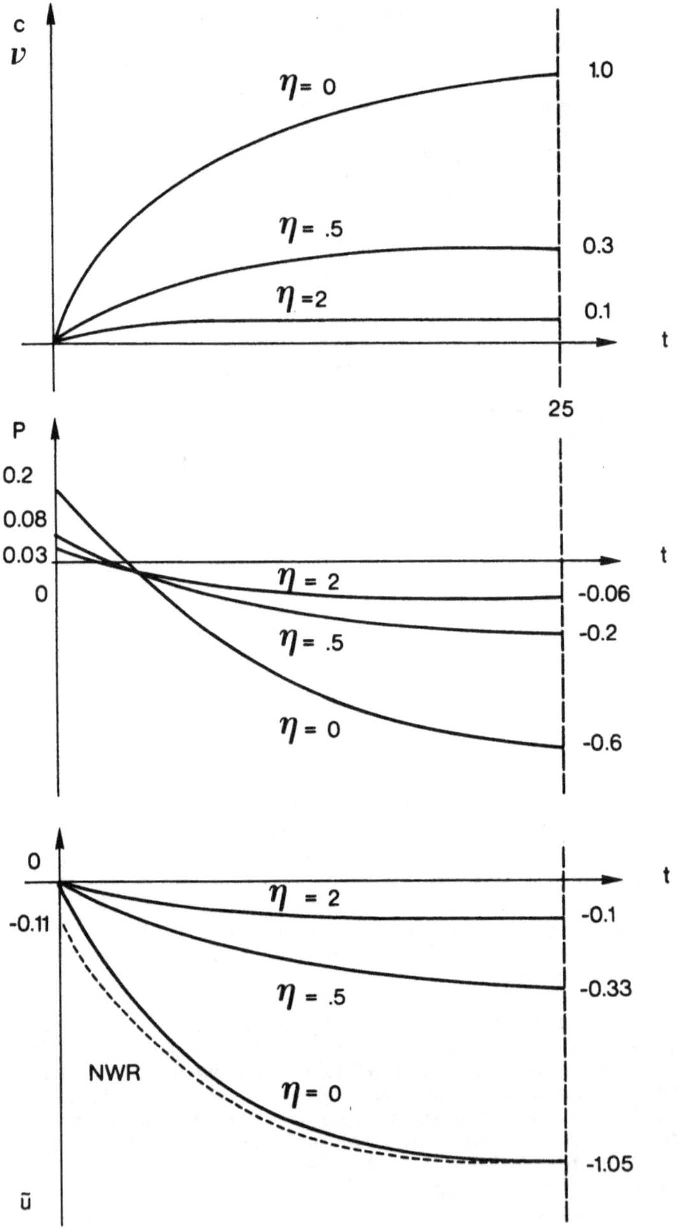

To absorb the additional labour real wages must decrease. A fall in nominal prices works in the wrong direction and therefore prices misadjust in the short run. Employment, output and consumption rise gradually as the real wage rate adjusts in the situation of an excess supply of labour. For $\eta = 0$ the full effect is realised, ultimately, as unemployment decreases by 1 per centage point ($\tilde{u}$ falls by 1.05%). When the labour market is less flexible ($\eta > 0$) the favourable outcomes of a positive supply shock are mitigated as appears from Figure 3. For $\eta = 2$ the results are even negligible. Supply-side policies aimed at a reduction of $\underline{u}_n$ do not work in an economy with a rather inflexible labour market. Of course, one should note that measures which contribute to a greater flexibility have a favourable effect on employment in the medium run and the long run. There may be policy measures which have a combined effect. In any case, supply-side policies require a careful analysis at the microeconomic level to begin with.

Returning to Figure 3 it can be concluded that under NWR there is a positive impact effect on employment, which can be attributed to the decline in the price level. In the medium run unemployment adjusts towards its new long-run equilibrium value ($u = \underline{u}_n = 0.04$).

5. Conclusions

As noted in the Introduction nominal wage rigidity may not be unreasonable for the US. A reduction in the long-run rate of unemployment of the US then requires supply-side measures affecting the natural rate. Fiscal expansion will have favourable effects in the short run at the expense of an increase in inflation. Assuming real wage rigidity the situation is different for Europe. Under real wage rigidity fiscal expansion has a lasting effect on unemployment, whereas the outcome of supply-side policies is uncertain. If the labour market is inflexible, measures aimed at a reduction in the natural rate of unemployment have little effect. Efforts should then be directed at improving labour market flexibility. In that sense there remains scope for a two-handed approach where the demand side and the supply side both receive due attention (e.g., Blanchard, et.al., 1985).

Extensions of the model are possible in different directions. Along more conventional macroeconomic lines account could be taken of capital accumulation, of non-tradeables and of more than one country. However, at least of equal importance may be extensions of the microeconomic foundations, especially with regard to labour demand, labour supply and wage setting (cf., Layard and Nickell, 1985).

References

[1] Argy, V. and J. Salop: "Price and output effects of monetary and fiscal policy under flexible exchange rates," IMF Staff Papers, 24(1977), 224-256.

[2] Blanchard, O., R. Dornbusch, J. Drèze, H. Giersch, R. Layard and M. Monti: "Employment and growth in Europe: A two-handed approach," Centre for European Policy Studies, 21(May 1985).

[3] Branson, W.H. and J.J. Rotemberg: "International adjustment with wage rigidity," European Economic Review, 13, 3(1980), 309-332.

[4] Casas, F.R.: "Efficient macroeconomic stabilization policies under floating exchange rates," International Economic Review, 16(1975), 682-698.

[5] England, R.: "A program for the solution of boundary value problems for systems of ordinary differential equations," Report PPN3/73, Culham Laboratory, 1976.

[6] Klundert, Th. van de and F. van der Ploeg: "Wage rigidity and capital mobility in an optimizing model of a small open economy," Discussion Paper No. 168, Centre for Economic Policy Research, London, 1987.

[7] Layard, R. and S. Nickell: "The causes of British unemployment," National Institute Review, (1985), 62-85.

[8] Lipton, D., J.M. Poterba, J. Sachs and L.H. Summers: "Multiple shooting in rational expectations models," Econometrica, 50(1982), 1329-1333.

[9] Ploeg, F. van der: "International interdependence and policy coordination in economies with real and nominal wage rigidity," Working Paper No. 986, Centre for Labour Economics, London School of Economics, 1987.

[10] Sachs, J.: "Wages, flexible exchange rates, and macroeconomic policy," Quarterly Journal of Economics, 95 (1980), 731-747.

[11] Sargan, J.D.: "Wages and prices in the U.K.," in Econometric Analysis for National Economic Planning ed. by P.E. Hart, G. Mills and J.K. Whittaker. New York: McMillan, 1964.

[12] Turnovsky, S.J.: "Domestic and foreign disturbances in an optimizing model of exchange rate determination," Journal of International Money and Finance, 1(1985), 157-171.

A CRITICAL VIEW ON THE ASSET MARKET APPROACH

BY THUSNELDA TIVIG

Universität Konstanz

1. Introduction

The more recent literature on the determination of flexible exchange rates is mainly of the monetary type, of which the monetary approach in a narrow sense, the stock approach, the portfolio balance approach and the asset market approach (AMA) are variants. The common base of these approaches can be summarized in four features:

(i) the interpretation of the exchange rate as a relative price of two currencies which, together perhaps with other prices, provides equilibrium in the international financial markets,

(ii) the consideration of more than one asset and of substitution possibilities between assets denominated in different currencies,

(iii) the formulation of asset demands as stock demands, and, at least tentatively, their microeconomic foundation by portfolio theory,

(iv) the emphasis on the role of expectations in the portfolio balance process and the determination of exchange (and interest-) rates.

With the definition of the exchange rate as the relative price of two currencies, the emphasis is shifted to the monetary sector, that is, on the supply of and demand for financial assets. As the latter are stocks, this view is called "stock approach" as opposed to the traditional analysis by means of equilibrium conditions referring to flow variables. The label "asset market approach" stresses the role of financial markets too, as compared to the customary emphasis on the real side of the economy. But whereas the monetary approach in a narrow sense concentrates on money markets only, the asset market approach explicitly takes into account asset markets, in which interest bearing, distinct assets are traded internationally. The notion "portfolio balance approach" finally underlines the fact that when several assets are available, wealth holders have to make portfolio decisions. Despite these distinctions, the literature largely uses the four notions synonymously.

2. A Critical Presentation of the Asset Market Approach

The AMA was initially designed for the analysis of capital movements and interest rate changes in a system of fixed exchange rates. It is based on the idea that private wealth holders allocate their wealth between the available domestic and foreign assets, the desired wealth structure depending generally on the expected returns of the assets, the transaction volume in the good markets and the level of wealth. Changes of these factors induce revisions of the desired portfolio structure and lead to price-and/or quantity movements. In a fixed exchange rates system there may be interest movements, in a flexible exchange rates system interest- and exchange rate movements. By quantity movements there are meant stock shifts which lead alone or in interaction with the price movements to restoration of portfolio equilibrium. These stock shifts may take either the form of a direct exchange of tradable assets between private domestic and foreign wealth holders, or they can come about because of the readiness of the domestic central bank to exchange foreign assets for domestic assets at a fixed exchange rate. The latter is valid for a system of fixed or managed exchange rates only, the former can occur in either exchange rate system provided there are at least two international tradable assets. The stock shifts imply gross capital movements which have solely a temporary stock adjustment character. They cease when the desired portfolio structure is reached. Thus short-run equilibrium requires portfolio balance and financial market equilibrium.

The private wealth holders decide not only on their portfolio structure, but also on their consumption level. If one takes into consideration that price movements imply, through valuation effects, changes in the level of wealth, the consumption and hence the savings decision is revised whenever the latter changes, e.g. with every exchange rate change. If, to the contrary, one assumes a constant level of wealth in the short run, the saving process occurs only in time, by accumulation of wealth through new savings. Long-run equilibrium is attained when all goods and asset markets clear, portfolio balance prevails and the current account is balanced.

The starting point of the AMA was the recognition that there had been periods, after switching to a flexible exchange rate system, with large international movements of financial capital and heavy exchange rate changes within relatively short spans of time. These facts raised interest in studying portfolio balance processes, which are characterized by simultaneous price- and quantity changes. Put differently, the AMA aims at studying exchange (and interest-) rate movements in the context of international financial capital movements, where the latter result either from reshuffling of portfolios or from current account imbalances. In the following it will be argued that the AMA provides an adequate framework for such an analysis, but that it has not been sufficiently exploited yet. The critics refer to two points: 1. the treatment of the simultaneity of price- and quantity movements in the short run, and 2. the interaction of international temporary stock shifts with sustainable flows of new savings along the adjustment path to a new long-run equilibrium. In his seminal paper, Kouri [11] describes the temporary portfolio balance process first as follows:

> "In the stock approach it is assumed that financial markets equilibrate very fast. Transaction costs are so small that it may be assumed that financial assets are always held in desired proportions. For example, if a market participant has too many sterling assets in his portfolio, he can exchange these assets for dollar assets in a very short time. For a large segment of the international short-term money market, this is probably a reasonable simplifying assumption. Revisions of existing portfolios are much larger in size than marginal additions to asset demands from new savings." (p. 281)

It is in no way clear from the text above that the author refers to a certain exchange rate system. But finally Kouri concludes:

> "Under fixed exchange rates portfolio equilibrium between domestic money and foreign assets at a given level of wealth is obtained through instantaneous capital inflows and outflows because the Central Bank supplies foreign assets at a fixed price. Under flexible exchange rates instantaneous portfolio equilibrium is obtained through changes in the valuation of assets - that is, through changes in the exchange rate." (p.300)[1]

Kouri's formal analysis corresponds to the distinction he made in the final part of his paper. His model is built up essentially of the equilibrium condition for the domestic money market

$$M = M^d(\pi Y, x, V); \qquad M^d_{\pi Y} > 0, \quad M^d_x < 0, \quad M_V > 0, \qquad (1)$$

and the differential equation for the wealth accumulation by new savings

$$\dot{F} = Y - C(Y, V/\pi); \qquad 0 < C_Y < 1, \quad C_{V/\pi} > 0, \qquad (2)$$

where M stands for the nominal stock of the domestic money, which is not tradable internationally, π is the exchange rate, which by the law of one price and setting the constant foreign goods price equal to one is equal to the domestic price P of the single international tradable good, M^d is the domestic nominal demand for domestic money, x is the expected rate of change of the exchange rate, Y is the real constant production income of domestics, V is the nominal wealth of private domestics, defined as $V = M + \pi F$, F is the real stock of foreign money owned by private domestics, and C is the real consumption demand of the private domestics.

For ease of exposition we abstract, unlike Kouri, from government activities. The model is closed by specifying an expectation process for the determination of x. For the purpose of this paper it is irrelevant what expectational form one chooses, as long as it does not pose stability problems; x could equally well be exogeneous. Note that because of the identity $P \equiv \pi$, real variables are expressed in terms of foreign currency or good units.

The system ((1),(2)) can be seen as an approximation by means of the moving equilibrium method of the system of differential equations

$$\dot{\pi} = \tau(M - M^d(\pi Y, x, V)); \quad \tau > 0, \quad \tau \to \infty \tag{1'}$$

$$\dot{F} = Y - C(Y, V/\pi). \tag{2}$$

Equation (1') describes the evolution of the exchange rate, which by assumption reacts instantaneously providing permanent equilibrium in the money market. In the above system, π is a fast and F a slow variable. The moving equilibrium method works as follows: system ((1'),(2)) is approximated - usually tacitly - by the system ((1),(2)), according to the postulated differences in the reaction speed of the endogeneous variables π and F. The equilibrium condition (1) serves to solve explicitly for the variable π as a function of F and the exogeneous variables x, Y and M

$$\pi = \pi(F; x, Y, M), \tag{3}$$

which is substituted into (2) yielding

$$\dot{F} = Y - C(Y, M/\pi(F; x, Y, M) + F). \tag{4}$$

The movement of the slow variable F is then studied under the assumption that the adjustment of the fast variable π to its respective short-run (temporary) equilibrium value has already occurred.

The above simplified exposition of the Kouri model reveals three points:

a. the verbally described exchange process of financial assets is not embedded in the formal structure of the model, i.e. there are no temporary stock shifts modelled as part of the portfolio balance process;

b. as no such stock shifts appear, one cannot identify a stock-flow interaction in the international movement of financial capital;[2]

c. short-run wealth changes by valuation effects are not captured, so the asset demands and the consumption demand do not change simultaneously.

In all asset market models there are only price- and no quantity adjustments modelled for the short run. Yet it would be wrong to conclude that this is an intrinsic feature of the AMA. The fact is that a precondition for short-run stock shifts is the existence of at least two international tradable assets. The point has been observed by several authors. So for example by Calvo and Rodriguez [3], who write:

> "The total stock of foreign currency is also given at any moment and not suspect to instantaneous jumps; the only way for this economy to alter the level of its foreign currency holdings is through an excess supply or demand of traded goods..."

and add in a footnote:

> "This follows from the assumption that foreign currency is the only internationally traded asset; if securities were included and allowed to be internationally traded then foreign exchange could be instantaneously acquired through sales of securities abroad." (p. 619)

Allen and Kenen [1] argue in a similar way:

> "If there is international trade in more than one asset...the stock version produces an instantaneous rearrangement of portfolios in response to a disturbance, with a gradual accumulation in holdings of all assets when the disturbance induces savings... When only one asset is traded internationally, by contrast, that asset can be exchanged only for a flow of goods; thus, even in the stock version of the model, any actual adjustment of portfolios is accomplished gradually as a result of saving."(p.70)

When reading the above passages one would expect to encounter stock shifts as part of the short-run portfolio balancing process in models with two international tradable assets. Most surprisingly, this is not the case.[3] The question is either simply not addressed, or, in a few papers, considered to be a methodological problem which cannot be solved in continuous time.[4] The models are usually specified in a way which does not make the neglect of stock shifts obvious. One variant is to specify the wealth evaluated at a long-run equilibrium exchange rate as the state variable. The dynamics of the system is then described in terms of this wealth variable, which does not "jump", and not in terms of the individual wealth components.[5] With this procedure, one "solves" two problems. First, the short-run exchange of assets between domestics and foreigners is not made explicit, as this exchange is neutral with respect to the level of wealth; it changes only the composition of wealth. Second, the simultaneity of wealth changes through valuation effects and through new savings is ignored. The latter problem would have to be considered in all models irrespective of the number of tradable assets, as long as they are denominated in different currencies. In Kouri's [11] model for example, the change in nominal wealth is not $dV/dt = \pi dF/dt$ but

$$\dot{V} = \dot{\pi} F + \pi \dot{F} = \dot{\pi} F + \pi B, \tag{5}$$

where B is the real current account balance.

The implication of neglecting $\dot{\pi} F$ in (5) is that the temporary portfolio balance process - which in the Kouri model is ensured by price adjustments only - is cut off from the saving process.

The methodological problem of modelling the stock-flow interaction adequately has been recognized by some authors in the 1975/76 literature, but has not been addressed afterwards up to the paper by Tobin and De Macedo [14]. So Frenkel and Rodriguez [7] write in their excellent paper on the AMA for fixed exchange rates:

> "...our analysis is complicated due to the fact that there are both adjustments that occur through time and adjustments that occur instantaneously in attaining a continuous portfolio balance. We have defined the capital account and the overall balance of payments in terms of changes in the holdings of securities and money ... that occur through time, therefore these definitions do not capture international transactions that occur discretely as those that are involved in the instantaneous exchange ... of cash balance for equities ... the expressions that pertain to the impact effects do not capture these initial stock adjustments." (p. 681, fn.)

Cooper [4] put the problem as follows:

> " ... formulation of the problem in terms of port-

folio equilibrium draws attention to the transitional dynamics, since exogeneous (or policy) changes lead to transitory stock adjustments in portfolios, plus continuing debt service payment.

Unfortunately, it is very difficult to model portfolio equilibrium and stock adjustment satisfactorily without shedding much of what is of interest for stabilization policy ... A complete model cannot avoid having both stock and flow equilibria, with explicit stock adjustment responses to disequilibria. This inevitably greatly complicates the model, and the results that are crucial for assessing the stabilization impact of monetary and fiscal measures depend critically on assumptions that are made about which markets clear instantaneously and which do not, and which markets clear in the short run by price adjustment as opposed to quantity adjustment. From the viewpoint of scientific analysis these 'dynamic' specifications, and tests of their validity, are of course desirable, however difficult they may be, for that is what economic stabilization is all about."(p.151)

Finally, Tobin and De Macedo [14] remark:

"The model we propose uses discrete rather than continuous time. The motivation is to include some of the effects of finite stock accumulation, which ... are not captured at all when asset markets are modeled as stock equilibrium at a point in time."(p.15)

There are at least three ways to avoid or solve the problem of modelling the stock-flow interaction in continuous time. The first is to choose a discrete time treatment, as Tobin and De Macedo do. The second is to postulate a slow portfolio adjustment process, as Karacaoglu and Ursprung [10] do. However, they do not give a solution for an adequate treatment of the wealth changes as the sum of valuation effects and new savings. Slow portfolio adjustment implies only that there are no instantaneous stock shifts; the exchange rate is as before a fast variable, so valuation effects occur. The third possibility is to apply the moving equilibrium method in the extended version of Witzel [16]. Witzel extended the method so as to analyze systems in which some endogeneous variables are no longer either slow or fast, but influenced in their time path by slow <u>and</u> fast movement components. Mathematically, a model in which such "mixed" variables appear is characterized by the usual treatment of fast or slow variables, and the formulation of differential equations for the mixed variables, in which the time derivative of other variables appears. In the simplified version of the Kouri model for example, the nominal wealth is such a mixed variable. Its time path is described by equation (5).

3. An Extended Asset Market Model

In what follows an extended Kouri model is presented. The structure of good markets is not changed. The extension refers to the asset structure of the model. I assume that the country is small in the good markets, but sufficiently large in financial markets for foreigners to find it attractive to hold domestic currency. This is in accordance with the view that when modelling an economy it may often be appropriate to give up the clear distinction between countries which are small or large in all markets and to differentiate between smallness or largeness of economies in a subset of markets. Real wealth of foreigners is then given by

$$V^* = M^*/\pi + F^* \tag{6}$$

where V^* is the real wealth of foreigners, M^* is the nominal stock of domestic currency held by foreigners, and F^* is the real stock of foreign currency held by foreigners.

All asset stocks are assumed to be positive. Asset demands are assumed to be linear homogeneous in wealth, the proportion of wealth held in different currencies being a function of x, the expected exchange rate change

$$M^d = m(x)V \quad ; \qquad m_x < 0, \tag{7a}$$

$$M^{*d} = m^*(x)\pi V^* ; \qquad m^*_x < 0, \tag{7b}$$

$$F^d = f(x)V \quad ; \qquad f_x > 0, \tag{7c}$$

$$F^{*d} = f^*(x)\pi V^* ; \qquad f^*_x > 0, \tag{7d}$$

where $F^d[F^{*d}]$ is the nominal domestic foreign demand for foreign currency, and M^{*d} is the nominal foreign demand for domestic currency. The expected momentary rate of change of the exchange rate is defined to be

$$x = \alpha \frac{\pi^e - \pi}{\pi^e} ; \qquad \alpha > 0, \tag{8}$$

where π^e is the expected long-run equilibrium exchange rate, the determination of which is specified below. Government expenditure is assumed to be fixed at the constant level of tax receipts. Asset supplies are exogeneous and fixed at M^s and F^s. Equilibrium in the money markets requires

$$M^s = M^d + M^{*d} \quad , \tag{9}$$

$$\pi F^s = F^d + F^{*d} \quad . \tag{10}$$

Note that portfolio equilibrium is not implied by money market equilibrium.[6] The trade balance is equal to disposable income Y_D minus private consumption C

$$B = Y_D - C(Y_D, V/\pi); \quad 0 < C_{Y_D} < 1, \quad C_{V/\pi} > 0, \tag{11}$$

where Y_D is constant.

The dynamics of the system can be described by the following system of differential equations

$$\dot{\pi} = \tau[M^s - (M^d + M^{*d})], \tag{12}$$

$$\dot{M} = \tau(M^d - M) + (1 - s)\pi B, \tag{13}$$

$$\dot{F} = -\frac{1}{\pi}\tau(M^d - M) + sB, \tag{14}$$

$$\dot{\pi}^e = a(\pi - \pi^e) + \gamma\tau[M^s - (M^d + M^{*d})], \tag{15}$$

$$\tau > 0, \quad \tau \to \infty, \quad 0 \leq s \leq 1, \quad 0 < a < \infty, \quad 0 \leq \gamma \leq 1.$$

Equation (12) says that exchange rate movements assure instantaneous rebalancing of the domestic money market. Equations (12) and (13) describe the changes over time of domestic holdings of domestic and foreign currency. The τ-terms reflect the stock shifts which are part of the instantaneous portfolio balancing process, whereas the s-terms give the accumulation of M and F through current account imbalances. Thus domestic wealth holders can accumulate foreign currency either by exchanging domestic currency for foreign currency or by running a trade balance surplus. Note that by summing up equations (13) and (14) one obtains the changes in wealth through new savings. The overall change in nominal wealth is composed of valuation effects and new savings

$$\dot{V} = \dot{M} + \dot{\pi}F + \pi\dot{F} = \dot{\pi}F + \pi B\ . \tag{16}$$

In what follows it will be assumed that s = 1 for simplicity. Equation (15) describes the formation of expectation by means of an adaptive-extrapolative process. If $\gamma = 0$ expectations are formed adaptively. If $\gamma = 1$ there are current value expectations held. The value of π^e which emerges from (15) is used to calculate x, the expected momentary rate of exchange rate change.[7]

The system ((12)-(15)) is approximated by means of the extended moving equilibrium method. The fast movement components are set equal to zero

$$M^s - (M^d + M^{*d}) = 0 \quad , \tag{17}$$

$$M^d - M = 0 \quad . \tag{18}$$

Equations (17) and (18) are used to solve explicitly for π and M, yielding $\pi(F, \pi^e)$, $M(F, \pi^e)$. These functions are then substituted into (14) and (15) yielding the reduced form of the approximation system, in which only F and π^e are endogeneous

$$\dot{F} = -\frac{1}{\pi(F, \pi^e)}\left(\frac{\partial M}{\partial F}\dot{F} + \frac{\partial M}{\partial \pi^e}\dot{\pi}^e\right) + B(\pi(F, \pi^e), M(F, \pi^e), F), \tag{19}$$

$$\dot{\pi}^e = a(\pi(F, \pi^e) - \pi^e) + \gamma\left(\frac{\partial \pi}{\partial F}\dot{F} + \frac{\partial \pi}{\partial \pi^e}\dot{\pi}^e\right) . \tag{20}$$

Temporary equlibrium which is characterized by portfolio balance for domestic and foreign wealth holders and equilibrium in all financial markets is thus described by equations (17) and (18),[8] while long-run equilibrium further requires $\pi^e = \pi$ and $B = 0$.

In order to check whether the approximation of the initial system of differential equations ((12)-(15)) by the reduced system ((19), (20)) is admissible, one has mainly to perform the stability analysis of the matrix of system ((19), (20)) and of the matrix of the fast subsystem.[9] It shows that both systems are locally stable if one assumes local currency preference, $m > m^*$.

4. Concluding Remarks

The AMA to flexible exchange rates took the existence of large international financial capital movements which result from reshuffling of portfolios, that is, which are not directly linked to current account imbalances, as a point of departure for analyzing temporary exchange rate fluctuations. However, in order for stock shifts to take place, at least two international tradable assets must exist. The dynamic analysis of models which explicitly incorporate stock shifts raises some problems when formulating the model in continuous time. Similar problems are posed by the inclusion of wealth changes through valuation effects. The inclusion of these effects helps - besides of closer approaching reality - to understand the concept of "incipient capital out- or inflows" which is rather obscure in the literature. The present paper suggests that these problems can be solved by applying the extended moving equilibrium method (Witzel [16]).

FOOTNOTES

1 Note that there are no <u>net</u> capital movements taking place in the case of stock shifts, as the change in the private domestic stock of foreign assets is equal in size but of opposite sign to the change in the reserve holdings of the domestic central bank, so the balance on capital account is zero.

2 There is, of course, a feedback from flows to stocks, as new savings result in higher stock holdings, but there is no interaction between current account induced capital flows and those capital movements which are the outcome of reshuffling of portfolios.

3 See Branson and Henderson [2], De Grauwe [5], Henderson and Rogoff [8],Ethier [6], Isard [9], who all have models with two distinct international tradable assets.

4 Most authors refer to Tobin and De Macedo [14] for this statement.

5 This is the procedure that Branson and Henderson [2] or Henderson and Rogoff [8] follow.

6 This will usually be the case with more than one asset traded internationally.

7 Note that only with $\gamma \neq 0$ expectations are revised simultaneously with the movement of π . Kouri [11] postulates that expectations are formed directly with respect to rates of change. In this case even with adaptive expectations one has instantaneous reactions in x, so the extended moving equilibrium method should be used, irrespective of how many assets are tradable internationally, as x is a mixed variable. For a detailed exposition of this point see Witzel [16] , pp. 129-139.

8 Equilibrium in the domestic money market is described by (17). From (17), (18) and the stock restriction $M^S = M + M^*$ one gets $M^{*d} = M^*$; this, together with the budget constraints $V = M^d + F^d$, $\pi V^* = M^{*d} + F^{*d}$ and the wealth definitions yields $F^d = \pi F$ and $F^{*d} = \pi F^*$; when further using the stock constraint $F^S = F + F^*$ one finally obtains $\pi F^S = F^d + F^{*d}$, which is the equilibrium condition for the foreign money market.

9 The calculations can be found in Tivig [13].

REFERENCES

[1] ALLEN, P.R. and P.B. KENEN: "Portfolio Adjustments in Open Economies: A Comparison of Alternative Specifications", Weltwirtschaftliches Archiv 11, (1976), 33-72.

[2] BRANSON, W.H. and D.W. HENDERSON: "The Specification and Influence of Asset Markets", in: Jones, R. and P.B. Kenen, eds., Handbook of International Economics, vol. 2 (1985), North Holland, 749-805.

[3] CALVO, G. and C.A. RODRIGUEZ: "A Model of Exchange Rate Determination under Currency Substitution and Rational Expectations", Journal of Political Economy, 85 (1977), 617-675.

[4] COOPER, R.N.: "Monetary Theory and Policy in an Open Economy", The Scandinavian Journal of Economics, 78 (1976), 146-163.

[5] DeGRAUWE, P.: "Macroeconomic Theory for the Open Economy" (1983), Gower.

[6] Ethier, W.: "Expectations And The Asset-Market Approach To The Exchange Rate", Journal of Monetary Economics, 5 (1979), 259-282.

[7] FRENKEL, J.A. and C.A. RODRIGUEZ: "Portfolio Equilibrium and the Balance of Payments: A Monetary Approach", American Economic Review, 65 (1975), 674-688.

[8] HENDERSON, D.W. and K. ROGOFF: "Negative Net Foreign Asset Positions and Stability in a World Portfolio Balance Model", Journal of International Economics, 13 (1982), 85-104.

[9] ISARD, P.: "Exchange Rate Determination: A Survey of Popular Views and Recent Models", Princeton Studies in International Finance, No. 42 (1978) Princeton University.

[10] KARACAOGLU, G. and H.W. URSPRUNG: "Slow Portfolio Adjustment And Exchange Rate Dynamics", unpublished working paper (1986).

[11] KOURI, P.: "The Exchange Rate and the Balance of Payments in the Short Run and in the Long Run: A Monetary Approach", The Scandinavian Journal of Economics, 78 (1976), 280-304.

[12] SCHLICHT, E.: "Die Methode der Gleichgewichtsbewegung als Approximationsverfahren", in: Helmstädter, E., ed., Neuere Entwicklungen in den Wirtschaftswissenschaften, Berlin (1978).

[13] TIVIG, T.: Flexible Wechselkurse aus der Sicht des Finanzmarktansatzes, unpublished dissertation, University of Konstanz (1986).

[14] TOBIN, J. and J.B. De MACEDO: "The Short-run Macroeconomics of Floating Exchange Rates: An Exposition", in: Chipman, J. and C.P. Kindleberger, eds., Flexible Exchange Rates and the Balance of Payments: Essays in Memory of Egon Sohmen, North-Holland (1980), 5-28.

[15] TURNOVSKI, S.: "Macroeconomic Analysis and Stabilization Policies", Cambridge University Press (1977).

[16] WITZEL, R.: "Eine Erweiterung der Methode der Gleichgewichtsbewegung mit Anwendungen", dissertation, Peter Lang Press (1985).

ON THE PROBLEM OF TIME-INCONSISTENCY IN THE THEORY OF STABILIZATION POLICY

Reinhard Neck
University of Economics, Vienna

1. INTRODUCTION

Starting with the fundamental contribution by Kydland and Prescott (1977), the possibility of time-inconsistent stabilization policies has received much attention in recent debates about macroeconomic theory and policy. Nevertheless, so far no consensus has been achieved with respect to the importance of this problem and its consequences for the theory of economic policy. In particular, the question whether "fixed rules" (e. g., Friedman's constant money-growth rule) instead of discretionary policy-making should be preferred in view of time-inconsistency, is still unresolved. In the present paper, the problem of time-inconsistency of stabilization policies is discussed within the framework of dynamic game theory. The crucial importance of strategic considerations as a cause of time-inconsistency has already been recognized by several authors, although so far many contributions to this topic still neglect insights obtained in the mathematical theory of dynamic games. We argue that time-inconsistency in the proper sense of the word is a property of a particular equilibrium solution concept, namely the global Stackelberg equilibrium solution; in our view, more severe problems may arise if subgame-perfectness instead of merely time-consistency is required as a desirable property of an equilibrium solution. However, there do exist solution concepts which exhibit this property, and the resulting policies may be either inefficient or efficient. Thus a shift of the theoretical discussion from the narrow time-inconsistency issue to the more general problem of designing efficient subgame-perfect stabilization policies in a non-cooperative environment seems to be appropriate. We give some suggestions on how such policies could look like within our game-theoretic framework.

2. A DYNAMIC GAME MODEL

Here we present a dynamic game model, which has been analysed in more detail in Dockner and Neck (1987a). We consider a deterministic scalar dynamic economic system in continuous time which is influenced by two decision-makers (players), to be called 1 and 2. Denoting time by t, the state variable by x(t), the control (instrument) variables of players 1 and 2 by $u_1(t)$ and $u_2(t)$, respectively, we restrict ourselves to a system described by the linear differential equation

$$\dot{x}(t) = ax(t) + bu_1(t) + cu_2(t) \tag{1}$$

with initial condition

$$x(0) = x_o > 0. \tag{2}$$

Players 1 and 2 have quadratic intertemporal objective functions, which they want to minimize, namely

$$J_1 = \frac{1}{2}\int_o^\infty \exp(-rt)[g_1x^2(t) + g_2u_1^2(t)]dt, \tag{3}$$

(4)

$$J_2 = \frac{1}{2}\int_0^{\infty}\exp(-rt)[m_1x^2(t) + m_2u_2^2(t)]dt, \tag{4}$$

respectively. That is, the objective functions are quadratic in the state variable, which is thus regarded as an economic "evil" by both players, and in the own instrument variable of each player. Extensions to time-varying systems, nonlinear systems, multivariable systems, non-quadratic objective functions, objective functions containing more arguments, finite-time-horizon problems, systems in discrete time, and more than two players are possible within the present conceptual framework, but usually do not allow analytical characterizations of the outcomes of the different solution concepts to be discussed in the next sections.

The problem of time-inconsistent stabilization policies has been usually discussed within the framework of macroeconomic models where private agents have rational expectations or perfect foresight. For this case, the above dynamic model can be interpreted as follows: Player 1 is the government with policy instrument $u_1(t)$ (e. g., money supply, government expenditures, taxes); player 2 is the aggregate private sector of the economy with forward-looking expectations concerning the behavior of the government, where $u_2(t)$ is a variable under the control of private agents (e. g., consumption etc.), based upon rational expectations about $u_1(t)$ and hence future values of $x(t)$. The state variable $x(t)$ may be the aggregate output gap, the rate of inflation, the rate of unemployment, or any other variable which is influenced over time by the evolution of both players' decision-variables and about which both players are concerned. Without too much loss of generality, all variables can be regarded as being measured as deviations from their "ideal" values (which may be constant or time-varying). Our dynamic game model can, however, also be interpreted for an economic system with a "passive" private sector, if there are more than two political decision-makers, for instance as a model of stabilization policies in a closed economy with government and central bank as policy-makers (e. g., Neck and Dockner (1987a)), or as a model for international policy coordination (e. g., Neck and Dockner (1987b)). Thus the problem of time-inconsistency, which can be analysed with this dynamic game model, can arise in those other areas, too.

In order to specify which strategies are available to each player, it is necessary to make some specific assumptions about the information pattern of the dynamic game, a point which is sometimes neglected in the economic literature on time-consistency. Roughly speaking, the information pattern of each player determines the information about the state variable available to him (her) at each point of time; players' strategies are then defined as mappings from the set of information patterns into the sets of feasible values of players' control variables. Thus, for each player the information set determines which information about the state he (she) will use when deciding about the action at time $t \geq 0$, as seen from the moment of planning the trajectory of decisions at $t = 0$. Important special cases are the open-loop information pattern, where the only information available to the player is the initial state $x(0)$, the feedback information pattern, which contains only the current state $x(t)$, and the closed-loop information pattern with memory, where all states $x(s)$, $0 \leq s \leq t$, are known at t. The latter information pattern involves memory, as past values of the state are recalled when they have been observed; the other two are memoryless. Open-loop strategies for player i may be written as

$$u_i = u_i[x(0),\ t],\ i = 1,2; \tag{5}$$

they are "path strategies", as their entire time path is decided upon once for all at t = 0. On the other hand, feedback strategies

$$u_i = u_i[x(t),\ t],\ i = 1,2, \tag{6}$$

may be called "decision rules", because they specify for each t what action to take, given the value of the state variable at t, x(t). Sometimes in the theory of stabilization policy, open-loop strategies are identified with "fixed rules" and feedback strategies with "contingent rules". "Fixed rules" in the sense of Friedman are indeed special cases of open-loop strategies, although in general strategies of the form (5) need not be constant over time.

Finally, we have to define the equilibrium solution concepts for our dynamic game. Among the non-cooperative solutions, where no binding agreements between the players are possible, we consider Nash and Stackelberg equilibria. A given pair of strategies u_1^*, u_2^* constitutes a Nash equilibrium solution, if

$$J_1(u_1^*,\ u_2^*) \leq J_1(u_1,\ u_2^*),\ J_2(u_1^*,\ u_2^*) \leq J_2(u_1^*,\ u_2) \tag{7}$$

for any admissible strategies u_1, u_2. This solution concept assigns symmetric roles to the two players; on the other hand, the Stackelberg equilibrium solution is asymmetric as it distinguishes between one player, the leader, who can enforce his (her) strategy on the other player, the follower, who reacts upon the leader's strategy. Denoting player 1 as the leader and player 2 as the follower, we can define the reaction function T^2 of the follower if we have for any admissible strategy of the leader u_1:

$$J_2(u_1,\ T^2(u_1)) \leq J_2(u_1,\ u_2)\ \forall u_2. \tag{8}$$

A given pair of strategies u_1^*, u_2^* then constitutes a Stackelberg equilibrium solution if for all admissible u_1

$$J_1(u_1^*,\ T^2(u_1^*)) \leq J_1(u_1,\ T^2(u_1)) \tag{9}$$

and

$$u_2^* = T^2(u_1^*). \tag{10}$$

Cooperative solution concepts start from the assumption that some binding agreement has been established by the players. How this agreement is brought about may be analysed by methods of game theory, too, but this is usually not a subject of cooperative game theory proper. There exists a great number of different cooperative solution concepts, and so far there is no consensus about their relative merits. However, at least in the two-players case, all of them require the property of Pareto-optimality or efficiency, which for our purpose will be used for defining the set of possible cooperative outcomes of our dynamic game. A pair of strategies u_1^*, u_2^* is efficient, if for all admissible strategies u_1, u_2

$$J_i(u_1^*,\ u_2^*) \leq J_i\ (u_1,\ u_2),\ i = 1,2. \tag{11}$$

The above definitions of the Nash and the Stackelberg equilibrium solution and of the Pareto-optimal solutions are valid both for static and for dynamic games, as they start from the normal form description of the game. For dynamic games, however, additional complications arise as these are usually formulated in an extensive-form description, where the evolution of the game, the information used by each player during the game, and to some extent the interactions between the players are explicitly shown. A differential game defined by the state equation (1) with initial condition (2), the objective functions (3) and (4), and some specific information pattern is given in extensive form. For a more rigorous analysis of this point, see Başar and Olsder (1982) or Başar (1986). Different information patterns lead to different equilibrium solution concepts within the classes of Nash and Stackelberg equilibrium solutions for dynamic games. Here we will consider only two of them, namely open-loop and feedback information patterns. The open-loop Nash equilibrium solution is the Nash equilibrium solution for a game with open-loop information pattern, i. e., both players determine their entire trajectory of actions only upon their knowledge of the initial state and of common knowledge (the state equation and the objective functions). The feedback Nash equilibrium solution is the Nash equilibrium solution for the feedback information pattern; its definition is based on dynamic-programming-type conditions for each player. Both the open-loop and the feedback Nash equilibrium solution are also Nash equilibrium solutions under closed-loop information patterns, because for dynamic games "informational non-uniqueness" holds: Adding to the players' informations does not destroy the equilibrium property of the Nash equilibrium of the original game, but may open up the possibility for further Nash equilibria. For Stackelberg equilibrium solutions the situation is different, however: What we have defined as the Stackelberg equilibrium solution corresponds to a global Stackelberg equilibrium solution for a dynamic game. If both players have open-loop information patterns, then the open-loop Stackelberg equilibrium solution is the global one. On the other hand, the feedback Stackelberg equilibrium solution cannot be obtained from the normal form of the game. Here the leader cannot commit himself (herself) for the overall game, i. e., for the entire time horizon, as is the case for the open-loop Stackelberg equilibrium solution, but he (she) can enforce his (her) strategy on the follower at each instant of time. As the feedback Stackelberg equilibrium solution of a dynamic game is equivalent to the feedback Nash equilibrium solution of a related game with an asymmetrical information pattern, it has an equilibrium property, too (cf. Başar and Haurie(1984)).

Next, we define the property of time-consistency for a dynamic game of the kind under consideration. An equilibrium pair of strategies (u_1^*, u_2^*) is said to be time-consistent if for every intermediate point of time $t_1 > 0$ it constitutes an equilibrium for the game that starts at $x^*(t_1)$ in t_1, where $x^*(t_1)$ is the value of the state variable that is reached at time t_1 when players have played u_1^*, u_2^* over the period $[0, t_1)$. Time consistency defined in this way means for an equilibrium solution that along the equilibrium path the continuation of the equilibrium solution remains an equilibrium. Some authors use an alternative definition for time-consistency, which is, however, much stronger and is equivalent to the notion of subgame-perfectness, which requires such as property for all subgames, not only along the equilibrium path. More precisely, an equilibrium pair of strategies u_1^*, u_2^* is said to be subgame-perfect if for every intermediate point of time t_1,(u_1^*, u_2^*)constitutes an equilibrium for every possible initial condition $x(t_1)$, provided only no-memory information patterns are taken into account. According to the more general notion due to Selten (1975), subgame-perfect equilibrium solutions for a game constitute an equilibrium solution for every subgame of the original game, where a subgame is a game that starts after a particular history has been played. Time-consistency is implied by subgame-perfectness, but not vice versa (Fershtman (1986)); therefore it seems useful to distinguish between these two concepts as distinct properties of an equilibrium solution for a dynamic game. Efficiency or Pareto-optimality, which we have introduced as definition of the set of cooperative solutions, may alternatively be seen as still another property of an equilibrium solution concept, although most non-cooperative equilibrium solutions do not fulfill this requirement.

3. OPEN-LOOP SOLUTION CONCEPTS

For the open-loop solution concepts of the dynamic game formulated in the previous section, we assume that both players consider open-loop strategies only (have open-loop information patterns). This can be interpreted to mean that both players commit themselves to certain courses of actions for the entire planning period, from which they will not deviate, even if that were in their short-run interest. Such commitments may be taken in isolation (in the non-cooperative case) or jointly as an agreement (in the cooperative case). It is important to note that such commitments are assumed a priori if we are considering open-loop solution concepts; they are not explained by the theory. If - for whatever reason - in a dynamic game of stabilization policies at least one player does not commit himself (herself) to a certain trajectory of strategic actions, then the open-loop solution concepts cannot be applied for that particular situation.

The open-loop Nash equilibrium solution for the two-person differential game under consideration can be obtained from necessary conditions of Pontryagin's maximum principle-type for each player. Omitting the time argument, we get as canonical system with costate variables p_1, p_2 for player 1 and 2, respectively:

$$\dot{x} = ax - (b^2/g_2)p_1 - (c^2/m_2)p_2, \tag{12}$$

$$\dot{p}_1 = - g_1 x + (r - a)p_1, \tag{13}$$

$$\dot{p}_2 = - m_1 x + (r - a)p_2, \tag{14}$$

with initial condition (2). At least if $a \leqq 0$, i. e., if the uncontrolled dynamic economic system is stable, the canonical system is unstable and exhibits the saddle-point property. The transversality conditions for the costate variables imply that the solution must not diverge. Hence the system (12) - (14) is reduced to its one-dimensional stable manifold to yield the equilibrium solution

$$x(t) = x_o \exp (s_1^N t), \tag{15}$$

where s_1^N ist the stable (negative) eigenvalue of the canonical system, given by

$$s_1^N = \frac{r}{2} - \sqrt{\frac{r^2}{4} - a(r - a) + g_1(b^2/g_2) + m_1(c^2/m_2)} \tag{16}$$

Strategies and equilibrium values of the cost functions (payoffs) for both players can be calculated in terms of the stable eigenvalue and the stable eigenvector of the canonical system; for details, see Dockner and Neck (1987 a).

Determining the open-loop Stackelberg equilibrium solution for the same game, with player 1 (the government) as the leader and player 2 (the private sector) as the follower, also leads to two optimal control problems, one of which is non-classical, however. In this case, the canonical system for the leader becomes:

$$\dot{x} = ax - (c^2/m_2)p_2 - (b^2/g_2)p_{11}, \tag{17}$$

$$\dot{p}_2 = - m_1x + (r - a)p_2, \tag{18}$$

$$\dot{p}_{11} = - g_1x + (r - a)p_{11} + m_1p_{12}, \tag{19}$$

$$\dot{p}_{12} = (c^2/m_2)p_{11} + a\,p_{12} \tag{20}$$

with initial conditions (2) and

$$p_{12}(0) = 0. \tag{21}$$

Here p_2 is the costate variable of the follower, whose dynamics (expressed by equation (18)) has to be taken into account by the leader as an additional dynamic constraint. Therefore the leader has two costate variables, one relating to the state variable (p_{11}) and the other relating to the costate variable of the follower (p_{12}). From the point of view of the leader, p_2 is a state variable, but it does not have a given initial condition. Instead, the necessary conditions for an optimum of the leader require that the corresponding costate variable satisfies condition (21).

As in the open-loop Nash equilibrium case, the canonical system exhibits "saddle-point stability" at least for $a \leq 0$. The open-loop Stackelberg equilibrium solution now becomes

$$x(t) = A_1\, x_o \exp(s_1^S t) + A_2\, x_o \exp(s_2^S t), \tag{22}$$

where s_1^S and s_2^S are the two stable eigenvalues of the four-dimensional canonical system (17) - (20), and A_1 and A_2 are constants which depend on the elements of the stable eigenvectors of that system. Again, the equilibrium payoffs and the control functions of both players can be calculated analytically. The stable eigenvalues are given by

$$s_{1,2}^S = \frac{r}{2} - \sqrt{\frac{r^2}{4} - a(r - a) + \frac{b^2 g_1}{2g_2} + \frac{c^2 m_1}{m_2} \pm \sqrt{\frac{b^2 g_1}{g_2}\left(\frac{b^2 g_1}{4g_2} + \frac{c^2 m_1}{m_2}\right)}}. \tag{23}$$

For the present discussion, it is important to note that the open-loop Stackelberg equilibrium solution is time-inconsistent. This can be seen most easily from condition (21). For $t_1 > 0$, in general we have, from the optimality conditions given by the solution of the leader's canonical system, $p_{12}(t_1) \neq 0$. However, if the leader had to solve his (her) problem anew at time t_1, this would result in a different solution involving $p_{12}(t_1) = 0$ from (21) instead of the continuation of the original open-loop Stackelberg strategy. Therefore the dynamic-programming principle of optimality does not hold for this time-inconsistent strategy of the leader. The same argument holds for an optimal control problem with a rational-expectations dynamic economic model, as the latter is structurally equivalent to an open-loop Stackelberg equilibrium with two players. In this case, p_2 may be interpreted as a rational-expectations dynamic variable for the private sector which is forward-looking and hence has a transversality instead of an initial condition. Then the open-loop strategy u_1 for the leader in our Stackelberg game can be interpreted as the "optimal", but time-inconsistent policy of the government (the leader). More generally, when private-sector agents are sensitive to the announcement of future policies, the structure of the economic model is not invariant to the policy rule of the government, and the state equations have to contain dynamic relations like that for p_2 in the present model.

Note, however, that the notion of the open-loop Stackelberg equilibrium solution presupposes the commitment of the leader to his (her) open-loop equilibrium strategy. Under the assumptions of this particular solution concept, there is hence no temptation to revise that policy, as a re-optimization at time points $t_1 > 0$ cannot take place due to the specific information pattern assumed. This points to the interpretation that optimal time-inconsistent government policies viz-a-viz a private sector with forward-looking expectations are either credible and will be executed (the "leader" keeps to its commitment), or otherwise the economic-policy problem is misspecified, as the possibility of re-optimization at later dates makes the "optimal" policy corresponding to the open-loop Stackelberg equilibrium strategy of the leader meaningless (the assumed commitment does not exist), or it may serve as a point of reference for comparisons with other strategies, at most.

Finally, the set of cooperative open-loop Pareto-optimal solutions for our dynamic game can be determined. Here the notion of a cooperative game requires a joint decision, i. e., an agreement or a bilateral commitment of both players from the start of the game. Open-loop Pareto-optimal solutions can be obtained from minimizing a weighted sum of the cost functions of both players, $\alpha J_1 + (1 - \alpha)J_2$, over all admissible pairs of open-loop strategies (u_1, u_2), subject to the dynamic model (1) - (2). This amounts to the solution of an optimal control problem for each $\alpha \varepsilon (0, 1)$ to trace out the Pareto-optimal frontier of the game in the payoff space. The necessary conditions result in the following two-dimensional canonical system now:

$$\dot{x} = ax - \{[b^2/(\alpha g_2)] + \{c^2/[(1 - \alpha)m_2]\}\}p, \tag{24}$$

$$\dot{p} = -[\alpha g_1 + (1 - \alpha)m_1]x + (r - a)p, \tag{25}$$

where p is the joint costate variable of the optimal control problem. Again this system has the saddle-point property, at least for $a \leqq 0$, and its solution implies

$$x(t) = x_o \exp (s_1^P t), \tag{26}$$

where

$$s_1^P = \frac{r}{2} - \sqrt{\frac{r^2}{4} - a(r - a) + [\alpha g_1 + (1 - \alpha)m_1] \left[\frac{b^2}{\alpha g_2} + \frac{c^2}{(1-\alpha)m_2}\right]} \tag{27}$$

is the stable eigenvalue of the canonical system (24) - (25). The control trajectories and the payoffs for both players can be determined as before, in this case for given α, of course.

As is well known from non-cooperative game theory, Nash and Stackelberg equilibrium solutions in general are not efficient. The same is true for our particular game, i. e., neither the open-loop Nash equilibrium solution nor the open-loop Stackelberg equilibrium solution belong to the set of open-loop Pareto-optimal solutions, apart from special values of the parameters of the economic model or the objective functions. This can be established from comparisons of the equilibrium values of the payoffs. Moreover, it is possible to compare the magnitudes of the stable eigenvalues for the three solution concepts, because they determine the speed of convergence of the state (the economic "evil", or the deviation of the target variable from its "ideal" value) towards its long-run equilibrium ($x = 0$). Denoting the eigenvalue for the uncontrolled system ($u_1(t) = u_2(t) = 0$ for all t) by $s_u = a < 0$, then the following relation can be established:

$$s_1^P < s_1^S < s_1^N < s_2^S < s_u. \tag{28}$$

This means that any Pareto-optimal solution will converge faster to the long-run equilibrium than either the open-loop Nash or the open-loop Stackelberg equilibrium solution. Both of these, in turn, improve upon convergence with respect to the uncontrolled system. Thus all three kinds of stabilization policies considered are in fact stabilizing and not destabilizing, as could have been suspected from the assertion of Kydland and Prescott (1977). Since a constant fixed rule of the Friedman-type would correspond to an uncontrolled system with respect to its convergence properties, this does not seem to be an appropriate answer to the time-inconsistency problem. Instead, both the open-loop Nash equilibrium solution and any Pareto-optimal solution are time-consistent, as they are derived from classical optimal control problems, and stabilizing in the sense of improving the convergence of the economic system to its desired state. Pareto-optimal solutions are efficient, in addition, hence they seem particularly attractive under a normative aspect; however, they require bilateral instead of unilateral commitment, which may be difficult to achieve in practice; thus, under a positive aspect, at first sight there still seems to exist a problem of implementing an "optimal" (which now means efficient) government stabilization policy.

4. FEEDBACK SOLUTION CONCEPTS

In contrast to the open-loop information pattern, where both players commit themselves to certain "path strategies", under the feedback information pattern players' strategies depend on the current value of the state variable $x(t)$ and require no pre-commitment at all. Such an information pattern is therefore adequate to model situations where neither player can be expected to keep to an announced policy; for example, it may be applied to model stabilization policies of a government which is not credible, which not committed to a "fixed rule", which exerts its discretionary power to influence the economic system at each moment, or which cannot expect to be in power for a longer period of time (although in the latter case a finite planning horizon would be more adequate). A feedback information pattern may, however, also be attractive as a model of private sector agents' behavior, because the private sector is in fact an aggregate of many "small" agents, who among themselves may not be able to coordinate their actions in order to commit their aggregate behavior.

Both the feedback Nash equilibrium solution and the feedback Stackelberg equilibrium solution consist of strategies which depend only on $x(t)$ and t, but not on previous states and particularly not on the initial state (they have no "memory"). The feedback Nash equilibrium solution is defined by the requirement that each player minimizes individually his (her) value function by the choice of his (her) equilibrium strategy at each point of time over the entire planning horizon, given the other player's equilibrium strategy, for all possible initial conditions. This implies that feedback Nash equilibrium strategies are subgame-perfect: In contrast to the open-loop Nash equilibrium strategies, at each $t_1 > 0$ both players will not only continue their equilibrium strategies if the game is along its equilibrium trajectory (time-consistency), but feedback Nash equilibrium strategies will also be adhered to if for some reason (either exogenous disturbances or the behavior of one player) the system is off its equilibrium trajectory.

The feedback Nash equilibrium solution is obtained from the solution of two partial differential equations of the Hamilton-Jacobi-Bellman type, which can be established by backward induction using the dynamic-programming principle of optimality. For our particular linear-quadratic differential game, using the current-value value functions makes these equations autonomous. Then the feedback Nash equilibrium strategies are given by linear feedback rules

$$u_1(t) = - (b/g_2)\, V_1\, x(t), \tag{29}$$

$$u_2(t) = - (c/m_2)\, V_2\, x(t), \tag{30}$$

which result in the state trajectory

$$x(t) = x_o \exp(s^F t) \tag{31}$$

with

$$s^F = a - (b^2/g_2)\, V_1 - (c^2/m_2)\, V_2. \tag{32}$$

The constants V_1 and V_2 are the real positive solutions to the system of algebraic equations

$$V_1(\frac{r}{2} - a) + V_1^2(\frac{b^2}{2g_2}) + V_1 V_2(\frac{c^2}{m_2}) = \frac{g_1}{2}, \tag{33}$$

$$V_2(\frac{r}{2} - a) + V_2^2(\frac{c^2}{2m_2}) + V_1 V_2(\frac{b^2}{g_2}) = \frac{m_1}{2}, \tag{34}$$

which are obtained from the current-value Hamilton-Jacobi-Bellman equations. Although V_1, V_2 can be calculated analytically in principle, this is very cumbersome because (33) - (34) leads to a pair of fourth-order algebraic equations, which do not yield meaningful analytical solutions; hence these two constants usually have to be determined numerically for particular values of the parameters.

The feedback Stackelberg equilibrium solution is also based on feedback information patterns for both players, but now the roles of the players are asymmetrical. This corresponds to a model where player 1 (the leader) announces his (her) decision rule, which depends on the current state x(t) and t, and player 2 (the follower) acts on the basis of the current state and the decision rule of the leader. Thus at any instant of time, the leader imposes his (her) action on the follower; the leader has an incremental informational advantage over the follower at each moment, but cannot commit himself (herself) for future time periods in a credible way, as is required for the open-loop Stackelberg equilibrium solution. A rigorous definition and analysis of this solution concept has been provided by Başar and Haurie (1984). For our particular dynamic game, the feedback Stackelberg equilibrium solution is identical to the feedback Nash equilibrium solution. This is a consequence of the particular objective functions, which we have assumed for our model, and does not hold for more general linear-quadratic games; in particular, if the objective functions contain terms of the form $u_1 \cdot u_2$, then the two solution concepts will in general imply different equilibrium strategies.

For the present purpose, it is important to stress the fact that the feedback Stackelberg equilibrium strategies are subgame-perfect, as are the feedback Nash equilibrium strategies. From this follows that they are time-consistent, too; however, subgame-perfectness is a stronger refinement of the respective (Stackelberg and Nash) equilibrium concept. This property allows the players to adapt their behavior at each moment of time, which is not only relevant for the strategic interaction between the players, but also in the presence of exogenous influences, particularly under uncertainty. The feedback Stackelberg equilibrium solution remains also stable (i. e., it gives the same feedback rules as strategies) if the dynamic system is subject to stochastic disturbances following a Markov process. The same is true for the feedback Nash equilibrium solution, which in the presence of stochastic Markovian perturbations even gives the informationally unique Nash equilibrium solution not only under a feedback information pattern, but also under a closed-loop information pattern with perfect state information. This stresses the robustness of the feedback equilibrium strategies with respect to deviations from equilibrium behavior, which is implied by their subgame - perfectness property. Since actual economic-policy models nearly always contain some stochastic elements, these results provide strong arguments for the use of feedback Nash or feedback Stackelberg equilibrium solutions to model strategic interactions between the government and the private sector, when non-cooperative behavior has to be assumed. The time-consistency of these solutions can be seen as a desirable by-product of their subgame-perfectness.

What does the foregoing discussion imply for the alleged conflict between optimality and time-consistency originally stated by Kydland and Prescott (1977)? As we have already remarked, if one takes a strictly game-theoretic point of view, this question seems somewhat ill-posed, because there is no unique concept of optimality for a government playing against a rational private sector. It is only useful to consider equilibrium strategies, and these and their properties depend upon the information pattern assumed. If we have an asymmetrical (Stackelberg) information pattern and players play either open-loop or feedback strategies, then either the time-inconsistent open-loop Stackelberg equilibrium strategy or the time-consistent and subgame-perfect feedback Stackelberg equilibrium strategy is "optimal" for the leader, respectively. The possibility of unilateral precommitment determines which of these two solution concepts is adequate in a particular economic model of interactions between the government and the private sector. One may, of course, compare the outcomes of the two different solution concepts and determine which of them leads to a higher payoff for the leader in order to compare the normative implications of "fixed rules" in the wider sense (open-loop equilibrium strategies) to contingent rules (feedback equilibrium strategies) from the point of view of the leader (the government); the same may be done for the follower (the private sector), and one may also compare the speed of convergence under the two different information patterns. The same comparison may be done for the open-loop and the feedback Nash equilibrium solutions.

In Dockner and Neck (1987a), this has been done numerically for some plausible values of the parameters of the model and the objective functions. It turns out that the feedback (Nash and Stackelberg) equilibrium solution shows the slowest convergence, and the leader has to be more active in the symmetric (Nash) solutions than in the open-loop Stackelberg solution, with correspondingly higher costs, whereas the reverse is true for the follower. Although this numerical example seems to confirm Kydland's and Prescott's conjecture that the leader would "prefer" an open-loop information pattern implying a time-inconsistent equilibrium strategy to a feedback information pattern implying a subgame-perfect equilibrium strategy (and to an open-loop Nash information pattern implying a time-consistent, but not subgame-perfect equilibrium strategy), Fershtman (1986) shows for another example that the leader's payoff in a feedback Stackelberg equilibrium solution may be better than his (her) payoff in an open-loop Stackelberg equilibrium solution. Therefore time-inconsistent fixed rules are not necessarily suboptimal as compared to time-consistent contingent rules. Unfortunately, at the moment we cannot give an analytical characterization of the classes of models where the leader is better off under an open-loop information pattern than under a feedback information pattern and vice versa. This problem is also an open one for other asymmetrical solution concepts proposed in the literature, where one player has an open-loop and the other a feedback information pattern; cf., e. g., Cohen and Michel (1986), or Fershtman's (1986) "perfect foresight feedback Stackelberg equilibrium".

What should be emphasized, however, is the purely hypothetical character of such comparisons. The logic of the game-theoretical approach implies that the information pattern is given to both players at the outset and cannot be chosen by either player. That means that either commitment is possible under some specific institutional conditions, or it is not possible, and these "givens" of the environment of the players determine which equilibrium solution concept has to be applied. Thus the normative problem is relegated to the question of institutionalizing certain information patterns and corresponding strategies, for instance by making self-commitments of the government credible and possible. This has to be analysed, however, apart from the outcomes of the different solution concepts. So far there is no general theory available which can answer this question; the important point, however, is that such a theory would have to start one step before strategic interactions within a dynamic model are investigated, as for the latter problem the information pattern has to be assumed to be given.

5. TIME-CONSISTENCY AND EFFICIENCY

The question of "suboptimality" of certain equilibrium outcomes of a dynamic game can be seen to be more general than the problem of time-inconsistency. From a game-theoretic point of view, the relevant problem is whether (time-consistent or time-inconsistent) non-cooperative equilibrium solutions are efficient in the sense of Pareto-optimality. Efficient outcomes can be expected for certain only in a cooperative game, where both players have to agree on their strategies. In special cases, such as that of the "command economy" of Hillier and Malcomson (1984), such efficient outcomes can be brought about by the government (one player) alone, but in general the non-cooperative solution concepts we have examined so far do not admit such a possibility; their outcomes are in general inefficient (suboptimal), as can be seen from the comparisons of the different speeds of adjustment between the non-cooperative and the open-loop Pareto-optimal solutions. However, this is in part due to the fact that we have considered only memoryless information patterns, where each player bases his (her) decisions at any time $t > 0$ on the initial state (open-loop) or the current state (feedback) and time only, but not on the states or the strategies which occurred at some intermediate points of time τ, $0 < \tau < t$. If we introduce memory strategies under an information pattern which includes past states and/or controls, then the situation becomes different, as totally new equilibrium solution concepts can be defined. For lack of space, here we only sketch the Nash equilibrium case; for related work on closed-loop Stackelberg equilibrium solutions and their applications to the problem of time-inconsistency, see, e. g., Carraro (1985) and the literature cited there.

In the Nash case, by introducing memory strategies the set of non-cooperative Nash equilibria is enlarged due to the informational non-uniqueness of Nash equilibrium solutions. In particular, if we allow for memory strategies where each player remembers deviations of the other player from a Pareto-optimal cooperative strategy in the past, he (she) may develop threats and retaliations in order to enforce outcomes corresponding to a Pareto-optimal solution without any binding agreement or unilateral commitment. This opens up the possibility of subgame-perfect Nash equilibrium strategies resulting in an efficient outcome, where each player is allowed to adapt his (her) control variable continuously, and a cooperative (i. e., jointly optimal) outcome may be nevertheless attained without explicit cooperation. Haurie and Pohjola (1987) have shown for a particular model how subgame-perfect non-cooperative equilibrium solutions in memory strategies for a differential game resulting in efficient outcomes can be obtained by combining an open-loop Pareto-optimal pair of strategies with a pair of threats defined as feedback Nash equilibrium strategies.

In our model, this means that at $t = 0$ the government and the private sector agree (explicitly or tacitly) to apply Pareto-optimal strategies for all $t > 0$, but this agreement is not binding. Instead, each player threatens to apply the non-cooperative feedback Nash equilibrium strategy as soon as one of the players does not keep to that Pareto-optimal strategy. Such threats are credible, since they imply subgame-perfect Nash equilibrium strategies, and they are effective if they make any unilateral change of action by any player unattractive. The latter requirement is fulfilled if the open-loop Pareto-optimal solution under consideration is preferred to the feedback Nash equilibrium solution by both players. For the model used in the present paper, this can be shown only numerically, but for several alternative models it is possible to show analytically that there exist values of α which make the Pareto-optimal solution strictly dominant with respect to the feedback Nash equilibrium solution. See, in particular, the derivations given for a model which has objective functions linear in the state variable (Dockner and Neck (1987b) and for a more general linear-quadratic model with a symmetric dynamic economic system modeling international policy coordination (Neck and Dockner (1987b)).

The basic idea is the use of "trigger strategies" which constitute a subgame-perfect non-cooperative Nash equilibrium, because at each $t > 0$ either at least one player has not played his (her) Pareto-optimal strategy for some non-empty interval of time, in which case both will apply the feedback Nash strategies for the entire planning horizon (a subgame-perfect equilibrium), or both have played their Pareto-optimal strategies in the past, and then neither of them has an incentive to deviate from this strategy in the future as it dominates the feedback Nash equilibrium strategy. Thus this theory leads to a self-enforcing and self-sustaining efficient outcome, which does not need any explicit pre-play communication. Further efficient subgame-perfect equilibria in memory strategies can be obtained by introducing the possibility of a finite retaliation period. The result is similar to the ones obtained by authors establishing the possibility of reputational mechanisms in repeated games between the government and the private sector enforcing time-consistent efficient government policies (Barro (1986), Backus and Driffill (1985)), although we use differential instead of repeated games.

What remains as an open problem, however, is the determination of the particular equilibrium solution to be established, because both in the repeated games and in the differential games with memory strategies, there usually exists a multiplicity of resulting equilibria. Here possibly some dynamic theory of bargaining will have to be applied, and hence some kind of explicit cooperation in the pre-play stage may be required. Another problem for further research is the extension of the theory of subgame-perfect efficient non-cooperative equilibria to cases where uncertainty and incomplete information of economic agents are present. Nevertheless, we think that the theory also in its present state of development can provide an attractive solution to the problem of time-inconsistency, as it shows that there exist non-cooperative equilibrium solutions which are not only subgame-perfect (and hence time-consistent), but also efficient or Pareto-optimal. Only under very special conditions they will result in constant controls for the government; from their structure, they are in fact contingent rules. Thus the problem of time-inconsistency cannot serve as a justification for recommendations of Friedman-type fixed rules.

6. CONCLUDING REMARK

In this paper the problem of time-inconsistent stabilization policies has been discussed within the framework of the theory of dynamic games. We have shown that time-inconsistent behavior arises in general if the interaction between government and private sector is modeled as an open-loop Stackelberg equilibrium. Assigning symmetric roles to both players, as is done in the open-loop Nash equilibrium, results in time-consistent government strategies, which still require some unilateral commitment and are not efficient. Feedback Nash and feedback Stackelberg equilibria are subgame-perfect and do not require any commitment; moreover, they are robust with respect to stochastic disturbances. Because the resulting outcomes of these equilibrium solutions are in general inefficient, they still leave us with the dilemma of time-inconsistency versus optimality. Efficient and subgame-perfect (and hence time-consistent) equilibria can be established, however, if memory strategies and the possiblity of threats and retaliations are included in the information pattern of the game. Further research will have to concentrate on refinements of these memory-strategy equilibria to rule out nonuniqueness typically associated with them.

ACKNOWLEDGEMENT

Many of the results used in this paper have been obtained by collaboration with Engelbert Dockner. The present author is, however, responsible for the interpretations given and hence for any errors and shortcomings.

REFERENCES

BACKUS, D., AND J. DRIFFILL (1985): "Inflation and Reputation," American Economic Review, 75, 530-538.

BARRO, R. J. (1986): "Reputation in a Model of Monetary Policy with Incomplete Information," Journal of Monetary Economics 17, 3-20.

BAŞAR, T. (1986): "A Tutorial on Dynamic and Differential Games," in Dynamic Games and Applications in Economics, ed. by T. Başar. Berlin: Springer-Verlag, 1-25.

BAŞAR, T., AND A. HAURIE (1984): "Feedback Equilibria in Differential Games with Structural and Modal Uncertainties", in Advances in Large

Scale Systems, ed. by J. B. Cruz. Greenwich CT: JAI Press, 163-201.

BAŞAR, T., AND G. J. OLSDER (1982): Dynamic Noncooperative Game Theory. London: Academic Press.

CARRARO, C. (1985): "Optimal Announcements and the Problem of Time Inconsistency," mimeo, Dipartimento di Scienze Economiche, Università degli Studi di Venezia.

COHEN, D., AND PH. MICHEL (1986): "How Should Control Theory Be Used by a Time-Consistent Government," CEPR Discussion Paper No. 141, London.

DOCKNER, E. J., AND R. NECK (1987a): "Time-Consistency, Subgame-Perfectness, Solution Concepts, and Information Patterns in Dynamic Models of Stabilization Policies," mimeo, Department of Economics, University of Saskatchewan, Saskatoon.

DOCKNER, E., AND R. NECK (1987b): "Zur Effizienz dezentralisierter Stabilisierungspolitik," in Operations Research Proceedings 1986, ed. by H. Isermann. Berlin: Springer-Verlag, 528-536.

FERSHTMAN, CH. (1986): "Fixed Rules and Decision Rules: Time Consistency and Subgame Perfection," mimeo, Hebrew University, Jerusalem.

HAURIE, A., AND M. POHJOLA (1987): "Efficient Equilibria in a Differential Game of Capitalism," Journal of Economic Dynamics and Control, 11, 65-78.

HILLIER, B., AND J. M. MALCOMSON (1984): "Dynamic Inconsistency, Rational Expectations, and Optimal Government Policy," Econometrica, 52, 1437-1451.

KYDLAND, F. E., AND E. C. PRESCOTT (1977): "Rules Rather than Discretion: The Inconsistency of Optimal Plans," Journal of Political Economy, 85, 473-491.

NECK, R., AND E. DOCKNER (1987a): "Conflict and Cooperation in a Model of Stabilization Policies: A Differential Game Approach," Journal of Economic Dynamics and Control, 11, 153-158.

NECK, R., AND E. DOCKNER (1987b): "Can the Gains from International Cooperation Be Secured without Policy Coordination?" Mimeo, University of Economics, Vienna, forthcoming in Ricerche Economiche.

SELTEN, R. (1975): "Reexamination of the Perfectness Concept for Equilibrium Points in Extensive Games," International Journal of Game Theory, 4, 25-55.

D. Empirical Models and Input-Output-Analysis

A COMPARISON OF SIMPLE BUSINESS CYCLE MODELS

Empirical Evidence for the Federal Republic of Germany *)

P. Kuhbier and J. Wolters
FU Berlin

1. INTRODUCTION

Investigating business cycles has a long tradition. The main interest focuses on two subjects: On the one hand, one looks for instruments apt to smooth business cycles (cyclical economic policy), on the other hand, the analysis and modeling of business cycles shall serve a better and earlier forecast of boom and depression periods. A lot of research activities concentrated especially on the latter aspect for quite some time in America in the National Bureau of Economic Research aiming at the discovery of leading and lagging business cycle indicators (e.g. Moore (1961), Burns/Mitchell (1967), Moore/ Shiskin (1967), Fels/Hinshaw (1968)). Such investigations have been resumed in the last two decades with growing intensity. Examples are the efforts made in order to develop a concept of a national budget that has no effect on the cyclical trend, and the search for business cycle indicators such as they have been proposed by various research institutes and the board of economic advisors.

In the most recent time, the analysis of structural changes and crises has pushed the interest in business cycles somewhat in the background. The discussion on business cycles itself concentrated on more basic questions such as proof of existence or ways of verification of business cycles. In fact, even nowadays causes of business cycles, their unequivocal describability as well as their existence are not undisputed. For this there may be several reasons: (1) Rarely, the economic process passes for a sufficiently long time span (20 - 30 years) without being influenced by structural disturbances which intermix with the cyclical components so that an analytic seperation is hardly or not at all possible. (2) Analogous problems arise because of the existence of seasonal fluctuations (which cannot be eliminated from the beginning, because one is forced to employ quarterly data in order to have sufficiently many degrees of freedom at one's disposal) as well as of trend components and disturbance terms. (3) Finally, how

---

*) The results reported in this paper are part of our work in the research project "Konjunkturstabilisierung" sponsored by the Deutsche Forschungsgemeinschaft, Bonn - Bad Godesberg. Grant number Wo 335/1-2. We like to thank Th. Pfoch for computational support.

good and to what an extent business cycles can be detected, largely depends on the analytic tools employed and/or on the types of models underlying the investigation.[1])

In the present article, we present a selection of results obtained in the above mentioned research project, which do not relate to the first subject of interest "controlling the economy" mentioned above, but to the latter subject "analysis and forecast of business cycles" or more generally speaking "analysis and forecast of cyclical and long run behaviour of macroeconomic variables".

In order to take to a certain extent into account the diversity of analytic tools and of possible ways to model the economic process, we investigated several approaches simultaneously: Two time series analysis approaches (one vectorautoregressive approach, one ARIMA-approach) and two econometric models (one model of the St. Louis equation type, one multiplier-accelerator model). Sample period is 1970.1 to 1984.4.

All models have been analyzed with respect to their forecast abilities (dynamic ex-post simulations for the time span 1970.1 until 1984.4, on the one hand, and ex-ante forecasts for the eight quarters 1984.4 until 1986.3, on the other hand). In order to draw appropriate comparisons with regard to the forecast abilities between time series analysis models and econometric models, all exogenous variables in the econometric models have also been modeled and forecasted by means of a univariate autoregressive-moving average processes so that all forecasts are "genuine" forecasts.

Moreover, all models have been investigated with respect to their dynamic characteristics (analysis of multipliers, eigenvalues, and power spectra) in order to reach a detailed insight into their cyclical and long run behaviour. The results presented here are less to be viewed as final results, but rather are to be valued as a contribution to the over-all discussion in this field.

In the following sections we will at first characterize the various models employed, then discuss their ability to forecast the economic process, and finally describe their dynamic properties. The article concludes with a summary.

2. MODELS

A difficult task in econometric modeling is the specification of a model. There are at least two different approaches which are widely used: On the one hand, econometric modeling relies on structural information obtained from economic theory and is tested against the observed data. On the other hand, time series analysis is employed to discover empirical regularities inherent in the observations.[2]) The use of time series models is mainly motivated by the hope to find an approximation to the data generating process. Here, we contrast it with traditional econometric models apt to generate business cycles.

2.1 ARIMA-Models

A simple way of describing a data generating process is to estimate a univariate time series model of the integrated autoregressive-moving average (ARIMA) type. For the variables which we are mainly interested in (YN, Y, PY), we used the traditional Box-Jenkins (1976) approach. Non-stationarities in the long run component and seasonality were removed by applying first and fourth order differences. The estimated model as well as some test statistics[3]) are shown below[4]).

YN ≙ Gross National Product in Current Prices

$$\underset{(2.0)}{(1 + 0.23L} + \underset{(2.0)}{0.22L^2})\Delta\Delta_4 YN_t = (1 - \underset{(2.9)}{0.41L^4} - \underset{(4.0)}{0.60L^5})u_t$$

$$SE = 3.82 \qquad Q = 8.30 \qquad \xi^*(7) = 2.45$$

Y ≙ Gross National Product in 1976-Prices

$$(1 + \underset{(1.3)}{0.16L})\Delta\Delta_4 Y_t = (1 - \underset{(3.7)}{0.52L^4} - \underset{(3.2)}{0.49L^5})u_t$$

$$SE = 3.86 \qquad Q = 7.79 \qquad \xi^*(7) = 3.08$$

PY ≙ Price Index of Gross National Product (1976 ≙ 100)

$$\Delta\Delta_4 PY_t = (1 - \underset{(3.4)}{0.46L^5})u_t$$

$$SE = 0.68 \qquad Q = 6.71 \qquad \xi^*(7) = 5.74$$

2.2 Vectorautoregressive Models

Estimation of a multivariate data generating process can be carried out by employing vectorautoregressive models as suggested by Sims (1980). The basic model contains five macroeconomic variables: Y ≙ Gross national product in 1976-prices (bill. DM) and YN ≙ Gross national product in current prices (bill. DM), resp.; PY ≙ Price index of gross national product (1976 ≙ 100); GAN ≙ Total government expenditures in current prices (bill. DM); M1N ≙ Money (currency and demand deposits) in current prices (bill. DM); PIM ≙ Price index of imports (1976 ≙ 100). The model has been estimated in two versions

VAR1: {Y, PY, GAN, M1N, PIM}
VAR2: {YN, PY, GAN, M1N, PIM}

by means of the following equations (S1, S2 and S3 are seasonal dummies):

$$z_{it} = a_0 + a_1 S1 + a_2 S2 + a_3 S3 + \sum_{j=1}^{5} \sum_{k=1}^{4} b_{jk} z_{jt-k} + u_{it}$$

$$(i = 1,\ldots,5)$$

i.e. each variable has been regressed on its own lagged values and on the lagged values of the four other variables (up to four lags have been included).

In the estimated equations we employed the levels of the variables, since we found that they were cointegrated. But we did not impose any restrictions with respect to the trend behaviour or the short and long run dynamics, i.e. parameterizing error correction mechanisms were not taken into account (see e.g. Engle/Granger (1987)).

The causal structure (in the sense of Granger-causality) is shown in the following table 1.

Table 1: Granger-Causality, F-Statistics, VAR1- and VAR2-Model

Lagged	Variables	Dependent Variables					
		Y	YN	PY	GAN	M1N	PIM
Y	VAR1	-	-	6.02**	3.38*	1.00	1.43
YN	VAR2	-	-	4.49**	2.50	1.02	0.95
PY	VAR1	0.32	-	-	0.38	1.99	0.91
	VAR2	-	1.88	-	0.37	1.06	0.50
GAN	VAR1	1.40	-	5.65**	-	0.76	1.80
	VAR2	-	2.03	4.45**	-	0.33	1.17
M1N	VAR1	2.52	-	3.80*	2.15	-	1.68
	VAR2	-	3.78*	3.36*	1.58	-	1.10
PIM	VAR1	2.98*	-	4.46**	2.65*	0.18	-
	VAR2	-	2.16	1.59	5.44**	0.36	-

It contains the values of the F-statistics ("*" ≙ significant for $\alpha = 0{,}05$; "**" ≙ significant for $\alpha = 0{,}01$), which test the hypothesis that all coefficients belonging to the variable heading row i are insignificant in the equation explaining the variable heading column j. The arrow schemes in figure 1 illustrate the causal structure:

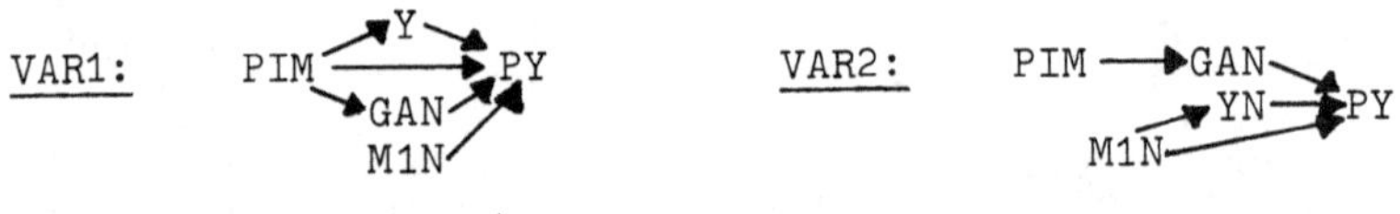

Figure 1: Structure of Granger-Causality, VAR1- and VAR2-Model

In both models the money stock and the index of import prices can be viewed as exogenous. The price index of gross national product is influenced by all other variables. Government expenditures and gross national product take a position in-between these two extremes.

2.3 Multiplier-Accelerator Model

One of the oldest approaches to explaining business cycles is the multiplier-accelerator model to which we try to find an empirical version. The estimated equations (OLS) as well as the corresponding test statistics are shown below.

YV ≙ Disposable Income (YVN * 100/PY)

$$YV_t = \underset{(2.3)}{9.43} - \underset{(7.5)}{15.01S1} - \underset{(11.5)}{16.02S2} - \underset{(12.4)}{14.66S3} + \underset{(7.9)}{0.51Y_t} - \underset{(3.7)}{0.33Y_{t-1}} + \underset{(7.8)}{0.72YV_{t-1}} + \hat{u}_t$$

$\bar{R}^2 = 0.99$ SE = 2.44 h = 0.43 Q = 11.42 $\xi^*(7) = 2.84$

KO ≙ Private Consumption in 1976-Prices

$$KO_t = \underset{(7.2)}{18.03} - \underset{(9.1)}{19.76S1} - \underset{(5.2)}{4.64S2} - \underset{(8.4)}{9.30S3} + \underset{(6.7)}{0.41YV_t} + \underset{(2.5)}{0.24KO_{t-1}} + \underset{(5.1)}{0.20M1_{t-1}} + \hat{u}_t$$

$\bar{R}^2 = 0.99$ SE = 1.56 h = 1.75 Q = 11.51 $\xi^*(7) = 2.41$

IP ≙ Private Gross Investment (in plant and equipment, inventory investment included) in 1976-Prices

$$IP_t = -\underset{(4.3)}{22.79} + \underset{(15.4)}{43.73S1} + \underset{(14.0)}{17.75S2} + \underset{(20.9)}{35.23S3} + \underset{(5.3)}{0.38Y_t} - \underset{(5.0)}{0.36Y_{t-1}} + \underset{(10.2)}{0.87IP_{t-1}} + \hat{u}_t - \underset{(1.9)}{0.30\hat{u}_{t-2}}$$

$\bar{R}^2 = 0.94$ SE = 2.67 h = -1.01 Q = 12.42 $\xi^*(7) = 0.73$

IM ≙ Imports in 1976-Prices

$$IM_t = -\underset{(8.3)}{35.17} + \underset{(5.7)}{3.73S1} + \underset{(8.1)}{4.51S2} + \underset{(5.0)}{2.90S3} + \underset{(8.7)}{0.24(Y + IM)_t} + \underset{(3.0)}{0.25IM_{t-1}} + \hat{u}_t + \underset{(3.5)}{0.50\hat{u}_{t-4}}$$

$\bar{R}^2 = 0.99$ SE = 1.49 h = -1.15 Q = 16.77 $\xi^*(7) = 1.73$

Y ≙ Gross National Product in 1976-Prices

$$Y_t = KO_t + IP_t + GA_t + EX_t - IM_t$$

For an explanation of test statistics see footnote 3.

It can be concluded that, compared to traditional standards, the results are very good. Also the dynamic multipliers (Figure 2) show the typical result: If government expenditures are permanently increased by one unit then the cumulated long run multiplier for real gross national product approaches one while the corresponding figure for an increase in the money stock is 0.3.

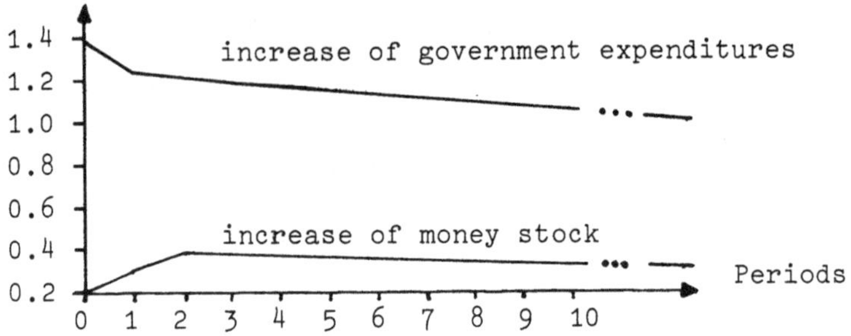

Figure 2: Cumulated Multipliers of the Multiplier-Accelerator Model for Real Gross National Product.

2.4 St. Louis Model

An alternative approach to an explanation of economic fluctuations in money, income, and inflation from a "monetarist's point of view" is the so-called St. Louis model (Andersen/ Carlson (1970)). The model is written in first differences. It basicly contains two stochastic equations for the gross national product (in current prices) and the price-index of gross national product, and, additionally, one identity (the demand pressure identity). In contrast to the usual approach, we did not apply polynomial distributed lags to describe adjustment processes and expectations, but rational lag distributions. This has the advantage that the residuals much better approximate white noise processes. OLS-estimates and values of some test statistics are shown below.

YN ≙ Total Spendings (Gross National Product in Current Prices)

$$\Delta YN_t = \underset{(1.6)}{4.56} - \underset{(1.0)}{3.36S1} + \underset{(0.5)}{1.91S2} + \underset{(1.9)}{7.20S3} - \underset{(2.8)}{0.27\Delta YN_{t-1}} - \underset{(3.5)}{0.31\Delta YN_{t-2}}$$

$$+ \underset{(2.4)}{0.25\Delta M1N_{t-3}} + \underset{(2.4)}{0.27\Delta M1N_{t-4}} + \underset{(4.5)}{0.77\Delta GAN_t} - \underset{(4.5)}{0.75\Delta GAN_{t-3}} + \hat{u}_t$$

$$\bar{R}^2 = 0.97 \quad SE = 2.99 \quad Q = 18.45 \quad \xi^*(6) = 1.84$$

PY ≙ Price Index of Gross National Product (1976 ≙ 100)

$$\Delta PY_t = \underset{(6.4)}{3.27} - \underset{(1.7)}{1.30S1} + \underset{(3.9)}{2.91S2} - \underset{(3.3)}{2.47S3} - \underset{(3.9)}{0.32\Delta PY_{t-1}} - \underset{(6.0)}{0.58\Delta PY_{t-2}}$$

$$+ \underset{(3.9)}{0.06D_t} - \underset{(5.7)}{0.07D_{t-3}} + \underset{(4.8)}{0.08D_{t-4}} + \underset{(2.0)}{0.09\Delta PIM_{t-4}} + \hat{u}_t + \underset{(2.9)}{0.57\hat{u}_{t-2}}$$

$$\bar{R}^2 = 0.96 \quad SE = 0.65 \quad Q = 19.60 \quad \xi^*(4) = 4.20$$

D ≙ Demand Pressure (Identity)

$D_t = \Delta YN_t - (YP - Y_{-1})_t$

For an explanation of test statistics see footnote 3.

The dynamic multipliers (figure 3) show the typical results for such models:

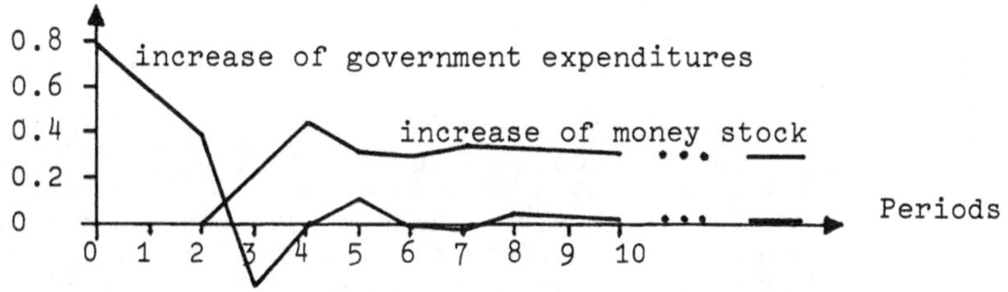

Figure 3: Cumulated Multipliers of the St. Louis Model for Gross National Product in Current Prices.

The cumulated long run multipliers of a permanent increase in government expenditure and in money stock with respect to gross national product in current prices (this time) are zero and 0.3, respectively [i.e. the latter multiplier is similar to the one in the multiplier-accelerator model].

2.5 Generating Processes for the Exogenous Variables

For all exogenous variables, an autoregressive-moving average relationship with seasonal dummies has been estimated. This has been done because we did not want to employ the actual values of the exogenous variables in our dynamic ex-post forecasts. Otherwise, there would have been no fair comparison between forecasts by means of econometric models and vector-autoregressive models. The results of the estimated generating processes for the exogenous variables are shown in table 2 at the end of the article.

3. FORECAST PERFORMANCE OF THE MODELS

We compare the forecasts obtained from all models for Y, YN and PY as far as possible. There are two types of forecasts: ex-post forecasts produced by means of dynamic simulations and ex-ante forecasts. The models have been estimated for the period 1970.1 - 1984.4. Hence, this is also the period for ex-post forecasts. Ex-ante forecasts can be obtained for the period 1984.4 - 1986.3. All forecasts are evaluated by means of their mean absolut percentage forecast errors and shown in table 3.

Table 3: Mean Absolute Percentage Forecast Errors.
Ex-Post Forecasts 1970.1 - 1984.4
Ex-Ante Forecasts 1984.4 - 1986.3

Variables / Models	Y		YN		PY	
	Ex-Post	Ex-Ante	Ex-Post	Ex-Ante	Ex-Post	Ex-Ante
ARIMA	22.82	1.54	2.00	0.98	8.31	0.97
VAR 1	1.53	1.42	-	-	0.78	0.81
VAR 2	-	-	1.20	1.04	0.74	0.91
Mult.-Acc.	2.24	0.74	-	-	-	-
St. Louis	1.82	1.85	1.69	1.51	1.21	0.46

The ex-ante forecast performance of the various models are very much alike. Minor differences are the following ones: (1) The St. Louis model is best in forecasting PY, but worst in forecasting the other variables. (2) The multiplier-accelerator model is best in forecasting Y (forecasts for PY and YN cannot be obtained from this model). (3) YN is best forecasted by a time series analysis model (the ARIMA-model being slightly better than the VAR2-model). See also figures 4 - 6. (All subsequent figures are at the end of the article.)

Also the ex-post forecasts are very similar to one another with one exception: The ARIMA-model does not appropriately forecast Y and PY, although their product YN is completely in line with the other ex-post forecasts[5]). Without taking the ARIMA-model into further consideration, the almost negligible differences between the ex-post forecasts of the other models are as follows: (1) The St. Louis model is, this time, best in no case and again worst in two cases (YN, PY). (2) The VAR-models are best in all cases. (3) The multiplier-accelerator model is worst in forecasting Y. But it may be emphasized once more that all differences are of not much importance. Figures 7 - 9 show typical results for Y, YN, and PY, respectively.

4. CYCLIC AND LONG RUN BEHAVIOUR OF THE MODELS

In order to get an idea with respect to stability and transient paths of the models, we computed all eigenvalues. Moreover, we derived the power spectra of the variables from the estimated models. This enables us to evaluate the cyclical properties created by the innovations of the models which is in line with the statement of Frisch (1933) that business cycles in stable linear dynamic economic models are created only by stochastic shocks superposed on the model solution.

Without loss of generality, a linear dynamic econometric model can be written as a system of first-order stochastic difference equations [6])

$$Ay_t = By_{t-1} + D(L)u_t \qquad (1)$$

with y being an m×1 vector of endogenous and stochastic exogenous variables, u being an m×1 vector of white noise residuals (the components of non-stochastic equations are zero), A and B being m×m matrices and D(L) a diagonal matrix with finite polynomials in the lag operator L as main diagonal elements:

$$d_j(L) = 1 + d_{j1}L + \dots + d_{jk_j}L^{k_j} \qquad j = 1, \dots, m.$$

This operator allows to treat moving average residuals of order k_j, j = 1, ..., m. Clearly, if there is no moving average process in the j-th equation, then $k_j = 0$. The m×m spectral matrix F(λ) of equation (1) can be written as

$$F(\lambda) = \frac{1}{2\pi}(A-Be^{-i\lambda})^{-1}D(e^{-i\lambda})\Sigma D'(e^{+i\lambda})((A-Be^{+i\lambda})')^{-1} \qquad (2)$$

$$i^2 = -1,\ 0 \leq \lambda \leq \pi,$$

Σ being the m×m contemporaneous variance-covariance matrix of the white noise residuals u_t. The main diagonal elements of (2) contain the power spectra of endogenous and stochastic exogenous variables of the model[7]). The off-diagonal elements correspond to the cross-spectra of the various variables. The power spectra describe the cyclical properties of the deviations of the variables from their deterministic components such as constant mean, seasonal dummies, or time trends.

In the frequency domain representation it is assumed "that the fluctuations of the underlying processes are produced by a large number of elementary cycles with different frequencies. The power spectrum gives the average relative contribution made by a small band of these elementary cycles to the variance of the process. If a cycle is very important for the mean-square variation of a process the power spectrum will exhibit a peak at the corresponding frequency band. The higher and narrower such a peak is, the more important and the more regular is the cycle. If the process is white noise, then no particular component dominates, and the spectral function is a constant over the frequency range." (Wolters, 1980, pages 14-15)

4.1 ARIMA-Models

The power spectra of the ARIMA-models refer to the first and fourth differences of the variables in question, i.e. $\Delta\Delta_4 Y$, $\Delta\Delta_4 YN$, and $\Delta\Delta_4 PY$. There is no indication for a long run component left and seasonal components seem to be almost completely eliminated as indicated by the trough near π/2. On the other hand, the power spectra of all three variables have maxima near frequencies 9π/40 and 5π/8 which correspond to periods of about 9 and 3 quarters, resp.. These periods do not point to the existence of a business cycle, but rather represent multiples of the seasonal component. Figures 10 - 12 show the power spectra of the ARIMA-models.

4.2 Vectorautoregressive Models

Both models (see table 4) have among their roots four roots which are approximately equal to: ± 1 and ± i, so that, after applying the differencing operator ($1-L^4$), their time paths would be stationary.

Table 4: Non-zero Eigenvalues of the VAR1-Model and VAR2-Model

VAR1-Model				VAR2-Model			
Real Part	Imaginary Part	Modulus	Period in quarters	Real Part	Imaginary Part	Modulus	Period in quarters
0.99		0.99		0.99		0.99	
-0.01	±0.98	0.98	4	-0.01	±0.99	0.99	4
-0.97		0.97		-0.97		0.97	
0.92	±0.18	0.93	32	0.88	±0.17	0.90	33
0.86		0.86		0.46	±0.66	0.81	7
-0.26	±0.77	0.81	3	-0.79	±0.14	0.80	2
-0.76	±0.11	0.77	2	0.78		0.78	
0.64	±0.42	0.76	11	-0.27	±0.73	0.78	3
0.45	±0.60	0.75	7	0.72	±0.19	0.74	24
0.04	±0.63	0.63	4	0.02	±0.73	0.73	4
0.54		0.54		0.50		0.50	
-0.44		0.44		-0.49		0.49	
-0.15		0.14		-0.02		0.02	

Moreover, this result shows that not all seasonal fluctuations in the data are captured by the included seasonal dummies. Two complex roots with a periodicity of four quarters together with periodicities of three and two quarters again point to seasonal fluctuations. This is further confirmed by steep peaks at $\pi/2$ and π in the power spectrum of ΔPY, ΔY and ΔYN, resp. (see figures 10-12). There are further roots with a periodicity of 7, 11, 24, 32 and 33 quarters. Although they could be multiples of the seasonal component, they may also point to a component of considerably lower frequency. Examining the power spectra of the models, one finds a peak at a frequency of $\pi/10$ which corresponds to a periodicity of 20 quarters.

4.3 Multiplier-Accelerator Model

Eigenvalues of the multiplier-accelerator model with fixed exogenous variables are shown in table 5.

Table 5: Non-zero Eigenvalues of the Multiplier-Accelerator Model

Real Part	Imaginary Part	Modulus	Period in Quarters
0.803		0.803	
0.780		0.780	
0.243	±0.073	0.254	22

They indicate that the homogeneous part of the model is stable. Trend components are caught up by the exogenous variables. There is one cyclic component of about 22 quarters. The power spectrum for ΔY (figure 10) gives no indication for a cyclic component of the corresponding frequency[8]). It rather exhibits periods of four and two quarters. Hence, it can be concluded that the model shows no clear evidence for a business cycle component.

4.4 St. Louis Model

This model, too, has stable eigenvalues (cf. table 6), taking the exogenous variables as fixed.

Table 6: Non-zero Eigenvalues of the Linearized St. Louis Model

Real Part	Imaginary Part	Modulus	Period in Quarters
- 0.241	±0.747	0.785	3
0.717		0.717	
- 0.135	±0.540	0.557	4
0.192		0.192	

Both complex eigenvalues indicate a periodicity of about three to four quarters. Also the power spectrum for ΔYN and ΔPY, the effect of the stochastic exogenous variables included (cf. figures 11 and 12), gives no evidence for longer cyclic components; it rather confirms periodicities of four and two quarters which are due to seasonal disturbances.

5. SUMMARY AND CONCLUSIONS

For the sake of comparing the forecast performance of economic models and their ability to capture cyclical movements in some major economic variables, we examined univariate ARIMA-models, low order vectorautoregressive models, and two simple econometric models of the multiplier-accelerator type and of the St. Louis type, i.e. on the one hand, we tried to model the time paths of the variables under consideration without any theoretical economic background, and, on the other hand, we used two different economic approaches, the Keynesian and the monetarist's view. With respect to the forecast performance of the models, we did not find major discrepancies, neither with respect to ex-post forecasts (with two exceptions) nor with respect to ex-ante forecasts. Furthermore, all models do reasonably well in describing past and future time paths of gross national product in constant and current prices as well as of the price index of gross national product. There are, however, crucial differences in their stochastic dynamic behaviour. Cyclic fluctuations are primarily due to seasonality which is not easily be dealt with, neither by taking fourth order differences (ARIMA-models) nor by means of constant seasonal dummies (all other models). Moreover, it is to be supposed that the dynamic structure of the investigated variables is more complex than it is usually modeled in a typical econometric approach by employing Koyck-lags.

From the behaviour of the models investigated, it may be concluded that the long run components look pretty much like random walks. Hence, no distinct business cycles can be found in the power spectra (except perhaps for the vectorautoregressive models), although the realizations of one or the other variable do not seem to exclude the possibility of their existence. These results suggest that shocks may rather have a permanent than a transitory effect on the economy.

FOOTNOTES

1) For a survey on theories of business cycles see Zarnowitz (1985).

2) For a survey on econometric and time series modeling see e.g. Wolters (1987).

3) Numbers in brackets are the absolute values of the t-statistics of the estimated parameters. SE is the standard error of regression. $\bar{R}^2$ denotes the determination coefficient corrected for degrees of freedom and h the value of the Durbin h-statistic (calculated in connection with some of the following models, e.g. the multiplier-accelerator model, see below). Q is the value of the Box-Pierce-Portmanteau statistic computed for the first 16 autocorrelation coefficients. $\xi^*(k)$ is Harvey's post sample predictive test statistic.

4) Estimation and simulation of all models has been carried out by means of Micro-TSP 5.0 on an IBM-PC-AT. L indicates the lag-operator, defined as $L^k x_t = x_{t-k}$, $k = \ldots -1, 0, 1, \ldots$ Data are taken from DIW, Sozialprodukt und Einkommenskreislauf, Berlin, Okt. 1986. M1N is taken from Monatsberichte der Deutschen Bundesbank. Potential output YP has been calculated by the authors.

5) Y is increasingly overpredicted and PY increasingly underpredicted. This may be a problem of initial values. But we did not pursue this any further.

6) For the following see e.g. Wolters (1980a).

7) The eigenvalues and the power spectra are computed by the program TRANSF (available on request). The elements of A, B and D are calculated by means of the estimated model parameters and the elements of Σ are estimated with the help of the estimated residuals, thereby assuming that the residuals of the stochastic exogenous variables and of the endogenous variables are independent. The assumption that the estimated parameters are appropriate substitutes for the true unknown parameters is very often used in analyzing econometric models.

8) Power spectra are computed by taking into account the generating processes of the exogenous variables.

REFERENCES

ANDERSON, L. C. and K. M. CARLSON: "A Monetarist Model for Economic Stabilization", Federal Reserve Bank of St. Louis Review, 52 (1970), 7-21.

BOX, G. E. P. and G. M. JENKINS: Time Series Analysis, Forecasting and Control, San Francisco: Holden Day, 1976, 2nd ed.

BURNS, A. F. and W. C. MITCHELL: Measuring Business Cycles, New York, 1967.

ENGLE, R. F. and C. W. J. GRANGER: "Cointegration and Error Correction: Representation, Estimation and Testing", Econometrica, 55 (1987), 251-276.

FELS, R. and C. E. HINSHAW: Forecasting and Recognizing Business Cycle Turning Points, New York, 1968.

FRISCH, R. (1933): "Propagation Problems and Impulse Problems in Dynamic Economics", in: Economic Essays in Honour of Gustav Cassel, London: New Impression, 1967, 171-205.

MOORE, G. H. and J. SHISKIN: Indicators of Business Expansions and Contractions, New York, 1967.

MOORE, G. H., Ed.: Business Cycle Indicators, New York, 1961.

SIMS, C. A.: "Macroeconomics and Reality", Econometrica, 48 (1980), 1-48.

WOLTERS, J.: "Business Cycle Stabilization Policies in a Small Econometric Model of the FRG", European Economic Review, 14 (1980), 9-43.

WOLTERS, J.: Stochastic Dynamic Properties of Linear Econometric Models, Berlin, Heidelberg, New York: Springer Verlag, 1980a.

WOLTERS, J.: "Ökonometrische Modelle bei Zeitreihendaten versus multivariate Zeitreihenmodelle - eine Übersicht", Statistische Hefte, 28 (1987), 1-25.

ZARNOWITZ, V.: "Recent Work on Business Cycles in Historical Perspective", Journal of Economic Literature, 23 (1985), 523-580.

Table 2: Generating Processes for Exogenous Variables. OLS Estimates 1970.1 - 1984.4

$EX \;\widehat{=}\;$ Exports in 1976-Prices

$$\Delta EX_t = \underset{(8.9)}{5.94} - \underset{(9.5)}{8.93}S1 - \underset{(5.0)}{4.69}S2 - \underset{(6.2)}{5.80}S3 + \hat{u}_t$$

$\bar{R}^2 = 0.60 \quad SE = 2.57 \quad DW = 1.95 \quad Q = 10.73$

$GA \;\widehat{=}\;$ Government Expenditures in 1976-Prices

$$\Delta GA_t = \underset{(26.5)}{5.15} - \underset{(25.5)}{12.40}S1 - \underset{(3.4)}{2.91}S2 - \underset{(8.6)}{3.76}S3 - \underset{(2.4)}{0.23}\Delta GA_{t-1} + \hat{u}_t + \underset{(7.5)}{0.93}\hat{u}_{t-4}$$

$\bar{R}^2 = 0.98 \quad SE = 0.73 \quad h = 0.37 \quad Q = 13.88$

$GAN \;\widehat{=}\;$ Government Expenditures in Current Prices

$$\Delta GAN_t = \underset{(23.4)}{16.77} - \underset{(24.0)}{25.02}S1 - \underset{(11.7)}{15.94}S2 - \underset{(10.3)}{17.05}S3 - \underset{(7.7)}{0.57}\Delta GAN_{t-1} - \underset{(4.3)}{0.32}\Delta GAN_{t-2}$$

$$+ \hat{u}_t + \underset{(7.5)}{1.09}\hat{u}_{t-4} + \underset{(4.3)}{0.55}\hat{u}_{t-8}$$

$\bar{R}^2 = 0.97 \quad SE = 1.96 \quad Q = 16.93$

$M1 \;\widehat{=}\;$ Money Stock (M1N × 100/PY)

$$\Delta M1_t = \underset{(6.6)}{6.33} - \underset{(9.4)}{12.76}S1 - \underset{(1.6)}{2.15}S2 - \underset{(3.9)}{5.27}S3 + \hat{u}_t - \underset{(2.4)}{0.33}\hat{u}_{t-5}$$

$\bar{R}^2 = 0.63 \quad SE = 3.73 \quad DW = 1.82 \quad Q = 8.83$

$M1N \;\widehat{=}\;$ Money Stock in Current Prices

$$\Delta M1N_t = \underset{(14.5)}{17.21} - \underset{(9.4)}{24.60}S1 - \underset{(5.3)}{11.28}S2 - \underset{(7.2)}{14.87}S3 - \underset{(2.7)}{0.32}\Delta M1N_{t-1} + \hat{u}_t - \underset{(3.5)}{0.55}\hat{u}_{t-5}$$

$\bar{R}^2 = 0.86 \quad SE = 4.57 \quad h = 4.35 \quad Q = 14.88$

$PIM \;\widehat{=}\;$ Price Index of Imports (1976 $\widehat{=}$ 100)

$$\Delta PIM_t = \underset{(2.8)}{0.87} + \underset{(2.7)}{0.33}\Delta PIM_{t-1} + \hat{u}_t$$

$\bar{R}^2 = 0.10 \quad SE = 2.12 \quad h = 0.72 \quad Q = 6.96$

$YP \;\widehat{=}\;$ Potential Gross National Product

$$\Delta^2 YP_t = -\underset{(3.1)}{0.03} - \underset{(2.4)}{0.22}\Delta^2 YP_{t-1} + \hat{u}_t + \underset{(7.0)}{0.91}\hat{u}_{t-4}$$

$\bar{R}^2 = 0.53 \quad SE = 0.06 \quad h = -1.09 \quad Q = 17.13$

$YP\text{-}Y_{-1} \;\widehat{=}\;$ Excess Supply (linearized)

$$(YP\text{-}Y_{-1})_t \;\widehat{=}\; -\underset{(4.6)}{13.29} + \underset{(3.9)}{18.18}S1 + \underset{(14.9)}{31.81}S2 + \underset{(4.3)}{9.01}S3 + \underset{(10.7)}{0.77}(YP\text{-}Y_{-1})_{t-1}$$

$$+ \underset{(4.6)}{0.37}\Delta YN_t + \underset{(2.7)}{1.09}\Delta PY_{t-1} + \hat{u}_t + \underset{(2.2)}{0.39}\hat{u}_{t-1}$$

$\bar{R}^2 = 0.89 \quad SE = 3.57 \quad h = 0.29 \quad Q = 10.50$

For an explanation of test statistics see table 1.

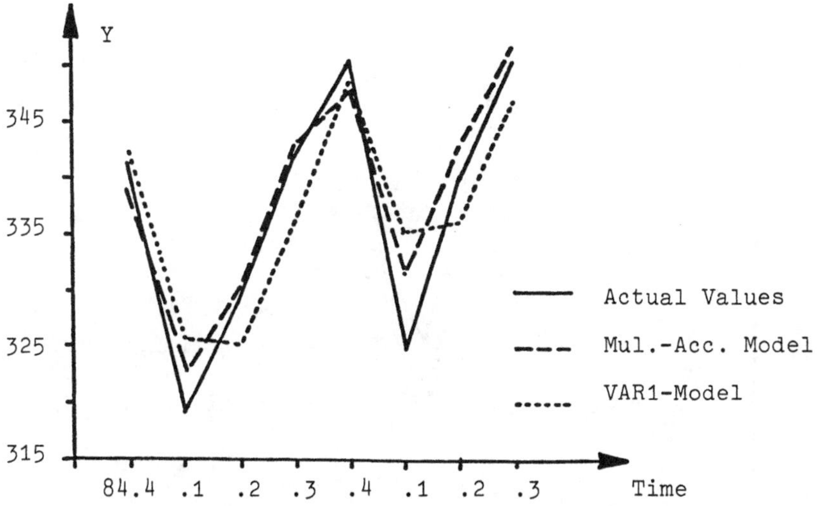

Figure 4: Alternative Forecasts of Gross National Product in Constant Prices

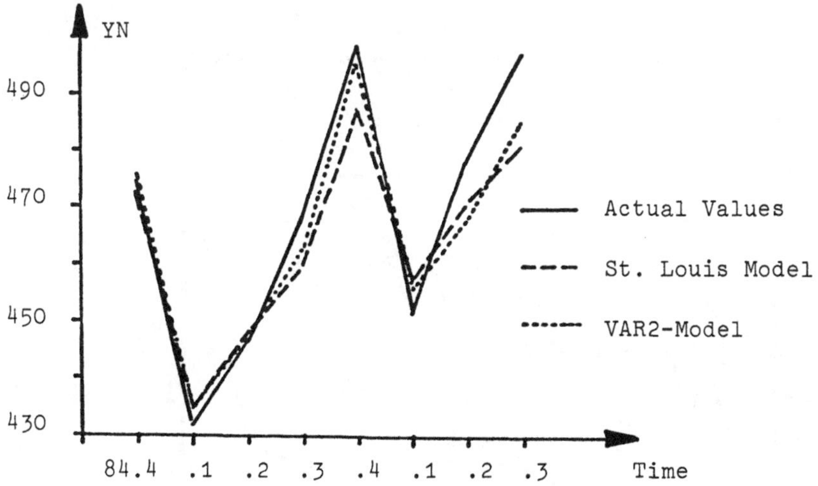

Figure 5: Alternative Forecasts of Gross National Product in Current Prices

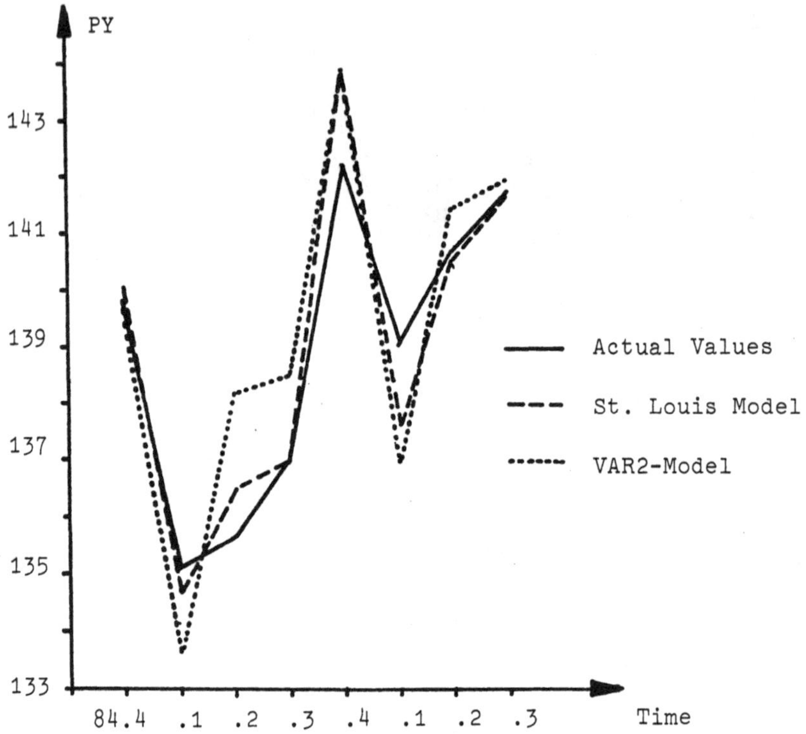

Figure 6: Alternative Forecasts of Price Index of Gross National Product

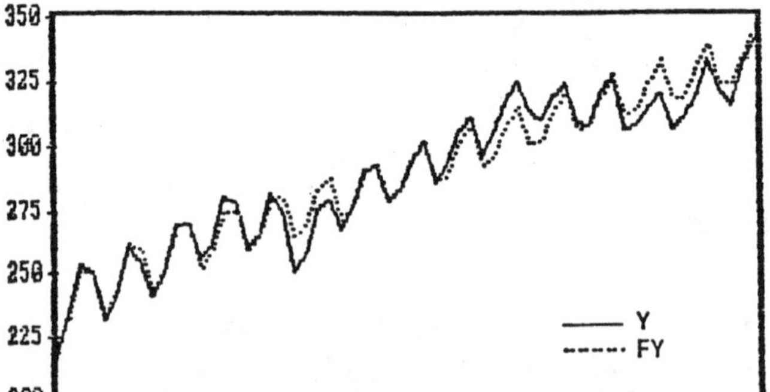

Figure 7: VAR1-Model, Actual and Ex-Post Forecasted Values of Y

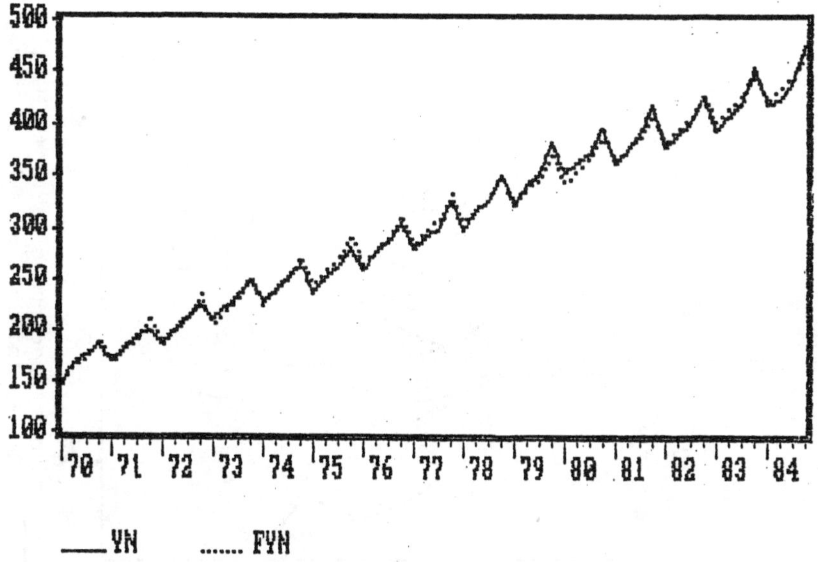

Figure 8: St. Louis Model, Actual and Ex-Post Forecasted Values of YN

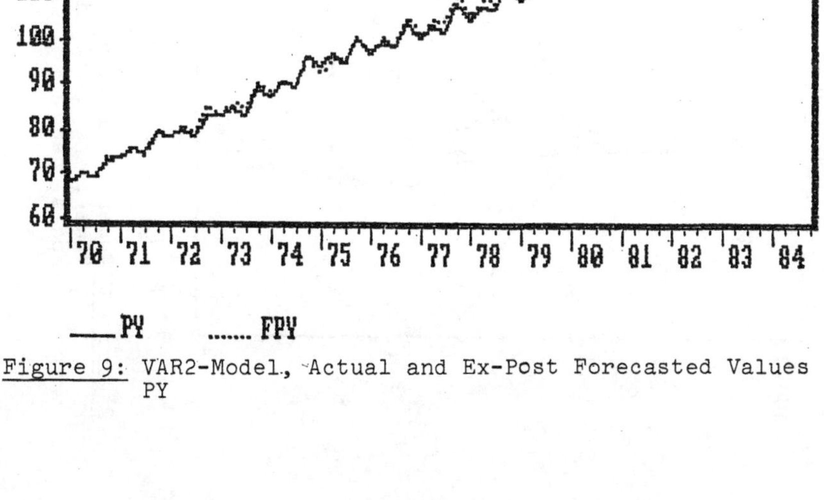

Figure 9: VAR2-Model, Actual and Ex-Post Forecasted Values of PY

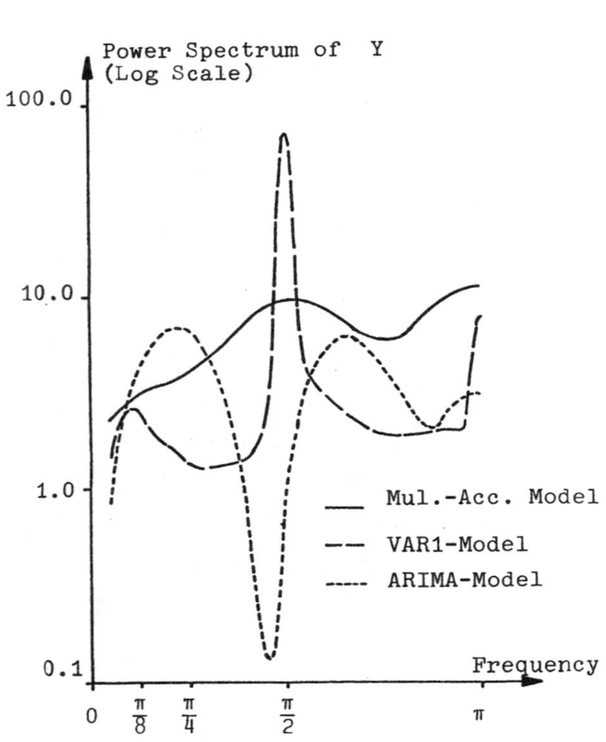

<u>Figure 10:</u> Power Spectrum of ΔY; ARIMA-, VAR1-, and Multiplier-Accelerator Model, resp.

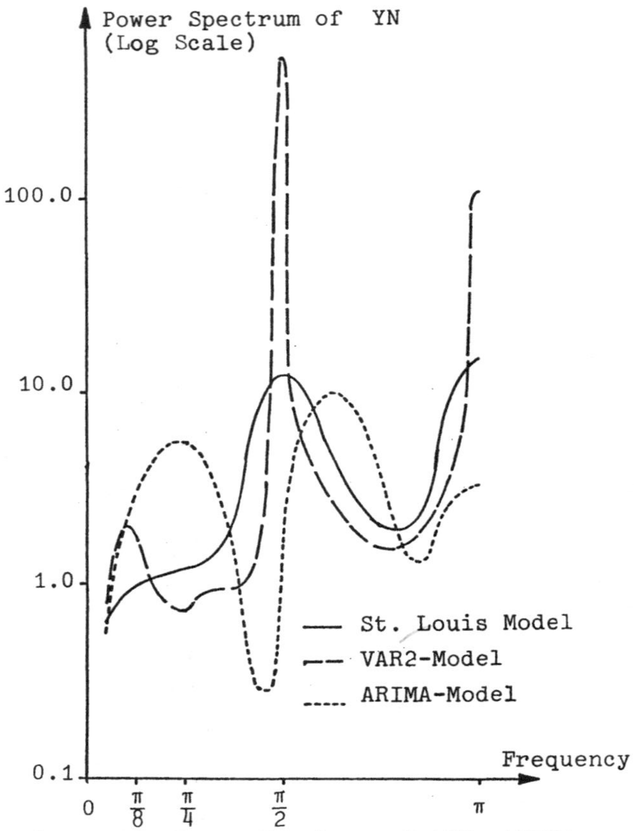

<u>Figure 11:</u> Power Spectrum of ΔYN; ARIMA-, VAR2-, and St. Louis Model, resp.

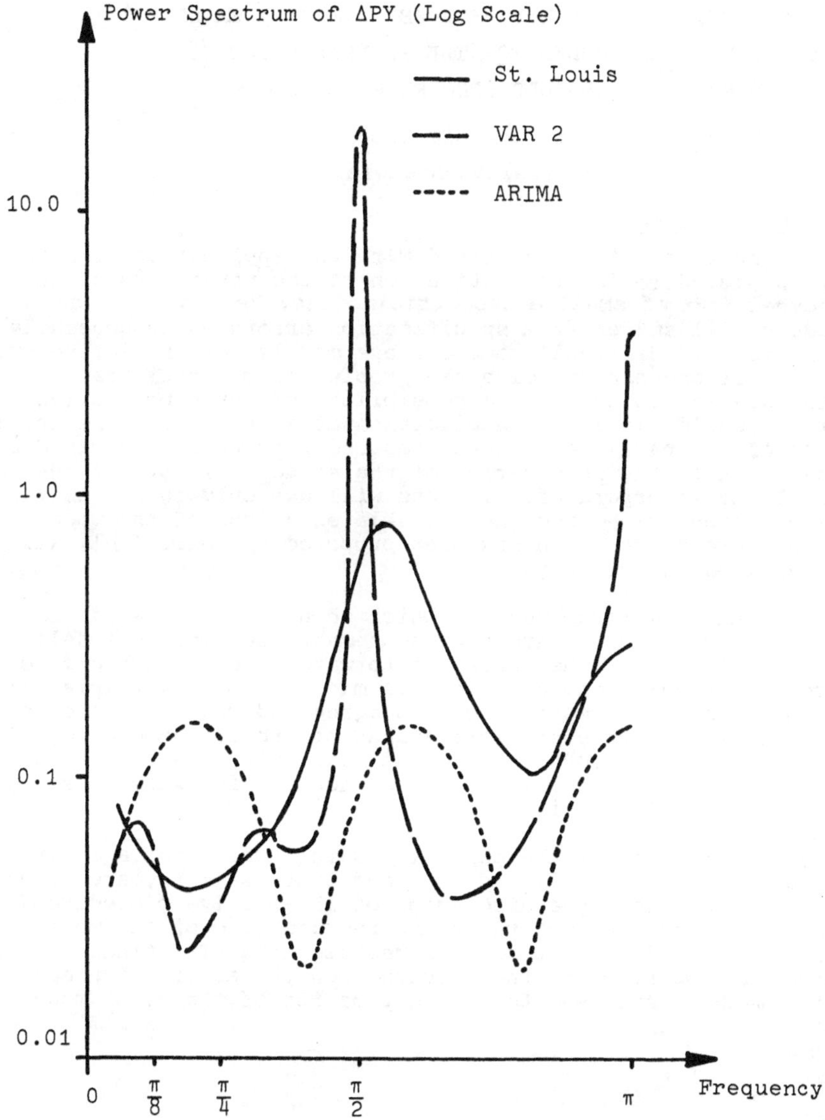

Figure 12: Power Spectrum of ΔPY; ARIMA-; VAR2-; and St. Louis Model, resp.

ON THE ESTIMATION OF THE REDUCED FORM OF A SMALL MACROECONOMETRIC MODEL FOR THE FEDERAL REPUBLIC OF GERMANY WITH POSSIBLE SPECIFICATION ERRORS

by Joachim Frohn
Universität Bielefeld

1. Introduction

This paper is mainly concerned with the application of a specific procedure for the estimation of the parameters of the reduced form of small econometric models. Very often such models will suffer from specification errors which generally will make it difficult to use these models for predictive purposes. If one cannot solve the problem of misspecification, a possible way to reduce the possibility of large predictive errors could lie in the application of a method for the estimation of the parameters of the reduced form where one can decide what part of the information of the structural form of the model should be incorporated, i.e. one will use only those restrictions which arise from the reliable equations of the model. Such a procedure which has been proposed by Court [1], will be applied in this paper.

The model, the parameters of which should be estimated, is a fourteen equations, interdependent macroeconometric model. It is a linear, rather straight forward model, which can be used for predictions of some main macroeconomic variables or as a possible "frame model" for disaggregated models, designed, for instance, for specific branches of the German economy. In order to be able to actualize the model within short periods of time, only such variables were included for which observations are easily available.

The organisation of the paper is as follows: After a short description of the model, the OLS- and TSLS-estimates are given. This is followed by a discussion of alternative procedures for estimating the parameters of the reduced form of the model on the basis of different sets of restrictions. The final section gives a comparison of the results when the alternative estimation methods are used for the estimation of the given model.

2. The model[1)]

It was the intention to specify equations for the most important variables of aggregated demand: private consumption, private investment, changes of inventories, and imports. Furthermore, at least one endogenous variable representing the general conditions on the supply side should be included. For this purpose, the level of utilization of capacity in the German production industry was chosen.

Because of the great importance of unemployment, the number of potential workers EP ("Erwerbspersonen") and the number of unemployed EL ("Erwerbslose") were also included as endogenous variables.

One main requirement during the specification analysis was, that all variables should "fit together"; therefore, for instance, instead of the variable "people officially registered as unemployed" ("Arbeitslose"), the variable "unemployed persons" ("Erwerbslose"), as reported by the German Statistical Bureau (Statistisches Bundesamt), was used in the model. As a consequence, labor productivity was measured in terms of potential workers minus unemployed persons.

The variables of governmental activity (consumption, investment, taxes, subsidies), money supply, and a representative interest rate were included as possible instruments.

For simplicity all equations were assumed to be linear; all variables were measured in real (1980-)terms.

The main criterion for the specification of the model was the predictive performance.

The model was specified and estimated on the basis of quarterly data from 1963,I to 1986,I (= 93 observation points). The predictive performance was checked on the basis of dynamic simulations of each equation, estimated from 1963,I to 1983,IV, thus leaving nine values to be predicted.

The resulting model consists of seven stochastic equations and seven definitions.

All the equations with the OLS- and the respective TSLS-estimates of the parameters, the multiple coefficients of the determination (R^2), the respective values of the Durbin-Watson-Statistics (DW), and the definition of the variables are given

---

1) I am very grateful to cand.rer.pol. R. Schröder for computational help.

below[1]:

Consumption function:

Meth.	Y_t^D	r_t	$C_{t-1}^{priv.}$	R^2	DW
OLS	0.1745 (4.89)[2]	-0.5383 (-2.51)	0.8265 (23.43)	0.99	2.38
TSLS	0.1547 (3.85)	-0.5253 (-2.44)	0.8457 (21.41)	0.99	2.44

Investment function:

Meth.	$\Delta C_t^{priv.}$	Y_{t-1}	$I_{t-1}^{priv.}$	$M1_t$	R^2	DW
OLS	0.3425 (3,60)	0.0895 (1.16)	0.7966 (13.62)	0.0382 (2.66)	0.95	2.01
TSLS	0.4874 (3.66)	0.0704 (0.89)	0.8131 (13.51)	0.0350 (2.38)	0.94	2.05

Inventory function:

Meth.	$\Delta C_{t-1}^{priv.}$	r_t	V_{t-1}	R^2	DW
OLS/ TSLS	0.4426 (3.25)	-0.4930 (-1.69)	0.4539 (4.96)	0.94	2.13

Import function:

Meth.	BSP_t	Imp_{t-1}	R^2	DW
OLS	0.1563 (4.08)	0.6586 (8.02)	0.99	2.89
TSLS	0.2191 (5.03)	0.5249 (5.63)	0.99	2.70

EL-function:

Meth.	Y_{t-1}	$Prod_{t-1}$	$Insolv_t$	Exp_t	R^2	DW
OLS/ TSLS	-13.95 (-3.67)	99269.2 (3.78)	0.7180 (22.04)	-14.19 (-5.16)	0.98	1.28

1) In all equations - with the exception of the equation for unemployed persons - seasonal dummies were used. In all equations an absolute term was included. The respective estimates are not reported here.

2) t-values in brackets

EP-function:

Meth.	$Bevw.1565_t$	EP_{t-1}	R^2	DW
OLS/ TSLS	0.0653 (2.15)	0.9412 (23.63)	0.93	1.95

γ-function:

Meth.	γ_{t-1}	ΔET_t	ΔBSP_t	$I^{priv.}_{t-1}$	R^2	DW
OLS	0.8334 (25.10)	0.0074 (5.92)	0.0532 (2.63)	-0.0387 (-2.42)	0.93	2.19
TSLS	0.8367 (24.47)	0.0017 (4.77)	0.0384 (1.62)	-0.0396 (-2.47)	0.93	2.16

Definitions:

$$BSP_t = C^{priv}_t + C^{st}_t + I^{priv}_t + I^{st}_t + V_t + Exp_t - Imp_t$$

$$Y^D_t = BSP_t - D_t - T^{ind}_t - T^{dirHH}_t + Z_t$$

$$Prod_t = BSP_t/ET_t$$

$$\Delta C^{priv}_t = C^{priv}_t - C^{priv}_{t-1}$$

$$ET_t = EP_t - EL_t$$

$$\Delta ET_t = ET_t - ET_{t-1}$$

$$\Delta BSP_t = BSP_t - BSP_{t-1}$$

The defintions of the variables:

C^{priv}: private consumption, Y^D: disposible income, r: long term interest rate, I^{priv}: private gross investment, γ: utilization of capacity of the German production industry, M1: money supply, V: inventories, BSP: gross national product, Imp: imports, EL: unemployed persons, Prod: productivity, Insolv: number of insolvencies, Exp: exports, Bevw1565: female inhabitants, aged 15 to 65, EP: potential workers, ET: employed persons, C^{st}: governmental consumption, I^{st}: governmental gross investment, D: depericiation, T^{ind}: indirect taxes, T^{dirHH}: direct taxes of households, z: subsidies. Sources of the data: All data were taken from publications of the Deutsches Institut für Wirtschaftsforschung [3], with the exception of γ (Ifo-Institut [4]), M1, r (Deutsche Bundesbank) [2], Insolv, Bevwl1565 (Statistisches Bundesamt [5]).

The following graphs show the fit and the results of the single equation simulation experiments.
The estimation results and the graphs make clear that one has to be aware of the possibility of specification errors at least with respect to the import equation and the equation for unemployed persons: In both cases the values of the Durbin Watson Statistic and the results of the prediction experiments are not satisractory. In two other equation - for investment and inventories - the predictive errors are rather large: Although the general structure of the development of the variables is predicted quite well, in several quarters the predictive errors are larger than 10 %.[1]

As it was not possible to solve these specification problems within the given set of variables and data the question was how such specification errors would effect the predictive performance of the model. To answer this question, R.H. Court's procedure of estimation of the reduced form was adopted; this method will be described briefly in the following section.

3. Estimation of the Reduced Form of the Model

Normally one can choose between two alternative estimation procedures for the reduced form of a model:

1. One can estimate the parameters without imposing the a-priori-restrictions given in a structural form (unrestricted estimation of the reduced form) by ordinary least squares. These OLS-estimates are consistent but suffer from the fact that only a very small part of the information incorporated in the structural form is used for the estimation of the reduced form.

---

1) There is a second specification error present in the model: In the equation explaining the number of unemployed persons, Insolv is used as an explanatory variable. Although this variable obviously should be regarded as an endogenous variable in the context of the model there is no explaining equation for this variable. To solve this problem one would have to find a reliable specification for this variable. We shall return to this problem briefly in the following section.

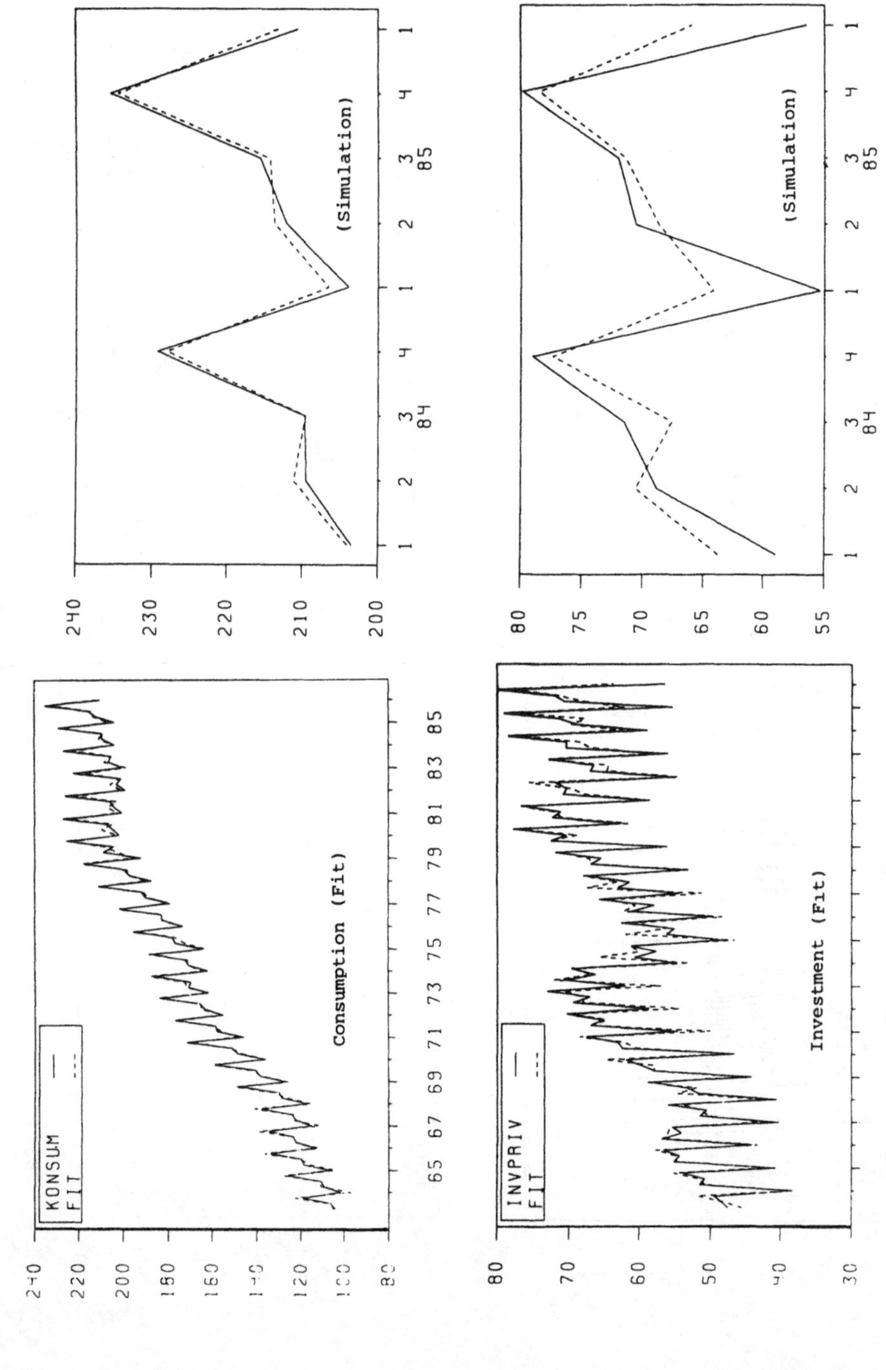
KONSUM
FIT
Consumption (Fit)
(Simulation)
INVPRIV
FIT
Investment (Fit)
(Simulation)

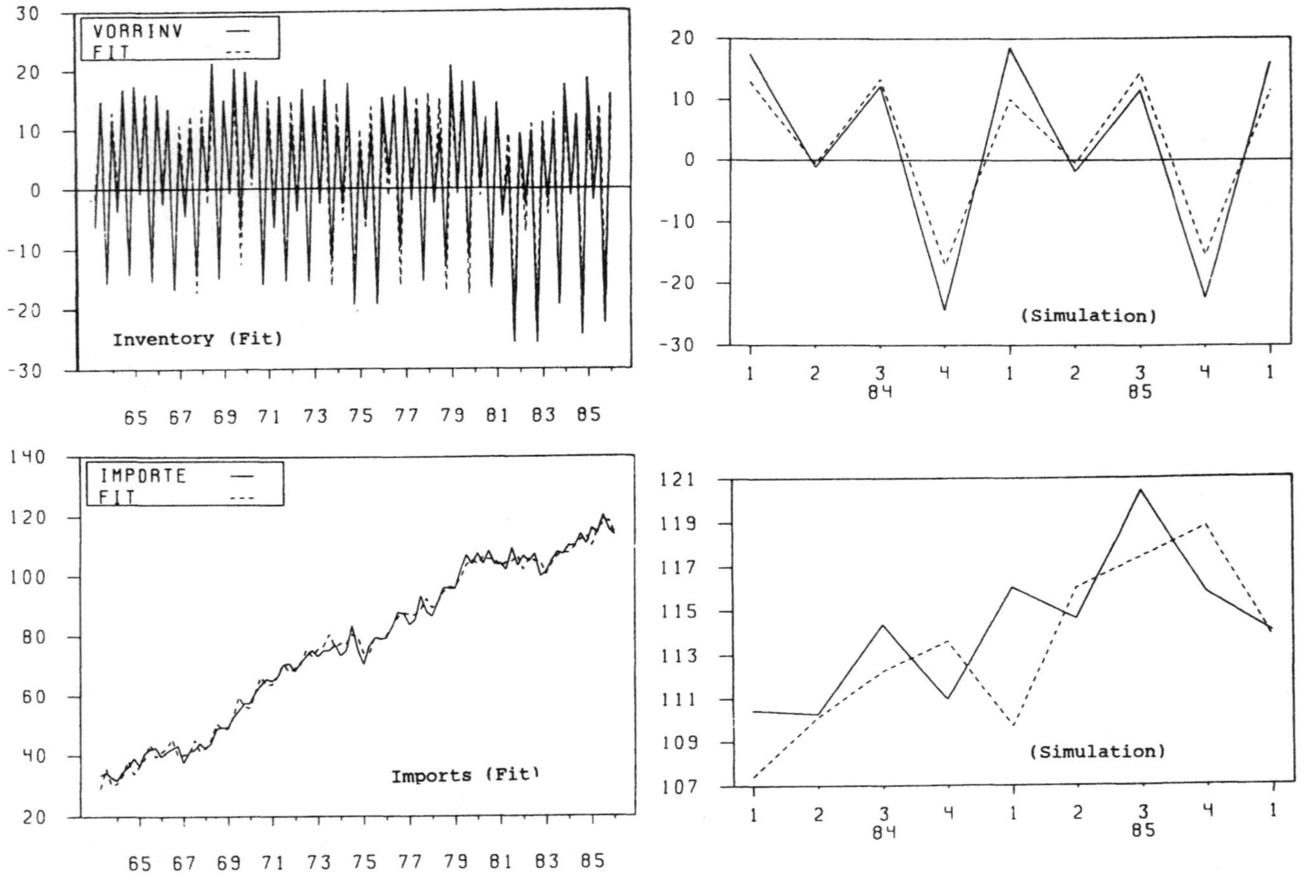
VORRINV
FIT
Inventory (Fit)
30
20
10
0
-10
-20
-30
65 67 69 71 73 75 77 79 81 83 85
20
10
0
-10
-20
-30
(Simulation)
1 2 3 4 1 2 3 4 1
84
85
IMPORTE
FIT
Imports (Fit)
140
120
100
80
60
40
20
65 67 69 71 73 75 77 79 81 83 85
121
119
117
115
113
111
109
107
(Simulation)
1 2 3 4 1 2 3 4 1
84
85

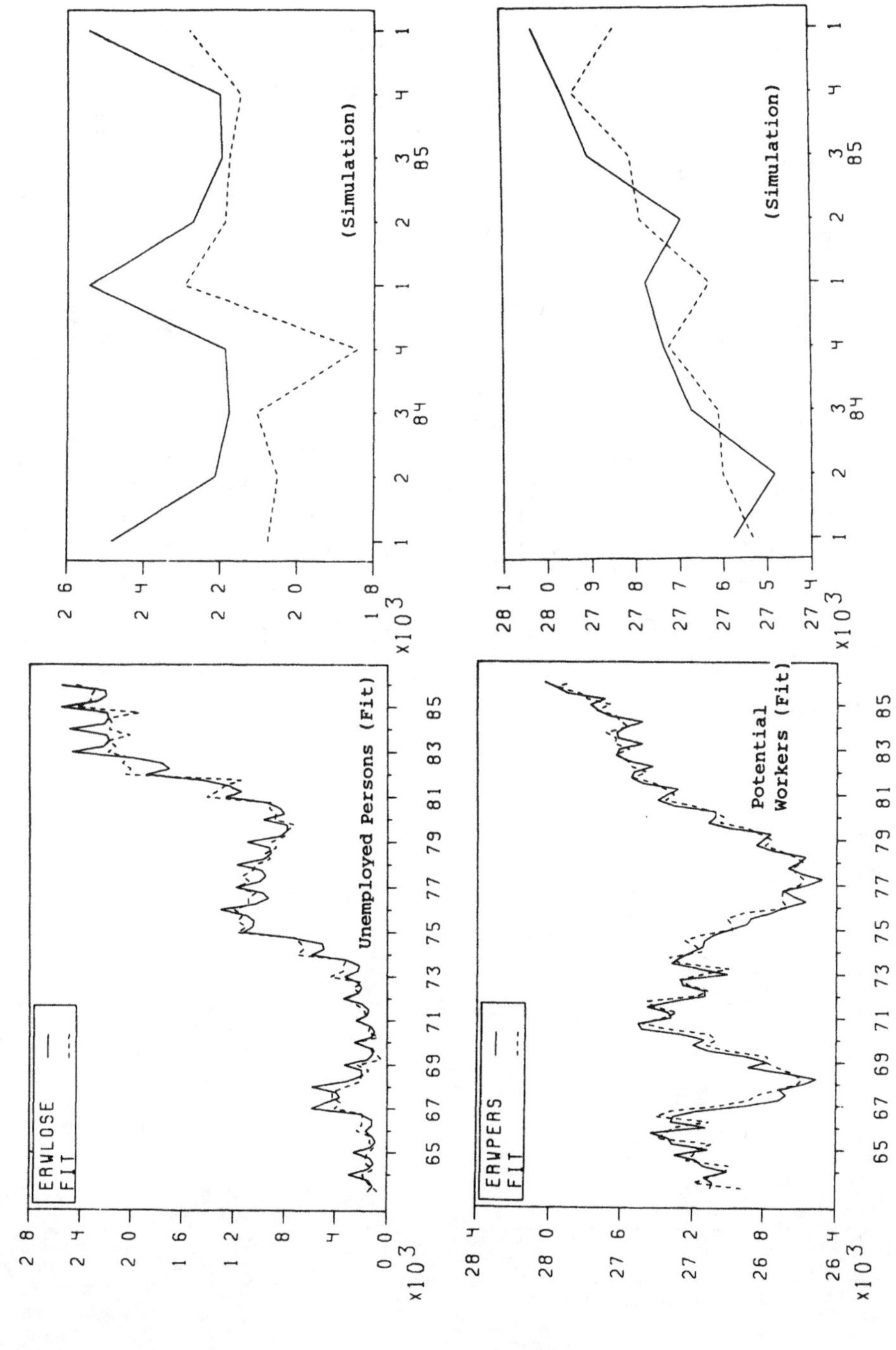
ERWLOSE
FIT
Unemployed Persons (Fit)
(Simulation)
ERWPERS
FIT
Potential Workers (Fit)
(Simulation)
x10 3
65 67 69 71 73 75 77 79 81 83 85
84
85

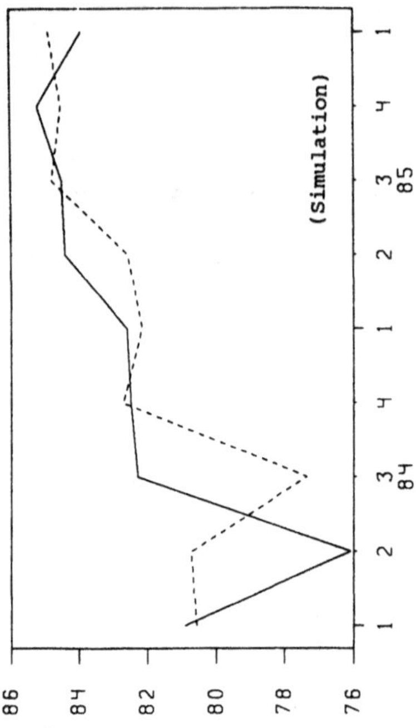
(Simulation)
86
84
82
80
78
76
1 2 3 4 1 2 3 4 1
84
85

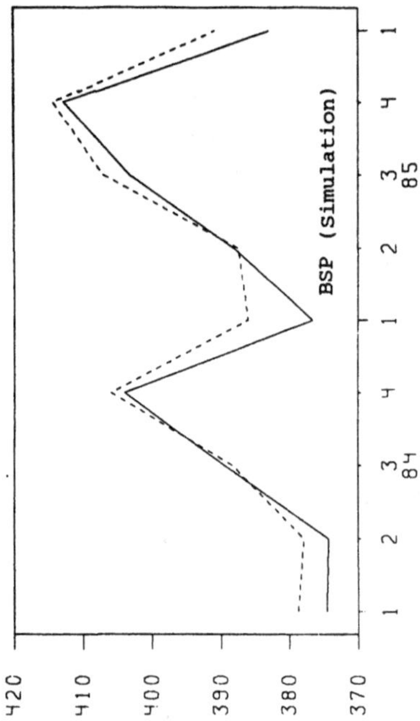
BSP (Simulation)
420
410
400
390
380
370
1 2 3 4 1 2 3 4 1
84
85

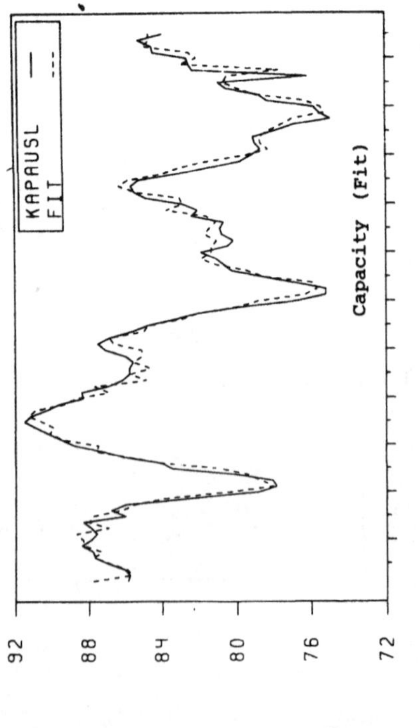
KAPAUSL
FIT
Capacity (Fit)
92
88
84
80
76
72
65 67 69 71 73 75 77 79 81 83 85

2. One can derive estimates for the parameters of the reduced form by taking adequate estimates of the parameters of the structural form, and then calculating $\hat{\Pi}$ according to $\hat{\Pi} = -\hat{B}^{-1}\hat{\Gamma}$ (derived reduced form estimation).

Obviously, the two procedures can be regarded as two extremes: The unrestricted estimation of the reduced form makes use of very little information of the a-priori-information of the model, the derived estimation of the reduced form will take into account all a-priori-information, including possible misspecifications.

R.H. Court now has proposed a procedure for the estimation of the reduced form, which allows to select specific parts of the structural form to be used within the estimation of the reduced form. His approach can briefly be described as follows:

We start from the structural form of a general interdependent model. Each structural equation can be written as follows:

(1) $y_i = Y_i\gamma_i + X_i\beta_i + u_i = Z_i\partial_i + u_i \qquad i = 1,2,\ldots,G$

with: y_i = "explained" endogenous variable of equation i
Y_i = other endogenous variables of the system included in question i
X_i = predetermined variables included in equation i
u_i = error term
γ_i, β_i = vectors of coefficients

The whole system:

(2)
$$\begin{bmatrix} y_1 \\ y_2 \\ \vdots \\ \vdots \end{bmatrix} = \begin{bmatrix} Z_1 & O & \cdots & O \\ O & Z_2 & \cdots & O \\ \vdots & \vdots & & \vdots \\ \vdots & \vdots & & \vdots \end{bmatrix} \begin{bmatrix} \partial_1 \\ \partial_2 \\ \vdots \\ \vdots \end{bmatrix} + \begin{bmatrix} u_1 \\ u_2 \\ \vdots \\ \vdots \end{bmatrix}$$

in obvious matrix notation:

(3) $y = Z\partial + u$

We assume: $E(u) = O$; $E(uu') = \Sigma \otimes I$.

Now, according to an evaluation of the estimates of the structural equations, only the following part of the structural form shall be used for the estimation of the reduced form:

(4) $y_s = Z_s\partial_s + u_s \qquad$ with $E(u_s) = O$; $E(u_s u_s') = \Sigma_s \otimes I$

The reduced form of the whole system is given as[1]:

(5) $$\begin{bmatrix} y_1 \\ y_2 \\ \vdots \end{bmatrix} = \begin{bmatrix} X & O & \dots & O \\ O & X & \dots & O \\ \vdots & \vdots & & \vdots \end{bmatrix} \begin{bmatrix} \Pi_1 \\ \Pi_2 \\ \vdots \end{bmatrix} + \begin{bmatrix} v_1 \\ v_2 \\ \vdots \end{bmatrix}$$

i.e.

(6) $y = (I \otimes X)\Pi$ with $E(v) = O$; $E(vv') = \Omega \otimes I$.

Court now proposes a simultaneous estimation of all equations belonging to the "reduced" structural form (4) and the complete reduced form (6). Therefore the following system should be estimated:

(7) $$\begin{bmatrix} y_s \\ y \end{bmatrix} = \begin{bmatrix} Z_s & O \\ O & I \otimes X \end{bmatrix} \begin{bmatrix} \partial_s \\ \Pi \end{bmatrix} + \begin{bmatrix} u_s \\ v \end{bmatrix}$$

with $$E\{\begin{bmatrix} u_s \\ v \end{bmatrix} \cdot [u_x' v']\} = \begin{bmatrix} \Sigma_s & \Phi' \\ \Phi^s & \Omega \end{bmatrix} \otimes I,$$

where $E(vu_s') = \Phi \otimes I$.

A 3SLS-estimation of system (7) leads to:

(8) $$\hat{\partial}_s = (z'[\Sigma_s^{-1} \otimes X(X'X)^{-1}X']z)^{-1}z'|\Sigma_s^{-1} \otimes X(X'X)^{-1}X'|y_s$$

(9) $$\hat{\Pi} = [I \otimes (X'X)^{-1}X']y_R - |\Phi\Sigma_s^{-1} \otimes (X'X)^{-1}X'|(y_s - Z\hat{\partial})$$

(10) $$\hat{Cov}(\hat{\partial}_s) = Q = (Z'[\Sigma_s^{-1} \otimes X(X'X)^{-1}X']Z)^{-1}$$

(11) $$\hat{Cov}(\hat{\Pi}) = (\Omega - \Phi\Sigma_s^{-1}\Phi') \otimes (X'X)^{-1} + JQJ'$$

wobei $J = [\Phi\Sigma_s^{-1} \otimes (X'X)^{-1}X']Z$

---

1) X: matrix of the observations of all predetermined variables of the model

Obviously, the 3SLS-estimation (8) of the parameters of the structural form and the estimation of the variance-covariance-matrix of the estimated parameters of the structural form (10) are identical with a 3SLS-estimation of the parameters of the "reduced" structural form (4) without taking into account the equations of the reduced form: As the equations of the reduced from are always exactly identified equations, and as the elimination of exactly identified equations from the structural form of a system will not change the 3SLS-estimates, this result is obvious.

On the other hand, the estimates (9) of the parameters of the reduced form according to this approach are different from the ordinary least squares estimates; and it can be shown, that - as far as asymptotic efficiency is concerned - the 3SLS-estimates (9) are superior to ordinary least-squares-estimates. Of course, this gain in efficiency can only be achieved if at least one of the structural equations is overidentified! If one is interested in the estimation of the reduced form only, one can obviously again eliminate all exactly identified equations from the structural form without changing the results.

A further comment might be of interest: If one combines the complete structural form and the reduced form for a 3SLS-estimation, it can be shown, that, because of the singularity of the variance-covariance-matrix of the combined system, parameter estimates of the parameters of the structural part of the total system cannot be calculated. On the other hand, the estimates of the parameters of the reduced form still can be determined, and it can be shown, that, in this case - of course not very surprising! - the parameter estimates of the parameters of the reduced form will coincide with the so-called derived reduced form estimation.

How can we now make use of the described estimation procedure in our case? According to the evaluation of the structural form, we should eliminate the four equations explaining the endogenous variables Imp, EL, I and V from the structural form[1]. Of course, these variables still should be included in the reduced form.

---

1) As far as the problem of the endogeneity of Insolv is concerned, if one cannot find a reliable specification for this variable one could include Insolv as one further endogenous variable in the reduced form without an equation for Insolv in the structural form. As so far the model has been simulated only within the observation period, no such attempt has been undertaken.

Thus, the two vectors y_s and y in our case will be defined as:

$$(12) \quad \begin{aligned} y_s' &= [C^{priv}\ E\ P\ \gamma] \\ y' &= [C^{priv}\ I\ V\ Imp\ EL\ EP\ \gamma]. \end{aligned}$$

The graphs of the following pages depict the results of historical simulations for 1985,2 to 1986,1 for the eight most important variables based on estimates for the reduced form of the model for the whole observation period, where the estimates have been calculated

1) as unrestricted reduced form estimates, -----
2) as derived reduced form estimates, -----
3) as "Court-estimates" on the basis of (12) -.-.-

An examination of the results show that of the 32 values to be predicted for the eight endogenous variables in 16 cases predictions based on Court's procedure came closes to the true values, whereas the other two methods came in first in only 9 (unrestricted estimation) or 7 cases (derived reduced form). The simulation on the basis of Court's estimates avoids extreme errors which occur in the case of the unrestricted estimates (imports) as well as in the case of derived estimates (capacity).

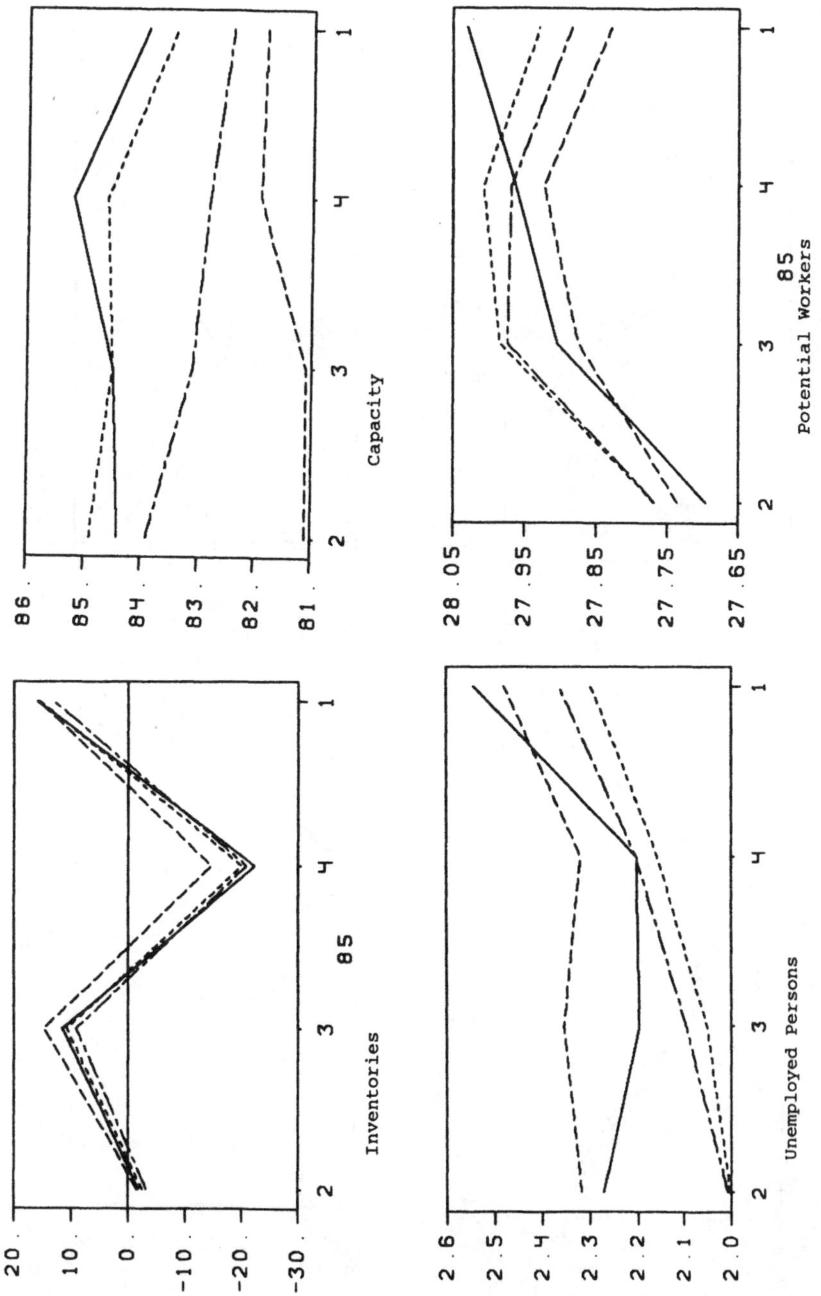
Inventories
Capacity
Unemployed Persons
Potential Workers
85
20
10
0
-10
-20
-30
86
85
84
83
82
81
2.6
2.5
2.4
2.3
2.2
2.1
2.0
28.05
27.95
27.85
27.75
27.65
2
3
4
1

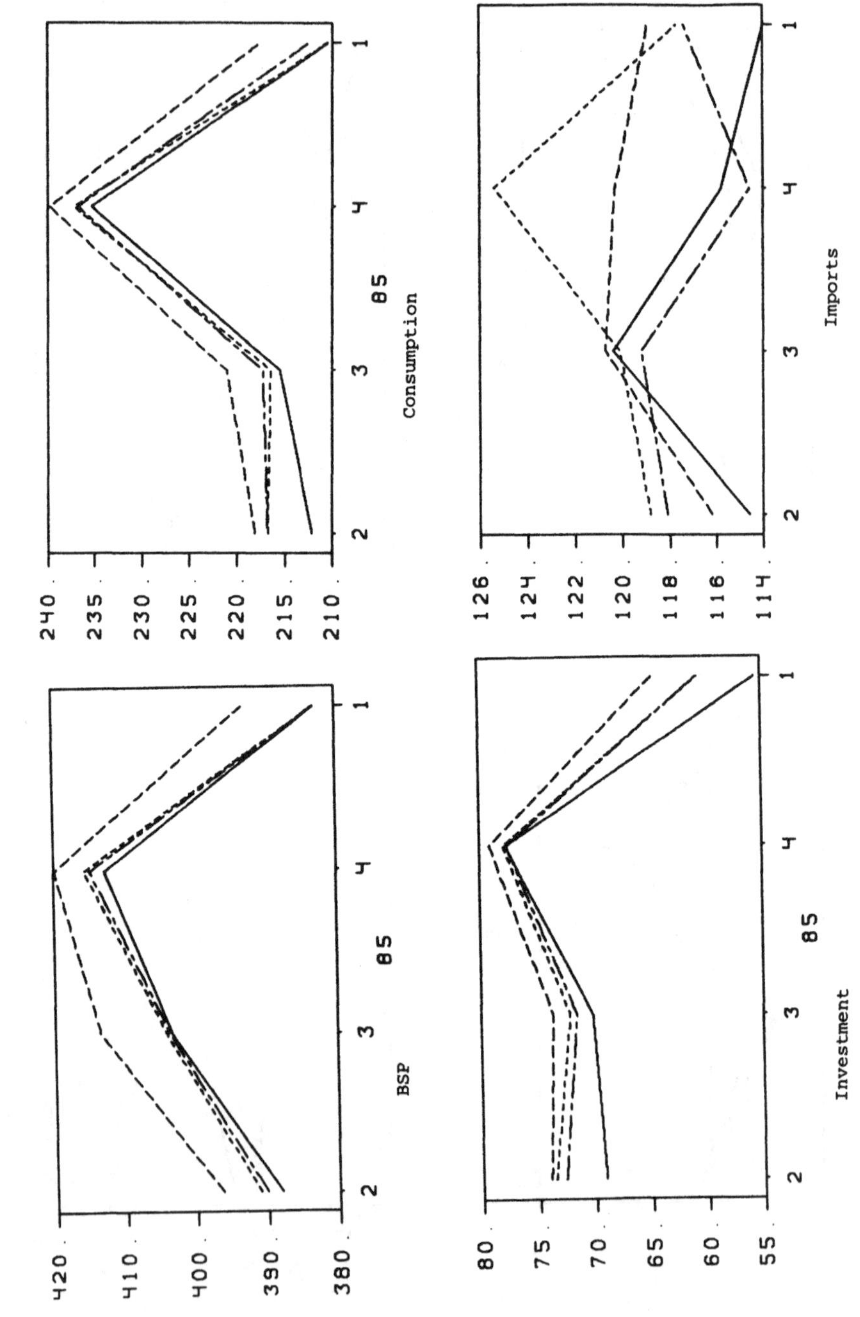
BSP
420
410
400
390
380
2
3
85
4
1
Consumption
240
235
230
225
220
215
210
2
3
85
4
1
Investment
80
75
70
65
60
55
2
3
85
4
1
Imports
126
124
122
120
118
116
114
2
3
4
1

4. Conclusion

The results show that one, of course, should try to improve the specification of the small model presented in this paper. But they also make clear that Court's procedure can be regarded as a proposing alternative if one wants to calculate predictions from a model which suffers from specification errors in the structural form.

Literature

[1] Court, R.H., "Efficient Estimation of the Reduced Form from Incomplete Econometric Models", Review of Economic Studies, 1973, p.411, ff.

[2] Deutsche Bundesbank, Frankfurt, Monatsberichte, several years.

[3] Deutsches Institut für Wirtschaftsforschung, Berlin, Vierteljährliche volkswirtschaftliche Gesamtrechnung für die BRD, several years.

[4] Ifo-Institut, München, Ifo-Konjunkturtest, several years.

[5] Statistisches Bundesamt, Wiesbaden, Statistische Jahrbücher, several years.

WAGE-PRICE DYNAMICS IN THE RWI-BUSINESS CYCLE MODEL

By Ullrich Heilemann
Rheinisch-Westfälisches Institut für Wirtschaftsforschung

Wage-price reactions have played a major role in the macroeconomic policy discussions in the FRG over the past 15 years. Many of the recent curative discussions of our employment problems are still centered around the wage-price nexus. To a large extent these debates were carried out theoretically or in very simple empirical terms. The message of macroeconometric models played only a minor role, but one of the main questions - how stable are wage-price reactions - wasn't even examined by most of them.

The following paper analyzes the problem of their intertemporal stability in the context of the single equation approach as well as in the context of a macroeconometric model. This will help to answer not only the question for the immediate effects of changes in the wage-price determinants but also for their macroeconomic consequences. The paper proceeds as follows: since the wage-price equations analyzed are taken from and the simulations are made with a model, an overview of this model is given first of all. Secondly, the wage determination is presented, followed by, thirdly, the price determination. Fourthly, to give an idea of the macroeconomic interactions of both sectors, the results of a restrained wage-policy-simulation are discussed. The paper ends with a summary and an evaluation of the results.

1. The RWI-business cycle model - an overview

The model to be used is the RWI-business cycle model. This is a medium-sized, short-term (quarterly) model which has been used for forecasting and simulation purposes since more than 10 years. It consists of about 40 stochastic equations and 75 definitions (HEILEMANN, MÜNCH 1984b). The demand-block of the model contains 30 equations, the distribution part 21 equations, the production-block 16 equations, while the price-part explains 12 variables and the government-sector is represented by 36 equations. Exogenous variables are the policy levers (social security-contribution rate, government construction outlays, short-term and long-term interest rates) on the one side and mainly internationally-determined variables (world trade, import-prices) on the other side. The theoretical base of the equations is - like in most applied econometric models - somewhat eclectical, comprising neoclassical elements as well as keynesian or monetarist ones. The general scheme of the model is, of course, keynesian resp. of the KEYNES-KLEIN-type. The reactions of the model may be labelled with respect to the role of demand and money and the stability of the pri-

Exhibit 1

Ex post forecasting accuracy[1] of the RWI-model 1973 to 1986

	RMSPE[2]				MAE[3]			
	Sample Period		1978-1 to 1979-4		Sample Period		1978-1 to 1979-4	
	I[4]	II[5]	I	II	I	II	I	II
GDP, origin								
Employed (x 1000)	0,5	0,2	0,3	0,3	86,7	44,9	179,4	53,4
GNP, demand, real								
Private consumption	1,1	2,0	0,7	2,0	1,4	3,3	1,1	3,0
Government consumption	1,4	1,3	1,5	0,8	0,7	0,8	0,8	0,5
Investment	2,4	3,3	1,1	2,1	1,2	2,0	0,6	1,4
Investment in machinery	4,2	4,7	2,0	3,0	0,7	1,1	0,5	0,8
Construction	2,7	3,8	1,3	2,2	0,8	1,3	0,4	1,0
Stocks, changes	88,2	96,0	27,1	80,3	2,1	2,4	1,4	3,5
Exports	2,6	2,6	1,8	1,8	1,8	2,3	1,3	1,1
Imports	1,2	2,8	1,7	2,3	1,5	2,2	1,0	1,9
GNP	2,7	0,9	9,7	0,5	2,7	2,7	1,8	1,6
Consumer prices, % change	0,8	0,5	0,4	0,3	0,7	0,4	0,4	0,2
GNP prices, % change	1,2	1,4	0,7	0,6	1,0	0,9	0,6	0,5
GNP, income distribution								
Gross-wage income	1,5	2,4	1,0	0,9	1,9	3,4	1,5	1,5
Gross profits	4,6	4,6	4,0	2,6	2,6	2,9	2,3	1,8
National income	1,7	2,0	0,9	1,0	3,4	3,8	1,8	2,1
Net-wage income	1,7	2,7	1,3	1,2	1,5	2,5	1,3	1,2
Net profits	7,7	5,7	7,7	3,6	3,2	2,8	3,5	1,9
Government Deficit	214,6	131,8	29,7	19,3	2,5	2,6	2,3	1,3

Author's computations.- 1) 8-quarter simulations within the sample period, if not noted other wise.- 2) Root-mean-square-percentage-error.- 3) Mean-average-error.-4) Sample period I: 1973-1 to 1982-4.- 5) Sample period II: 1976-3 to 1986-2.

vate sectors generally as post-keynesian (DAVIDSON 1981; CHICK 1983). The formation of expectations is in most cases of the adaptive or "weak rational"-type (ECKSTEIN 1983). - Sample periods of the model are usually the last 40 quarters of the data base available. For the present study, however, two versions of the model with the sample period 1973-1 to 1982-4 resp. 1976-3 to 1986-2 have been used. The older version has been widely examined (HEILEMANN 1985). The parameters have been estimated with OLS. Ex post and ex ante-performance of the model seem to be rather satisfying up to now.

Exhibit 1 presents some measures of forecasting accuracy for the most important variables of the two model versions. The simulations are (dynamic) 8 quarter solutions over the complete sample period as well as the period 1978-1 to 1979-4 which is the base of the comparative analysis referred to later on.

The results indicate a satisfying forecasting accuracy and give so some cause to believe in the simulation results.

2. The wage sector

The wage explanation follows the so-called augmentation-hypothesis (Aufstockungshypothese). This means a two-step approach with, firstly, the explanation of the negotiated wages and, secondly, on this basis the explanation of the actual wages resp. the gross wage income per employee (R. SCHMIDT 1972).

The explanation of the negotiated wage rate follows the well-known enlarged PHILLIPS-approach with

$$TLGHJW_t = A_1 + A_2\ PCPJW_{t-1} + A_3 \frac{1}{4} \sum_{i=1}^{4} AL_{t-i} + u_t \qquad (1)$$

with
TLGHJW: index of negotiated hourly wages, change rates against previous year;
PCPJW : consumer price-index, change rates against previous year;
AL : registered unemployed (x 1000);
u : error term.

Unemployment and prices explain up to 80% and on the average at least 65% of the observed development. Compared with the highly reputated explanations of private consumption this seems to be not too bad. Nevertheless, it still leaves some room for the influence of other factors, such as general union aggressiveness, strike activity, level of unionization or political factors. Though it cannot reasonably be denied that

Exhibit 2

Stability of the negotiated wages explanation[1].

1961-3 to 1986-2[2]

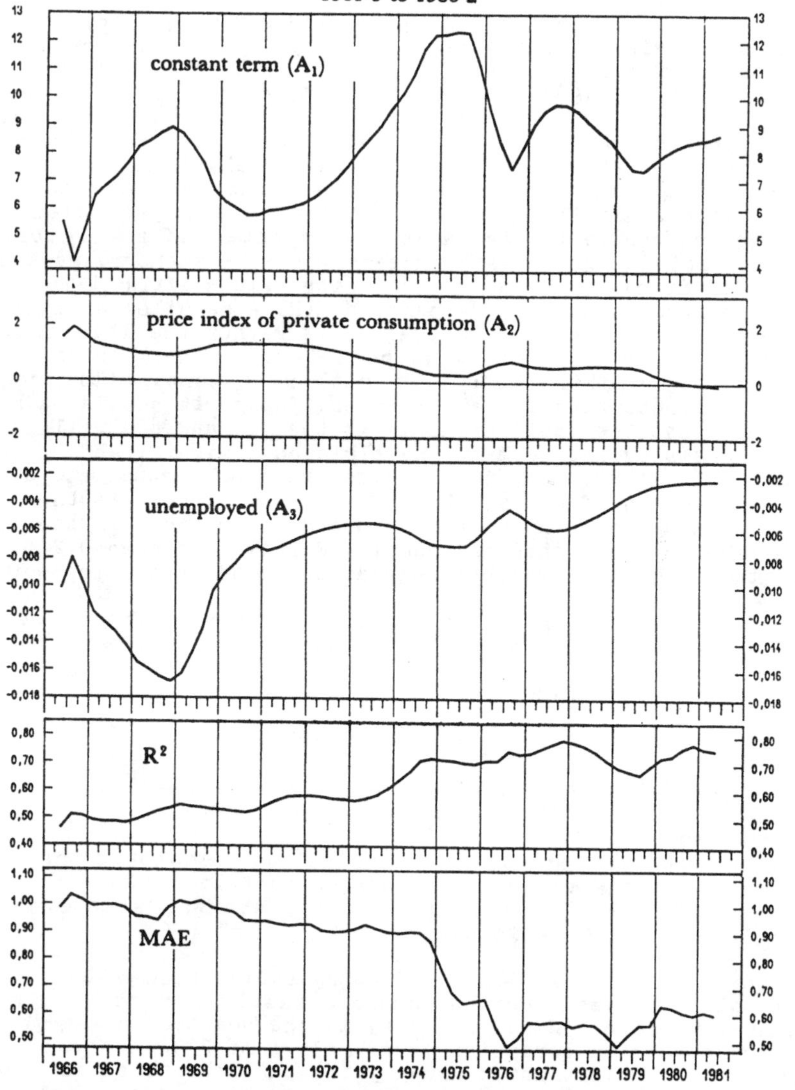

Author's computations.- [1]Changes against previous year.- [2]Each estimation covers 40 quarters, the results have been assigned to the 20th quarter of each sample period.- MAE: mean averaage error.

these factors played some role in the West German wage-setting processes of the past, they seem to have been only of temporary or episodical importance.

To evaluate the formal and substantial quality of this equation, their intertemporal stability was examined by moving regressions for a 40 quarter "window" (WOLL et alii 1976; HEILEMANN and MÜNCH 1984a) (Exhibit 2). Bearing in mind the well-known statistical and econometric difficulties of the PHILLIPS-approach like autocorrelation, multicollinearity, single equation bias etc. the results of this analysis may be summarized as follows: (1) up to 1972 the constant term and the price term show the greatest influence (ß-coefficients). Whether the constant term can be interpreted as an indicator of productivity growth as suggested by some authors, may be a matter of dispute, though at least its values after 1970 seem to be contradicting this hypothesis. (2) A detailed analyis of the price coefficient shows that he is rather flexible in the short run. The question must be left open whether this indicates bad specification or a rather rational attitude in times of fluctuating inflation rates. Though it should be noted, too, that the coefficient is in the 70s and 80s well below 1. (3) Unemployment seems to have played a significant role as a wage determinant only in the first half of the 60s and again in the second half of the 70s. Of special interest is the fact that this parameter shows a continuing diminution. In the face of a rise in the number of unemployed from less than 0.2 m. to more than 2 m. this was probably an unavoidable development.

The explanation of the change rate in the gross wage income per employee is based on the change rate of the negotiated wages and the GNP growth as an indicator of the wage drift and of the labour time component:

$$BLGAJW_t = A_1 + A_2\ TLGHJW_t + A_3\ (0.6\ BSP80JW_t + 0.3\ BSP80JW_{t-1} + 0.1\ BSP80JW_{t-2}) + u_t \qquad (2)$$

with
BLGAJW : gross wage income per employed person, change rates against previous year;
BSP80JW: GNP in 1980 prices, change rates against previous year.

Exhibit 3 shows a satisfactory explanation of the gross wages per employee over the complete sample period. The most important factor (ß-coefficient) is the negotiated wage rate. The corresponding parameter takes values of 0.9 at the end of the sample period, while usually the parameter lies somewhere between 0.7 and 0.8. Correspondingly, the influence of the GNP is reduced. As GNP serves as an indicator of the wage drift, this shift reflects a relaxation of the labour market during this period.

Exhibit 3

Stability of the gross-wage per employed explanation[1]

1961-3 to 1986-2[2]

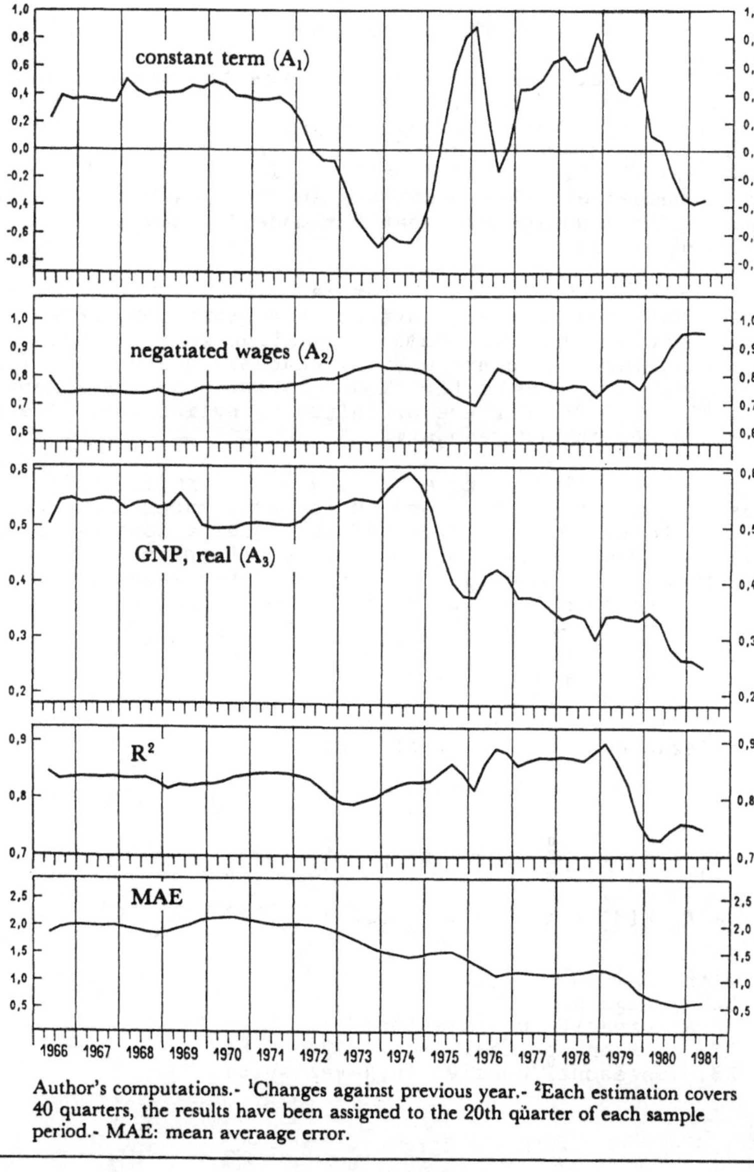

Author's computations.- [1]Changes against previous year.- [2]Each estimation covers 40 quarters, the results have been assigned to the 20th qüarter of each sample period.- MAE: mean averaage error.

Measured by Chow-tests, CUSUM- and CUSUM-square-tests (not shown here) these shifts do not seem to be of major importance, even though the explanatory power shows some noticeable break ins.

3. The price sector

The need to explain prices in models like the present one is threefold: firstly, any diagnosis of the short-term development requires information about the price development; secondly, the price sector is required to bridge the demand side with the distribution side; finally, prices play an important role in a number of other equations and relying here on exogenous estimations would mean that the model's results would be conditioned too much.

The price sector of the RWI-model explains the price level for private consumption, government consumption, investment in machinery and tools, housing construction, business construction, government construction, changes in stocks and exports. The price levels for fixed investment, investment in construction and for GNP are explained by definitions. The import prices are set exogenously.

All(stochastic) price equations of the model follow more or less the same concept, based on the well-known mark-up hypothesis; in earlier versions of the model a monetary aggregate (M2) played a role, too, but in recent years it lost its significance. The wage pressure (negotiated wages minus productivity) serves as an indicator of the domestic costs. The supply/demand conditions are represented by the capacity-utilization rate (actual minus potential output). Since the FRG is integrated to a large extent in world trade, import prices are a third factor in price determination. Assuming a distributed-lag-reaction of prices and incorporating the seasonal dummies leads e.g. for the price index of private consumption to

$$PCP_t = A_1 + A_2 \frac{1}{4} \sum_{i=1}^{4} LDR + A_3 \frac{1}{4} \sum_{i=0}^{3} KAPA_{t-i} + A_4 PIM_t + A_5 DS1 + A_6 DS2 + A_7 DS3 + A_8 PCP_{t-1} + u_t \tag{3}$$

with
LDR : wage pressure;
KAPA: capacity utilization;
PIM : import prices, index, 1980=100;
DS. : seasonal dummies (0,1-variables);

Exhibit 4

Stability of the price explanation[1]

1962-2 to 1986-2

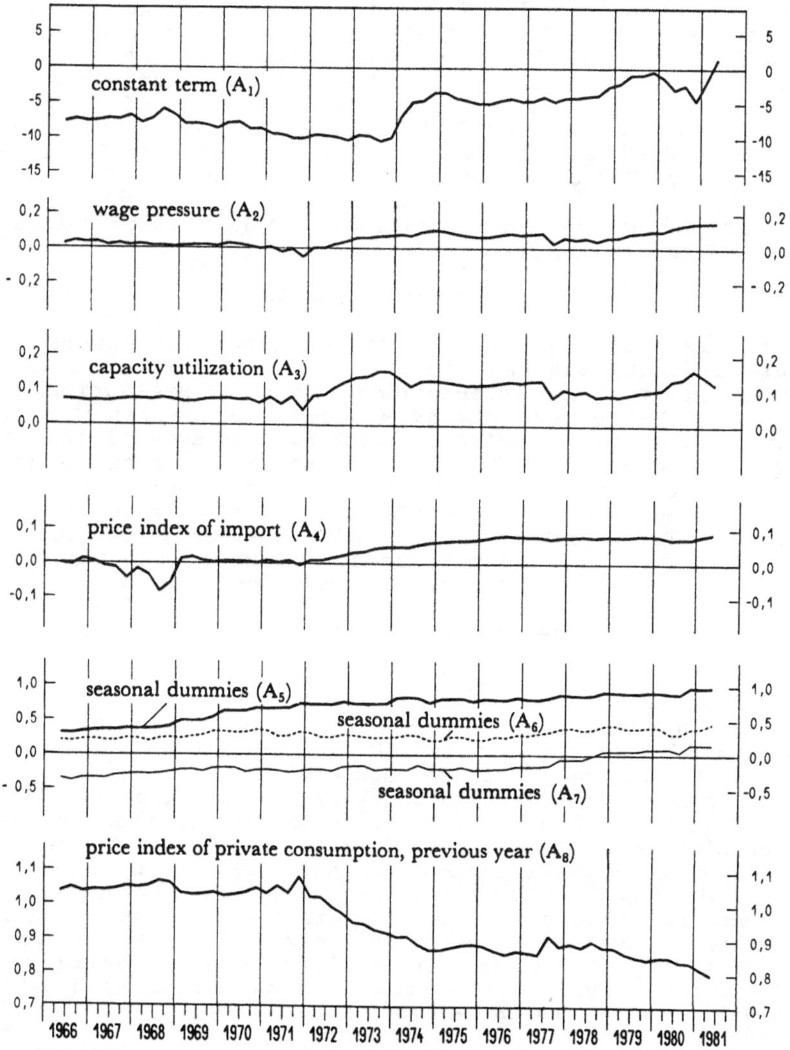

Author's computations.- [1]Each estimation covers 40 quarters, the results have been assigned to the 20th quarter of each sample period.

RWI ESSEN

Exhibit 4 shows the coresponding results of the moving regressions-approach. As the RWI-model has been in use since 1977 onwards, the results prior to 1972 have the character of backcasts, presented to facilitate the comparison with the more actual results (HEILEMANN and MÜNCH 1984a). Judged by the t-statistics and the R^2 (not shown here), the results are not too bad. Neglecting the lagged endogenous variable, import prices and wage pressure are the most important variables (ß-coefficients), while capacity utilization is of minor importance. An interesting feature is the increasing adjustment velocity: the mean lag $(1/(1- A_8))$ is reduced from 5 quarters in the early 70s to 2 quarters in the early 80s.

Again, neither Chow-tests nor CUSUM- and CUSUM-square-tests indicate structural shifts.

4. Wage-price dynamics

The single equation approach yields important information about direction and magnitude of the various influences. He ignores, however, the interactions within the wage-price sector and with the rest of the economy. To analyze these effects, a more comprehensive approach such as a multi-equation model has to be employed. With the transition from the (single) equation level to the system level, however, it becomes more difficult to get a clear picture from the model's causal structure. To facilitate this, exhibit 5 gives a schematic representation of the most important links of the wage-price sector in the RWI-business-cycle model. The important employment reaction (which is not a part of the wage price section) follows the well-known concept of an implicit production function.

To analyze the wage-price dynamics at the system level, it seemed appropriate to simulate a restrained wage policy (for similar exercises cf. LÜDEKE 1980, MARTIENSEN 1984, JAHNKE 1985, BUCHER and ROSSI 1985 and WALLIS (ed.) 1986). To do so, the negotiated wages had to be exogenized first of all and were fixed then to 1% point below the actual development over the complete simulation period (1978-1 to 1979-4). The effects of this on gross wage income are not exactly the same. But since GNP growth will be reduced, too, the differences are very small. Of course, the exogenization of the (nominal) wages means that second round (price-wage-)effects are ignored. But these effects are much smaller than the primary effects and thereby the analysis is much facilitated. - The (nominal) interest rates were left unchanged.

As already mentioned, the simulations are performed with the same model estimated over different sample periods. The differences between the reactions of this two versions should indicate shifts in the model reactions. Unfortunately, this approach makes it difficult to separate changes in the various parts of the model. The alternative would be to use only one

Exhibit 5

Wage-price interactions in the RWI-business cycle-model[1]

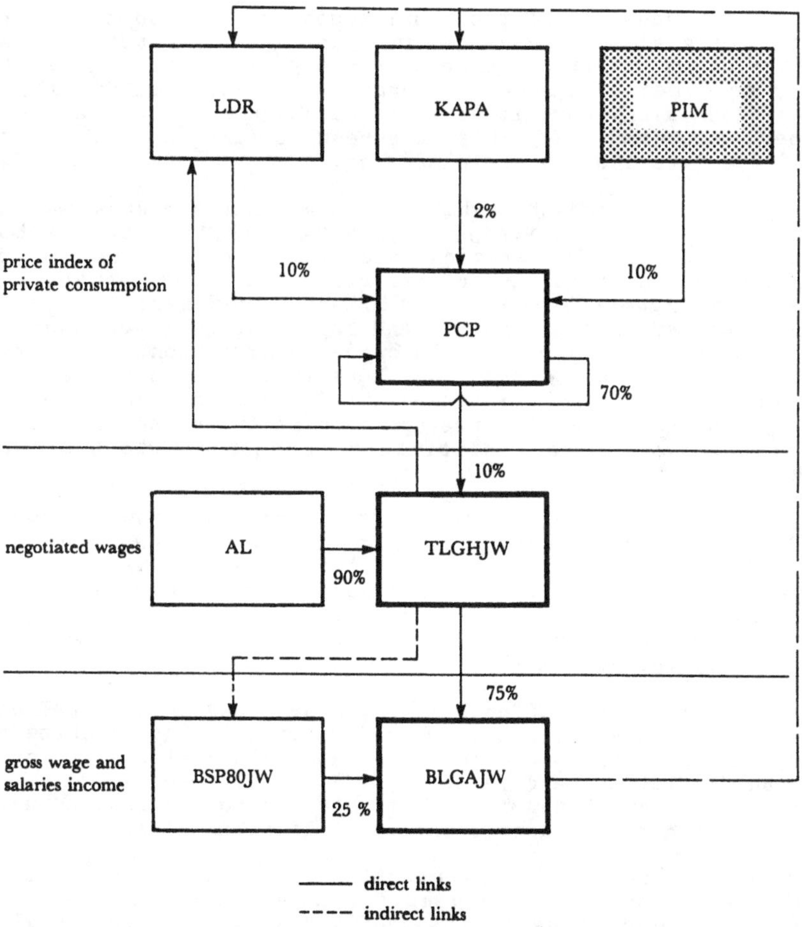

Author's computations.- [1]The %-data represent the comparative influences (ß-cofficients,%) of version II (sample period 1976-3 to 1986-2).

version and exchange the equations of interest. This, however, neglects the empirical relations between the various equations and would reduce the simulations to pure "pragmatic numerology" (BRUNNER). - The selection of the two sample periods reflects a compromise between the actuality, a priori knowledge about differences in the reactions of the single equations (exhibit 2, 3, 4) and the need for a common simulation period.

The discussion of the results concentrates on the macroeconomic goals of growth, employment, price stability and distribution. (Whether there exists a goal of external stability in times of flexible exchange rates is a matter of dispute, and in any way, it would be difficult to define one in numerical terms.) Firstly, the results for the model version II (1976-3 to 1986-2) are discussed.

As to GNP-growth, exhibit 6 shows that a restrained wage policy works contractively, beginning with the 5th quarter, however, with a diminishing rate. The main cause of the contraction is the reduction of private consumption, enhanced by the accelerator-caused reduction of investment, while the growth-caused rise of imports and the growth/wage-cost-caused reduction of exports work in opposite direction. It is worth noting that the wage-cost reductions in the short run are not of such a magnitude to compensate for demand reductions even though the habit-persistance-approach of the consumption function employed already stabilizes consumption to a a large degree.

In the first year these effects dominate the employment effects to a large extent. The reductions of the wage rate certainly reduces wage costs, but the reduced GNP leaves the wage costs unchanged. A significant rise of employment can only be detected in the second year. Compared with the high expectations often held of these effects, here they turn out to be comparatively small.

The strongest effects - of course a personal judgement - can be registered with the price development. Though these are very small in the beginning, too, at the end of the 8th quarter the inflation rate is about 0.4% points lower than in the base-line simulation. This is mainly due to the reduced wage-pressure and the reduced capacity utilization rate.

As to the income distribution the results show that the wages and salaries are reduced for about the same amount as wages. Correspondingly, the profit shares rise. The accelerated wage-income reduction in the second year is a consequence of the further shrinking (compared with the base-line simulation) of macroeconomic activity since the employment effects are rather moderate.

Exhibit 6

Effects of a restrained wage policy[1] in the RWI-model[2], version II
1978-1 to 1979-4, differences to the base line solution
% changes against previous year

	1978 1 q.	1978 2 q.	1978 3 q.	1978 4 q.	1979 1 q.	1979 2 q.	1979 3 q.	1979 4 q.	1978	1979	78/79
GDP, origin											
Employed (x 1000)	-3,0	-7,8	-2,4	4,0	7,1	9,2	20,5	32,6	-2,3	17,3	7,5
GNP, demand, real											
Private consumption	-0,4	-0,6	-0,6	-0,5	-0,7	-0,9	-0,9	-0,8	-0,5	-0,8	-0,7
Government consumption	0,0	0,0	0,0	0,0	-0,1	-0,1	-0,1	-0,1	0,0	-0,1	-0,1
Investment	-0,2	-0,2	-0,3	-0,1	-0,3	-0,3	-0,3	-0,1	-0,2	-0,3	-0,2
Investment in machinery	-0,4	-0,6	-0,9	-0,3	-0,8	-1,1	-1,1	-0,6	-0,5	-0,9	-0,7
Construction	0,0	0,0	0,0	0,1	0,0	0,1	0,2	0,2	0,0	0,1	0,1
Stocks, changes	-0,1	0,1	-0,1	-0,1	-0,1	-0,0	-0,0	-0,0	-0,4	-0,1	-0,5
Exports	0,2	0,3	0,3	0,2	0,4	0,5	0,4	0,4	0,2	0,4	0,3
Imports	-0,3	-0,5	-0,6	-0,5	-0,7	-0,9	-0,9	-0,9	-0,5	-0,8	-0,7
GNP	-0,1	-0,2	-0,2	-0,1	-0,2	-0,3	-0,2	-0,2	-0,2	-0,2	-0,2
Consumer prices,% change	0,0	0,0	-0,1	-0,1	-0,2	-0,2	-0,3	-0,4	0,1	-0,3	-0,2
GNP prices,% change	-0,1	-0,2	-0,3	-0,4	-0,5	-0,7	-0,7	-0,9	-0,2	-0,7	-0,5
GNP, income distribution											
Gross-wage income	-0,9	-1,0	-1,0	-0,9	-1,8	-1,8	-1,8	-1,7	-0,9	-1,8	-1,4
Gross profits	1,2	1,1	0,6	0,6	1,4	1,1	0,7	0,5	0,8	0,9	0,9
National income	-0,3	-0,4	-0,5	-0,5	-0,8	-1,0	-1,1	1,1	-0,4	-1,0	-0,7
Net-wage income	-0,8	-0,9	-0,8	-0,8	-1,6	-1,6	-1,5	-1,4	-0,8	-1,5	-1,2
Net profits	1,5	1,3	0,8	0,7	1,8	1,4	0,8	0,5	1,0	1,1	1,1
Government Deficit	-0,4	-0,5	-0,4	-0,2	-0,4	-0,4	-0,3	0,2	-1,5	-1,3	-2,8

Author's computations.- 1) 1 % point reduced growth of negotiated wages.- 2) 8-quarter simulations.

Exhibit 7

Effects of a restrained wage-policy[1]

Differences from the base-line simulation, changes against provisions year

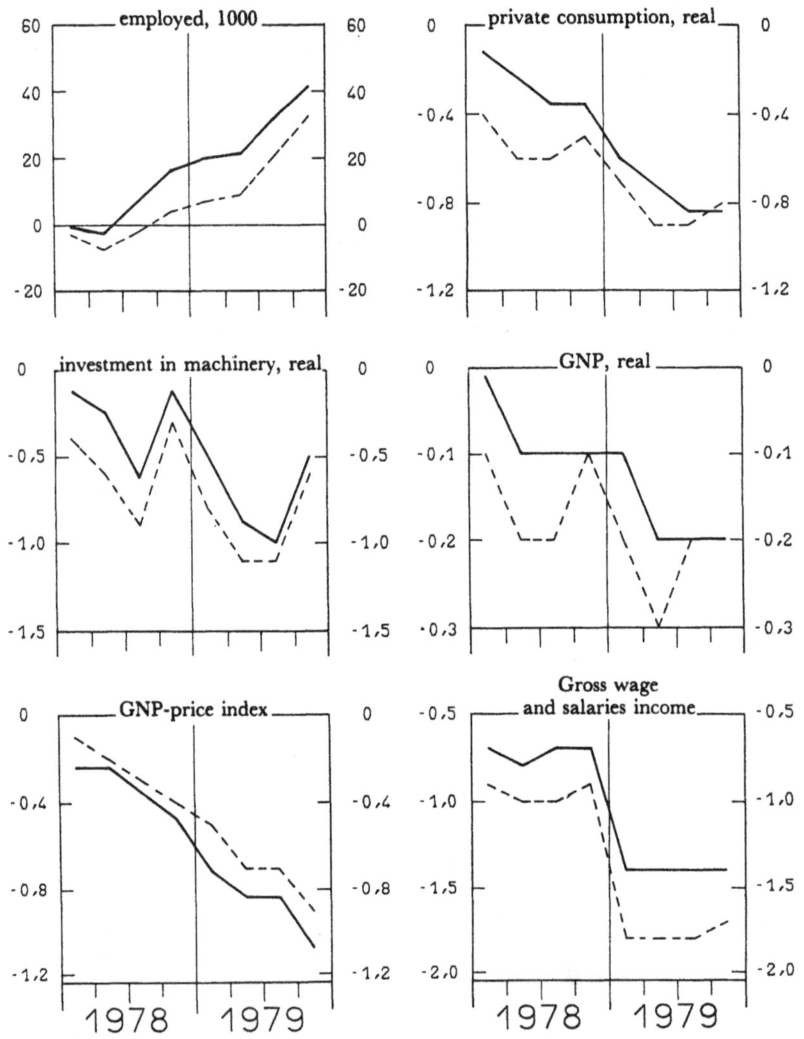

Author's computations.-
[1]For details cf. text.

How do these reactions compare with the effects of the older model version (1973-1 to 1982-4)? Firstly, the effects in general do not seem to be too different (Exhibit 7). This indicates at least some stability in the reactions of the West German economy or at least those of the model used. As to specific reactions, it can be said that the growth reduction of version I is not as large as that of version II, especially in the first year. This is mainly due to the lower short run marginal propensity to consume in version I (mpc = 0.2 vs. 0.7 in version II). The employment effects are somewhat stronger which can be attributed to both a larger demand and cost elasticity of the employment function. The price reactions of the two model versions are nearly identical.

5. Summary

Bearing in mind the various reservations made, the wage-price reactions in the RWI-business cycle model and their changes may be characterized as follows:

(1) the intertemporal stability analysis of the wage and price equations showed some important changes in the parameters of the determinants. But few indicators of large and sudden parameter shifts were found.

(2) Growth, prices and employment react comparatively slowly on reduced wage growth. Especially for economic growth it takes a long time till the originally contractive process is reversed. Only prices seem to react immediatedly in the right direction.

(3) Basically, the same can be said about the magnitude of the reactions: usually it takes 8 quarters until significant effects can be detected.

(4) A comparison of these results with those for the earlier periods reveals that the dynamics have increased.

(5) Marginal propensity to consume (or demand in general) is a pivotal turnaround for the contractive growth effects of a restrained wage policy. In this respect the lower mpc of the older version is more favourable to such a policy. Furthermore, the results reveal the important role of expectations since, only if it is possible to overcome the consumption-caused growth reductions by stabilising or augmenting the mpc or otherwise, a wage reduction will result immediately in higher employment.

Rheinisch-Westfälische Institut
für Wirtschaftsforschung, Essen

References

A. Bucher and V. Rossi, Analysis of the stabilisation mechanisms of macroeconomic models: a comparison of the Eurolink models. "Economic papers", 39, Brussels 1985.

V. Chick, Macro Economics after Keynes. Cambridge, Mass. 1983.

P. Davidson, Post Keynesian Economics. "Public Interest", Special Issue 'The Crisis in Economic Theory', New York, vol.20 (1980), pp.151-173.

O. Eckstein, The DRI-Model of the U.S. Economy. New York 1983.

U. Heilemann, Zur Prognosepraxis ökonometrischer Modelle. "Zeitschrift für Wirtschafts- und Sozialwissenschaften", 105.Jg. (1985), S.683-708.

U. Heilemann and H.J. Münch (a), The Great Recession: A Crisis in Parameters? In: P. Thoft-Christensen (ed.), System-Modelling and Optimization. Proceedings of the 11th Conference 1983. (Lecture Notes in Control and Information Sciences, vol.59.) Berlin 1984, pp.117-129.

U. Heilemann und H.J. Münch (b), Einige Bemerkungen zum RWI-Konjunkturmodell. In: G. Langer, J. Martiensen und A. Quinke (Hrsg.), Simulationsrechnungen mit ökonometrischen Makromodellen. München 1984, pp.355-385.

W. Jahnke, Simulation verschiedener Strategien zur Verringerung der Arbeitslosigkeit. Paper presented to the Annual Meeting of the "Statistische Gesellschaft", Bonn, September 27, 1985.

D. Lüdeke, Zur empirischen Relevanz der Lohn-Preis-Spirale und der Lohn-Investitionsaversion. In: Empirische Wirtschaftsforschung. Festschrift für R. Krengel aus Anlaß seines 60. Geburtstages. Hrsg. v. J. Frohn und R. Stäglin. Berlin 1980, pp.213-235.

J. Martiensen, Simulationsexperimente mit makroökonometrischen Modellen für die Bundesrepublik Deutschland. In: G. Langer, J. Martiensen und A. Quinke (Hrsg.), Simulationsexperimente mit ökonometrischen Makromodellen. München 1984, pp.9-75.

R. Schmidt, Kurzfristige Prognosefunktion für die Tarif- und Effektivlohnentwicklung in der Bundesrepublik Deutschland. "Die Weltwirtschaft", vol.1972, pp.237-264.

K.F. Wallis (ed.), M.J. Andrews, P.G. Fisher, J.A. Longbottom, and J.D. Whitley, Models of the UK economy. A Third Review by the ESRC Macroeconomic Modelling Bureau. Oxford 1986.

A. Woll. B. Faulwasser und B.-T. Ramb, Zur Zeitstabilität des Phillips-Theorems. "Kredit und Kapital", vol.9 (1976), pp.293-316.

Short-term Time Structures in the Input-Output Analysis

by

Ulrich Meyer

Universität Bamberg

Content

Figures, Table

1. Introduction

Input-output models are used to analyse the inter-industry transactions of an economy. The analysis aims at the inter-industry relationships of quantities (or values) which result from intermediate and - in dynamic input-output analysis - from investment deliveries between sectors. The time structure which results from production, delivery and reaction lags has until now scarcely been analysed.

Dynamic input-output models, of course, have a time aspect in describing the possibility for growth of an economy over time. The time component in these models is due to the fact, that in every period the sectoral capacities allow for levels of production greater than intermediate and final demand - leaving room for investments (with a gestation lag of one period in most models) to expand the capacities. Apart from this there is no time structure in the models, i.e. it is assumed that production of a good and production of all the necessary inputs for that good take place simultaneously.

Dynamic input-output models have rarely been empirically applied.[1] The most often empirically applied model is the static open input-output model. It serves to compute the effect of a change in final demand on sectoral gross production levels and therefore on employment, capacity utilization etc. In this context, the temporal distribution of effects is also often of interest, as well as the comparative static "total effect". For example, in the case of additional government final demand as an anticyclical programme, the interest lies not only in the induced production as a whole, but also in the question in which time periods the effects occur, i.e. the time profile of these effects. The time structure of the effects results from various lags, for instance reaction lags during decision making within firms concerning changes in production and orders of intermediate goods, or from the fact that production itself requires time.

This last aspect is now being discussed on theoretical terms under the title "Input-Output Analysis with Distributed Activities".[2] For that purpose production relationships are described under explicit consideration of the time dimension of production: Each input coefficient a_{ij}, which gives the total input requirement of good i in the production of good j, is split into a series of input coefficients $a_{ij}(0)$, $a_{ij}(-1)$, $a_{ij}(-2)$, ... where $a_{ij}(-s)$ is the quantity of good i necessary as input s periods before the production of one unit of good j. (So, the sum $\Sigma_s\, a_{ij}(-s)$ corresponds to the "old" input coefficient a_{ij}.) Similarly, in containing production capacities, $b_{ij}(-s)$ are the capital coefficients describing which investment in good i is necessary in period t-s in order to expand capacity of sector j in period t by one unit.[3] By using the mathematical term "convolution product", instead of the matrix product, this production technique can be described in a model which corresponds formally to the static (only intermediate coefficients) resp. to the dynamic input-output model and which can be analysed using similar methods.

This interesting approach will not be discussed further in this paper, firstly due to the difficulties regarding necessary empirical data and secondly because other aspects appear more important for the short term time structural relationships. The approach developed here resembles the attempt of taking lags into account, as was undertaken a few years ago by Schnabl/Holub [5] by use of the qualitative input-output analysis.

Within the framework of qualitative input-output analysis, so-called distances between the sectors can be defined. To that purpose, all direct and indirect supply relations between the sectors are considered and the distance e_{ij} from sector i to sector j is defined as the shortest length of such a "supply path" between sector i and sector j. If these distances can be interpreted as time distances, then the distance e_{ij} shows how quickly a sector i is affected by a change in production in sector j via intermediate demand. However this (demand orientated) approach suffers from the fact that it supposes one uniform time lag for all sectors between a change in demand for the output of a sector and the resulting input demand of this sector. In addition, the loss of information from failing to consider quantitative input-output relationships appears too great for the achieved results to be regarded as economically usable - for example, to find out which sectors are particularly suitable as a direct starting point for an anticyclical programme, because the indirect effects originated by them quickly pass through the whole economy.

The approach outlined here is based on the static input-output model, in as much as investment demand is taken as exogenous. The approach describes the time structure of the adaption of sectoral production to demand; final demand is treated as exogenous and intermediate demand as endogenous. The adaption of the level of production is governed for each sector by its specific demand and stock situation.

2. Time sequence of effects caused by a change in final demand

A change of final demand can cause two entirely different effects on sectoral production. With regard to an increase in final demand this shall be demonstrated for a simple economy with a linear production structure, for which the deliveries between sectors S1, S2, S3 and final demand F can be graphically illustrated as follows:

$$(S1) \xrightarrow{x_1^s} (S2) \xrightarrow{x_2^s} (S3) \xrightarrow{x_3^s} (F) \;,$$

where x_i^s is delivery of good i.

Case 1: "Backwards effect" Firstly, sector 3 increases its production (x_3↑) as a result of additional demand for its products. Sooner or later, this leads to increasing (intermediate) demand for good 2, so that production in sector 2 also increases (x_2↑), which finally affects sector 1 as well (x_1↑). In case 1, the influence of final demand on sectoral production runs in the opposite direction to the delivery flows.

Conditions for the working of the backwards effect are sufficient production capacities (primary inputs), of course, but also sufficient stocks of input goods. Only then all the sectors can increase their production without the input suppliers having to increase their output first. If, contrarily, there are no stocks of intermediate goods, then the chain of effects runs in accordance with the following case 2.

Case 2: "Forwards effect" The increased final demand for good 3 (x_3^d↑) leads to an increase in the intermediate demand for good 2 (x_2^d↑) and then for good 1 (x_1^d↑). As a result of this, production in sector 1 rises. After the increase in deliveries to sector 2, the production can rise here, and finally in sector 3, too. So we have a backward moving "wave of orders" first, and then the effect proceeds in the direction of the production structure.

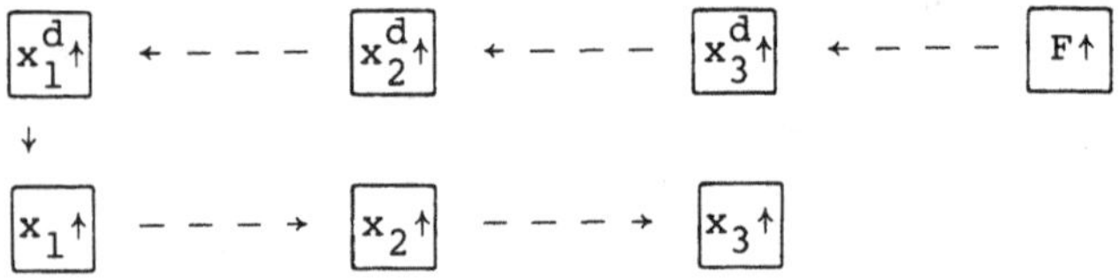

In the case of the backwards effect, the speed at which the effect spreads depends on:
- how quickly firms react to an increase in demand by deciding to raise production levels (instead of satisfying additional demand from output stocks, if possible),
- how quickly production can be raised and
- how quickly the firms, which now produce more, develop additional intermediate demand themselves (which essentially depends on input stock levels).

The speed of the forwards effect on the other hand depends essentially on the length of the production process (including delivery time).

In reality both effects will generally appear together and will overlap one another. But there exists an asymmetry, concerning positive and negative final demand impulses: the forwards effect, whose speed is limited by the time consumption of the production process, can only be efficacious in the case of an **increase** in final demand (and stock clearance), whereas the backwards effect can appear with both positive and negative changes in final demand.

If the production structure is interdependent, the chains of effects are more complicated. The overall effect, with regard to its time structure, can be thought, however, as a compound of many different forwards and backwards effects which overlap and influence each other.[4]

These interrelated effects shall be built into a model below, which, on the basis of empirical data, describes the temporal structure of the quantitative effects of changes in final demand on each sector. The above remarks show that the sectoral stock levels are of decisive importance for the speed of diffusion of a demand shock. Low stocks of input goods (and the impossibility of obtaining goods quickly from suppliers) slow down the speed of diffusion of a (positive) final demand shock, since a production increase, according to the forwards effect has to start in the sectors furthest away from final demand. On the other hand, low stocks of output goods have, according to the backwards effect, an accelerating effect, since an increase in demand necessitates a rapid rise in production in order to satisfy demand.

In a model with n sectors, output stocks o_j of sectors j are to be described as vector $o=(o_j)$ of length n. The description of input stocks is more complicated: for each sector j, stocks s_{ij} of all the input goods i=1,...,n must be considered. Such detailed statistical information is not as yet available. A model which describes the development of input stock levels, moreover, would have to take into account the varying storage possibilities of different inputs (e.g.: electricity and steel). Therefore, we shall not include input stocks explicitly, but construct a model only for such situations where there are no limits on production from the input side. That is, we only take into account the demand orientated backwards effect. This procedure would be appropriate for situations with unused capacities in the business cycle.

3. A model to describe the backwards effect

The basic hypothesis of the model is shown by the following equation:

$$x_j(t)-x_j(t-1) = d_j(x_j^d(t)-x_j(t-1)) + v_j(o_j^*(t)-o_j(t)). \qquad (1)$$

The firms in sector j change their production x_j from period t-1 to period t with respect to the demand situation $(x_j^d(t)-x_j(t-1))$ and the (output-)stock situation $((o_j^*(t)-o_j(t))$. The **reaction-to-demand coefficient** $d_j>0$ describes which part of the excess of current demand over previous production (which can be negative) causes a rise (or a fall) in production in the current period t. Correspondingly, the **reaction-to-stock coefficient** $v_j>0$ describes to what extent a difference between the stock level $o_j^*(t)$, desired for period t, and the actual stock level $o_j(t)$ causes additional (or less) production in period t.

For clarity's sake, equation (1) shows a time lag of only one period. The empirical application in the next section uses the slightly more general equation

$$\begin{aligned} x_j(t)-x_j(t-1) &= d_j(0)\,(x_j^d(t)-x_j(t-1)) + \sum_{s=1}^{S_{dj}} d_j(s)\,(x_j^d(t-s)-x_j(t-s)) \\ &\quad + v_j(0)\,(o_j^*(t)-o_j(t)) + \sum_{s=1}^{S_{vj}} v_j(s)\,(o_j^*(t-s)-o_j(t-s)). \end{aligned} \tag{1'}$$

In (1') it is assumed that the adaption of production possibly depends, in the sense of distributed lags, on the demand and stock situations in previous periods (at most period $t-S_{dj}$ respectively $t-S_{vj}$).

The change in stocks of output goods (measured at the start of the period) is the difference between production and demand:

$$o_j(t)-o_j(t-1) = x_j(t-1)-x_j^d(t-1). \tag{2}$$

The basis for equation (2) is the assumption that either there are always sufficient stocks to satisfy demand at current production levels (then $o_j(t)>0$ for all t), or that unsatisfied demand will be transfered into the next period as a backlog of orders (=negative stock).

For the desired stock level $o_j^*(t)$, we assume that it is proportional to demand, with $g_j>0$ as **stock-level coefficient.**

$$o_j^*(t) = g_j\, x_j^d(t). \tag{3}$$

The demand x_i^d for goods produced in sector i consists of intermediate demand x_{ij}^d by sectors j=1,...,n and final demand f_i:

$$x_i^d(t) = \sum_{j=1}^{n} x_{ij}^d(t) + f_i(t), \qquad (4)$$

where intermediate demand of sector j

$$x_{ij}^d(t) = a_{ij}\ x_j(t) \qquad (5)$$

results from production $x_j(t)$ according to the input-coefficient a_{ij}.

Equation (5) means that production in period t leads to corresponding intermediate demand during **the same period.** In the empirical application the more general equation

$$x_{ij}^d(t) = b_{ij}a_{ij}x_j(t) + (1-b_{ij})a_{ij}x_j(t-1) \qquad (5')$$

is assumed, meaning that a share (b_{ij}, $0 \leq b_{ij} \leq 1$) of inputs necessary for production leads to intermediate demand without delay whereas the remaining part ($1-b_{ij}$) of input requirement leads to intermediate demand one period later.

Equations (1) to (5) can be summarized in a system of linear difference equations with 2n equations

$$\begin{pmatrix} x(t) \\ o(t) \end{pmatrix} = \begin{pmatrix} M^{-1}(I-\hat{d}-\hat{v}(I-A)) & -M^{-1}\hat{v} \\ I-A & I \end{pmatrix} \begin{pmatrix} x(t-1) \\ o(t-1) \end{pmatrix} + \begin{pmatrix} M^{-1}((\hat{d}+\hat{v}\hat{g})f(t)+\hat{v}f(t-1)) \\ -f(t-1) \end{pmatrix}, \qquad (6)$$

where $\hat{d}$, $\hat{v}$, $\hat{g}$ represent diagonal matrices of the respective sectoral coefficients and where M indicates the matrix $I-(\hat{d}+\hat{v}\hat{g})A$ (I being the unity matrix of order n).

The system of difference equations (6) describes the development of production levels x(t) and stock levels o(t) over time, on the basis of given starting values x(0), o(0). This development depends on the reaction-to-demand coefficients d_j, on the reaction-to-stock coefficients v_j and on the stock-level coefficients g_j. The development is very complex; in order to gain an insight into the interrelation of effects, we shall consider the case n=1. Equation (6) then becomes

$$\begin{pmatrix} x(t) \\ o(t) \end{pmatrix} = \begin{pmatrix} \frac{1-d-(1-a)v}{m} & -\frac{v}{m} \\ 1-a & 1 \end{pmatrix} \begin{pmatrix} x(t-1) \\ o(t-1) \end{pmatrix} + \begin{pmatrix} \frac{(d+vg)f(t)+vf(t-1)}{m} \\ -f(t-1) \end{pmatrix} \quad (7)$$

where all variables are scalars, where a is the aggregate input coefficient, and where $m=1-(d+vg)a$. In this aggregate version, the model resembles the Metzler's inventory cycle model.

If the matrix from (7) is represented by $\begin{pmatrix} AB \\ CD \end{pmatrix}$ then the condition for stability for (7) is:[5]

$$S_1 = 1 - (AD-BC) > 0 \quad \text{and}$$
$$S_2 = 1 + (AD-BC) + (A+D) > 0 \quad \text{and}$$
$$S_3 = 1 + (AD-BC) - (A+D) > 0.$$

The insertion of the coefficients from (7) for A, B, C and D leads to $S_3 = v(1-a)/m$, which means that stability can only hold when $m>0$. Therefore, S_1 and S_2 yield the necessary and sufficient stability condition:

$$B_1 = (1-a)d - agv > 0 \quad \text{and}$$
$$B_2 = B_1 + 2(1-d) - (1-a)v/2 > 0.$$ [6]

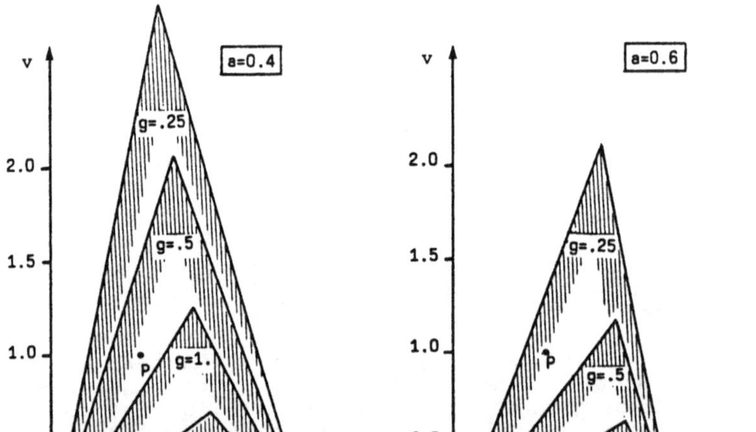

Fig. 1: Areas of Stability

Figure 1 depicts the areas of stable parameter constellations with respect to the reaction coefficients v and d for various stock-level coefficients g and two different input coefficients a. (Example to use the diagram: the combination P = (d,v) = (0.5, 1.0) is stable for a=0.4 when g=0.5 but unstable when g=1.0. It is stable for a=0.6 when g=0.25 but unstable when g=0.5.)

Other things being equal, a large stock-level coefficient g has a destabilizing effect in that it endogenously overstrengthens exogenous changes in demand. A similar argument holds for the input coefficient a. The shape of the stability areas shows that the reaction-to-demand coefficient d must lie within a certain interval: if adjustment to demand is too low, an increasing gap appears over time between demand and production (and also between desired and actual stocks); too great an adjustment leads to continued overreactions.

These statements concerning stability can be transferred to each individual sector in a disaggregated model. One has to notice, however, that for each sector i the intermediate demand $\Sigma_j a_{ij}x_j$ of all sectors takes the place of intermediate demand ax in the aggregated model, so that the adaption processes of the individual sectors influence each other.

4. Application of the model

In this section, results of a model for the Federal Republic of Germany with 56 endogenous sectors are described. As described in the appendix, the coefficients of the model are computed on the basis of monthly statistics from the "Statistisches Bundesamt" (statistical office of the FRG) as far as possible. For some coefficients plausible estimates had to be made because of missing empirical data. All coefficients can be found in the table in the appendix. Because of insufficient data and of some problems associated with estimating, this attempt can only be seen as a first step towards empirical application. So, the following results are mainly for demonstration.

In the model, the temporal effect of an exogenous increase in final demand in period 1, on an economy previously in a stationary equilibrium, is described. In terms of the model-exogenous vector of final demand that means:

$$f(t) = \bar{f} \qquad \text{für } t = \ldots,-2,-1,0 \quad \text{and}$$

$$f(t) = \bar{f}+\Delta\bar{f} \qquad \text{für } t = 1,2,3,\ldots$$

As the model is linear, we can take $\bar{f}=0$ without loss of generality. Therefore, the calculated values $x_j(t)$, $x_j^d(t)$,

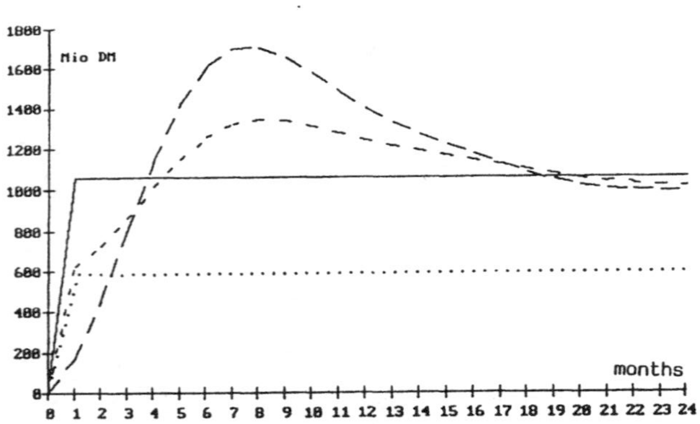

fig. 2.1: Whole economy (j=1,...,56)

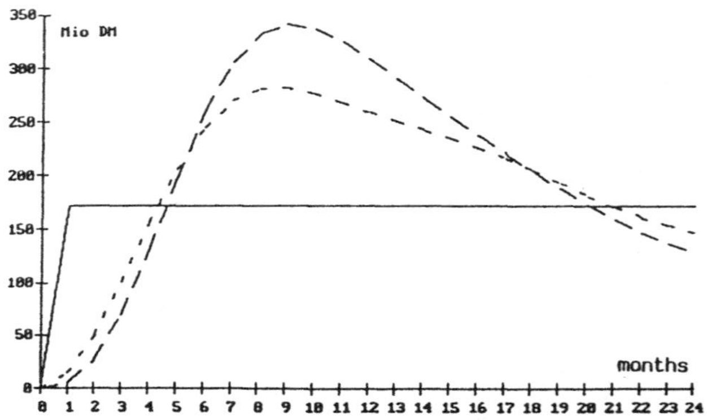

fig. 2.2: Basic and producer goods industry (j=8,...,17)

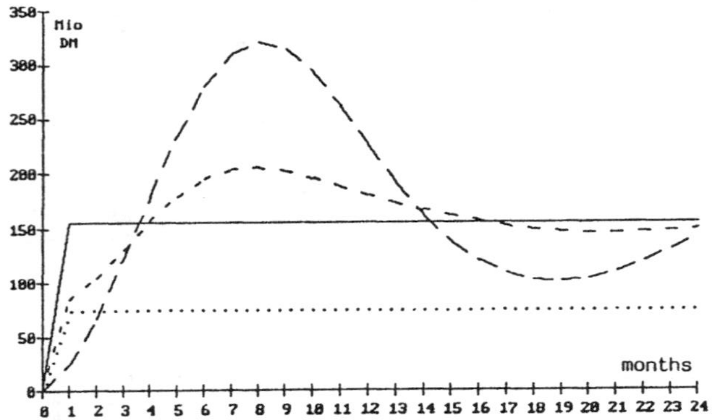

fig. 2.3: Capital goods industry (j=18,...,27)

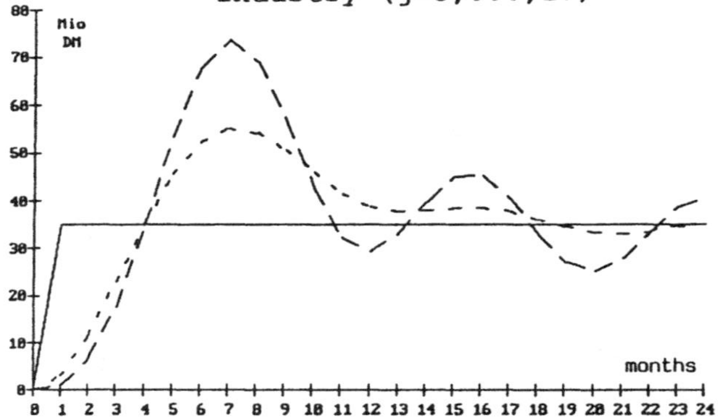

fig. 2.4: Consumer goods industry (j=28,...,36)

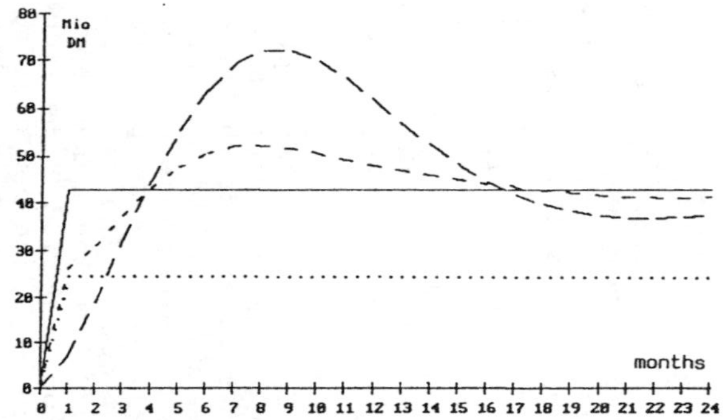

fig. 2.5: Electrical engineering industry (j=25)

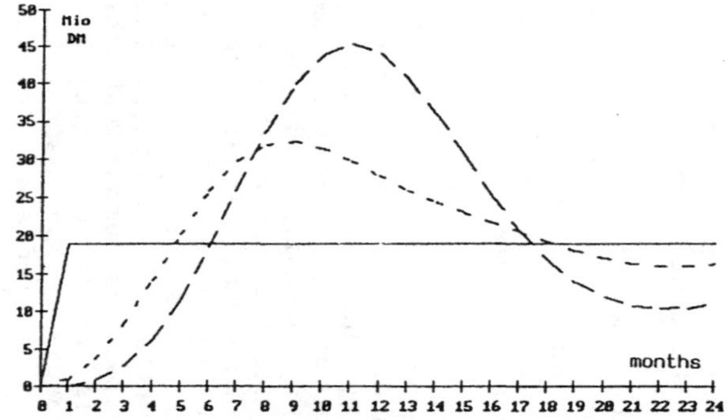

fig. 2.6: Chemical industry (j=8)

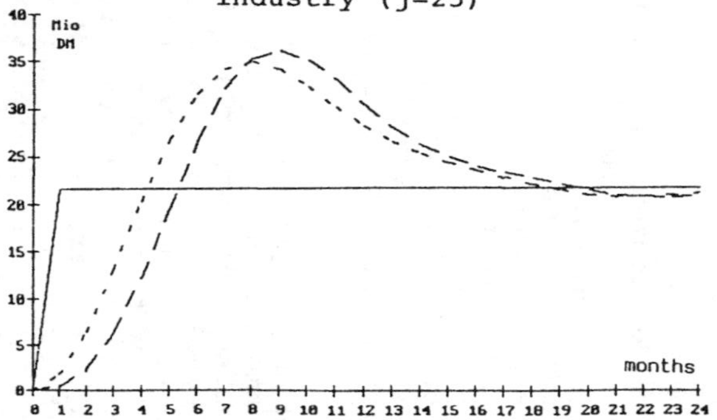

fig. 2.7: Other transport (j=46)

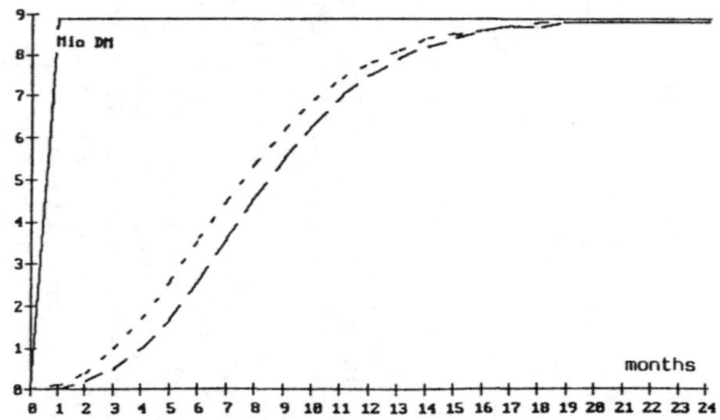

fig. 2.8: Credit institutions (j=47)

demand (x_j^d): ------ final demand ($\Delta\bar{f}_j$): .... (zero for fig. 2.2, 2.4, 2.6, 2.7 and 2.8)
production (x_j): —— —— new equilibrium ($\Delta\bar{x}_j$): ————

$o_j(t)$, $o_j^*(t)$ for t>0 are to be interpreted as additional production, demand levels etc. induced by the demand impulse $\Delta\bar{f}$.

As additional demand, $\Delta\bar{f}$, the 1975 programme for construction and other investment strengthening of the German government has been taken.[7] The monthly sectoral expenditure totals and the comparative static effect $\Delta\bar{x}=(I-A)^{-1}\Delta\bar{f}$ are laid down in the last two columns of the table in the appendix.

As an example the pattern of production and demand over time for several economic sectors is shown in figures 2.1 to 2.8, which also include $\Delta\bar{f}$ (when it does not equal zero) and $\Delta\bar{x}$. Figure 2.1 shows for the whole economy that the increase of demand in the first period is somewhat greater than the exogenous demand impulse. In the following periods, there is a further increase and then finally an oscillation towards the new equilibrium level. Production follows demand with a temporal delay and shows (for the whole economy as well as for the most sectors) a significantly greater overshooting over the new equilibrium level and greater oscillating movements. The reason for this is that the differences between demand and production lead to changes in stock levels, which strengthen the fluctuations in production. This is especially the case at the start of the process, where the rise in desired stock levels, caused by the exogenous rise in demand, leads to stock deficits.

The economic development, as shown in Figure 2.1, is the sum of the developments of the parts of the economy which are by no means synchronised, as shown in figures 2.2 to 2.4. The following four figures represent single chosen sectors: A slower increase in demand and production can be seen, especially at the start, in the sectors which are only indirectly affected by the final demand shock ($\Delta\bar{f}_j=0$; here: Chemical industry, transport, credit institutions). Furthermore, the service sectors show smaller differences between demand and production than the producing sectors because of the impossibility of (output) stock keeping.

For each of the 56 sectors in Figure 3, the black bar represents the number of periods which go by, from the onset of additional demand until production reaches or exceeds the level of the new equilibrium $\Delta\bar{x}_j$. These times are based on the above application of the model. If the initial values were altered, so that, other things remaining equal, there were positive stock levels at the start, then the effect of the demand impulse would be slower, as the white bars show. The times represented by them result, if for each sector in period 0 the stock level is as high as the desired stock level at the new equilibrium ($o_j(0)=g_j\Delta\bar{x}_j$ for j=1,...,40). Moreover (what is not shown here) are the oscillations towards the new equilibrium of smaller amplitudes in this case, due to less overshooting.

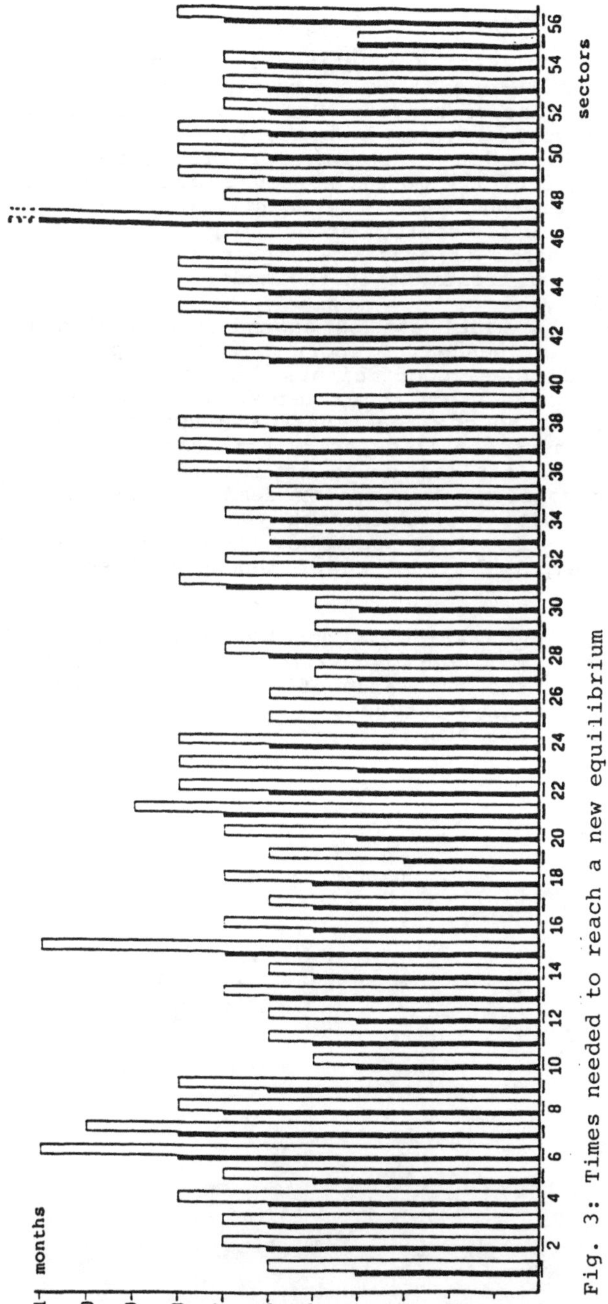

Fig. 3: Times needed to reach a new equilibrium

5. Concluding remarks

The empirical application described is supposed to demonstrate the working of the model in principle; it does not claim to give a true picture of reality. For that purpose the underlying data material would have to be better prepared, with additional information, in order to obtain more reliable estimates for the coefficients of the model. Moreover, those sectors such as coal-mining, ship-building, the construction sector and utility industries, whose output stock keeping is virtually also determined by other aspects than those mentioned here, should receive a special treatment in the model.

In addition, the model can be expanded by including further factors which influence the short-term development of production. The stock-level coefficients g_j, for instance, could be modelled as dependent on the interest rate and/or on the economic situation and the reaction-to-demand coefficients d_j as dependent on the sectoral capacity utilization. And, of course, the Keynesian multiplier could be included in the model (by means of endogenizing consumption demand), instead of letting the Leontief multiplier determine the structure of the model alone. Necessary for that would be, however, a detailed analysis of the temporal structure of income induced consumer demand.

The sphere of application is considerably expanded when production is not modelled as being only demand determined in accordance with the backwards effect but when the input side of production is also included in the model. By doing so, the forwards effect described above could be described and, more generally, the accelerator effect of changes in final demand i.e., the net investments which are induced in the adjustment process (through capacity bottlenecks respectively overcapacities) and which influence the process themselves.

Appendix: Data sources and coefficients of the model

The input coefficients a_{ij} were computed from the input-output tables of 1980 for the Federal Republic of Germany (Statistisches Bundesamt, Fachserie 18, Reihe 2), whose 58 endogenous sectors were aggregated to 56 sectors, j=1,...,56 (by combining sectors 7 and 8 and sectors 38 and 39).

The coefficients d_j, v_j and g_j for the manufacturing industry sectors **(j=8,...,40)** were computed using data provided by the Statistisches Bundesamt on a monthly basis (especially Fachserie 4, various Reihen) between January 1979 and December 1983 (60 months; t=1,...,60). For that purpose the yearly sectoral production values were converted into monthly production values $x_j(t)$ with the help of monthly net production indicies. Order receipts were used as an indicator for demand $x_j^d(t)$ (monthly indicies, using the hypothesis $\Sigma_{t=1}^{60} x_j^d(t) = \Sigma_{t=1}^{60} x_j(t)$, are converted into absolute totals). Monthly sectoral stock levels were calculated by extrapolating the stock level at the start of the year by means of monthly production (+) and monthly turnover (-).

The stock-level coefficient $g_j = o_j^*(t)/x_j^d(t)$ was determined as the relationship between average (realized) stock level and average monthly demand: $g_j = \Sigma_{t=1}^{60} o_j(t) \,/\, \Sigma_{t=1}^{60} x_j^d(t)$. (In the same way g_j was determined for j=3,...,7). The disired stock level $o_j^*(t)$ could then be computed as $g_j x_j^d(t)$. The reaction-to-demand and the reaction-to-stock coefficients d_j and v_j were determined with previously seasonally adjusted production and demand data by using regression analysis. For that purpose, equation (1') was estimated for each sector j for the values S_{dj}=0,1,2,... and S_{vj}=0,1,2,... by the ordinary least squares method, with the structure of the parameter vectors ($d_j(0)$,..., $d_j(S_{dj})$) and ($v_j(0)$,...,$v_j(S_{vj})$) exogenously given as linear decreasing. Out of the estimated alternatives, the coefficients were chosen in accordance with statistical criteria and plausibility.

The data for the computations described above are not available for sectors **j=1,...,7**. The coefficients d_j and v_j (for j=1,2 also g_j) were therefore established "by hand" in accordance with plausibility.

In the service sectors, **j=41,...,56** there is no (output) stock keeping; that means g_j=0 and v_j=0. Due to the shortage of suit-

No. (j)	Industry	g_j	Coefficients $d_j(s)$ s=0	s=1	s=2	s=3	s=4	Coefficients $v_j(s)$ s=0	s=1	s=2	s=3	s=4	$\Delta\bar{f}$ MioDM	$(I-A)^{-1}\Delta\bar{f}$ Mio. DM
1	Agriculture	1.00	.17	.13	.10	.07	.03	.05	.04	.03	.02	.01	7.7	11.1
2	Forestry and fisheries	1.00	.17	.13	.10	.07	.03	.05	.04	.03	.02	.01	1.1	5.1
3	Electricity and heating	0.04	.40	.27	.13	.00	.00	.10	.07	.03	.00	.00		9.6
4	Gas supply	0.19	.40	.27	.13	.00	.00	.10	.07	.03	.00	.00		3.1
5	Water supply	0.03	.40	.27	.13	.00	.00	.10	.07	.03	.00	.00		0.9
6	Coal mining	0.92	.10	.08	.06	.04	.02	.03	.03	.02	.01	.01		5.9
7	Other mining	0.92	.10	.08	.06	.04	.02	.03	.03	.02	.01	.01		1.1
8	Chemical industry	1.03	.13	.11	.08	.05	.03	.03	.02	.02	.01	.01		18.9
9	Mineral oil industry	0.60	.11	.07	.04	.00	.00	.10	.07	.03	.00	.00		14.9
10	Synthetics	0.76	.33	.22	.11	.00	.00	.15	.10	.05	.00	.00		21.2
11	Rubber processing	1.08	.30	.24	.18	.12	.06	.13	.11	.08	.05	.03		1.6
12	Mineral processing	0.74	.40	.27	.13	.00	.00	.15	.10	.05	.00	.00	0.3	58.8
13	Ceramics	1.25	.13	.09	.04	.00	.00	.03	.02	.01	.00	.00		2.8
14	Glass and glasswares	0.99	.21	.16	.12	.08	.04	.10	.08	.06	.04	.02		3.2
15	Iron and steel production	1.45	.14	.09	.05	.00	.00	.05	.03	.02	.00	.00	1.0	37.6
16	Non-ferrous metals	1.29	.27	.18	.09	.00	.00	.07	.05	.02	.00	.00		7.6
17	Founding	1.06	.26	.17	.09	.00	.00	.13	.09	.04	.00	.00		5.9
18	Cold rolling mills	0.72	.28	.22	.17	.11	.06	.08	.07	.05	.03	.02		12.8
19	Steel & light metal const.	3.51	.10	.07	.03	.00	.00	.05	.03	.02	.00	.00	36.3	46.4
20	Mechanical engineering	2.74	.15	.10	.05	.00	.00	.05	.03	.02	.00	.00	15.6	31.6
21	Office machine production	1.07	.11	.08	.04	.00	.00	.05	.04	.02	.00	.00		0.5
22	Road vehicle construction	0.80	.16	.11	.05	.00	.00	.07	.05	.02	.00	.00		4.3
23	Ship-building	7.73	.04	.03	.01	.00	.00	.03	.02	.01	.00	.00	1.2	1.3
24	Air and space construction	4.24	.07	.05	.02	.00	.00	.05	.03	.02	.00	.00		0.1
25	Electrical engineering	1.91	.19	.13	.06	.00	.00	.04	.03	.01	.00	.00	24.5	42.7
26	Precision eng., optics	1.18	.19	.15	.11	.08	.04	.13	.10	.08	.05	.03	0.4	1.6
27	Ironware and metalware	1.20	.26	.21	.16	.10	.05	.09	.07	.05	.03	.02	0.3	13.7
28	Musical instruments, toys	0.85	.22	.17	.13	.09	.04	.11	.09	.06	.04	.02		0.2
29	Wood processing	0.76	.31	.25	.19	.13	.06	.11	.09	.07	.04	.02		11.5
30	Wood manufacturing	0.72	.28	.22	.17	.11	.06	.21	.16	.12	.08	.04		12.3
31	Cellulose/paper production	0.77	.17	.12	.06	.00	.00	.08	.05	.03	.00	.00		2.2
32	Paper/cardboard products	0.78	.27	.21	.16	.11	.05	.10	.08	.06	.04	.02		3.6
33	Printing	0.38	.32	.25	.19	.13	.06	.24	.19	.14	.09	.05		3.4
34	Leather industry	0.68	.22	.14	.07	.00	.00	.15	.10	.05	.00	.00		0.1
35	Textile industry	1.57	.33	.27	.20	.13	.07	.11	.09	.07	.04	.02		1.4
36	Clothing industry	0.86	.19	.13	.06	.00	.00	.08	.05	.03	.00	.00		0.2
37	Food industry	0.31	.27	.18	.09	.00	.00	.09	.06	.03	.00	.00		5.0
38	Tobacco industry	0.46	.30	.20	.10	.00	.00	.10	.07	.03	.00	.00		0.6
39	Construction	3.60	.16	.13	.10	.07	.03	.02	.02	.01	.01	.00	314.9	319.9
40	Inside construction	0.35	.25	.16	.08	.00	.00	.13	.09	.04	.00	.00	173.5	184.4
41	Wholesale, trade agencies	0.00	.27	.21	.16	.11	.05	.00	.00	.00	.00	.00	2.6	23.5
42	Retail trade	0.00	.27	.21	.16	.11	.05	.00	.00	.00	.00	.00		1.5
43	Railways	0.00	.27	.21	.16	.11	.05	.00	.00	.00	.00	.00	0.4	4.0
44	Shipping and ports	0.00	.27	.21	.16	.11	.05	.00	.00	.00	.00	.00		1.2
45	Mail and communication	0.00	.27	.21	.16	.11	.05	.00	.00	.00	.00	.00		6.2
46	Other transport	0.00	.27	.21	.16	.11	.05	.00	.00	.00	.00	.00	0.1	21.6
47	Credit institutions	0.00	.27	.21	.16	.11	.05	.00	.00	.00	.00	.00		8.9
48	Insurance	0.00	.27	.21	.16	.11	.05	.00	.00	.00	.00	.00		5.0
49	Property rental	0.00	.27	.21	.16	.11	.05	.00	.00	.00	.00	.00		3.7
50	Hotels and restaurants	0.00	.27	.21	.16	.11	.05	.00	.00	.00	.00	.00		6.4
51	Education, science	0.00	.27	.21	.16	.11	.05	.00	.00	.00	.00	.00		4.5
52	Health & veterinary serv.	0.00	.27	.21	.16	.11	.05	.00	.00	.00	.00	.00		0.7
53	Other services	0.00	.27	.21	.16	.11	.05	.00	.00	.00	.00	.00	10.3	53.5
54	Government	0.00	.27	.21	.16	.11	.05	.00	.00	.00	.00	.00	1.7	6.3
55	Social services	0.00	.27	.21	.16	.11	.05	.00	.00	.00	.00	.00	0.6	0.6
56	Private organizations	0.00	.27	.21	.16	.11	.05	.00	.00	.00	.00	.00		0.5
	Sum												592.5	1057.2

Sectors and Coefficients of the Model, Final demand impulse $\Delta\bar{f}$ and the comparative static effect $(I-A)^{-1}\Delta\bar{f}$

able demand indicators, d_j could not be estimated and was set to the value 0.8 in the sum for all sectors, distributed to the coefficients $d_j(0)$ to $d_j(4)$.

The shares b_{ij} (see equation (5')) were set to 0.9 for the sectors i=3,4,5 and for all other sectors i to 0.5 (with j=1,...,56).

Footnotes

1 Besides the difficulty in finding reliable data for the sectoral capital coefficients this is certainly due to the instability of such models. See [3 or 2, p. 71 ff] for instance.

2 See ten Raa [8; 9] or Pörting [4] for instance.

3 With regard to capital coefficients this approach has been used before, see Volkonsky [10] or Meyer, U. [2, p. 43 ff] for instance.

4 Besides these effects, induced (positive or negative net) investments could appear as a further effect of a change in final demand. We shall not consider this aspect below - firstly because these effects are less short-term and secondly because otherwise it would be necessary to consider investment reaction times, investment gestation times and capacity utilisation.

5 See Gandolfo [1, p. 59] for instance.

6 The condition $m>0$ is contained in B_1 for $d\leq 1$ and in B_2 for $d>1$.

7 "Programm zur Stärkung von Bau- und anderen Investitionen". The sectoral expenditure totals are taken from [6] and have been converted to fit underlying sectoral division used here. In accordance with the approximate duration of the programme, the totals have been divided by 16 to obtain monthly values.

References

[1] Gandolfo, G.: Economic Dynamics: Methods and Models. Amsterdam 1985

[2] Meyer, U.: Dynamische Input-Output-Modelle. Königstein/Ts. 1980

[3] Meyer, U. and J. Schumann: Das dynamische Input-Output-Modell als Modell gleichgewichtigen Wachstums. Mit einem Anwendungsbeispiel für die Bundesrepublik Deutschland. Zeitschrift für die gesamte Staatswissenschaft 133 (1977), 1-37

[4] Pörting, H.: Verteilte Aktivitäten in der Input-Output-Analyse. Dissertation Münster 1987 (fortcoming)

[5] Schnabl, H. and H.W. Holub: Qualitative und quantitative Aspekte der Input-Output-Analyse. Ansatzpunkte zu einer Neuorientierung der Konjunkturpolitik. Zeitschrift für die gesamte Staatswissenschaft 135 (1979), 657-678

[6] Stäglin, R.: Multiplikatorwirkungen des Konjunkturprogramms von 1975: Anwendung des um den Keynes'schen Multiplikator erweiterten Input-Output-Modells. DIW, Beiträge zur Strukturforschung, Heft 45. Berlin 1976

[7] Statistisches Bundesamt:

Ausgewählte Zahlen für die Bauwirtschaft

Fachserie 4, Reihe 2.1: Indizes der Produktion und der Arbeitsproduktivität, Produktion ausgewählter Erzeugnisse im Produzierenden Gewerbe

Fachserie 4, Reihe 2.2: Auftragseingang und Umsatz im Verarbeitenden Gewerbe, Auftragseingang und Auftragsbestand im Bauhauptgewerbe

Fachserie 4, Reihe 4.2.1: Beschäftigte, Umsatz und Investitionen der Unternehmen und Betriebe im Bergbau und im Verarbeitenden Gewerbe

Fachserie 4, Reihe S.7: Neuberechnung der Produktionsindizes und des Index der Arbeitsproduktivität auf Basis 1980

Fachserie 18, Reihe 1: Konten und Standardtabellen

Fachserie 18, Reihe 2: Input-Output-Tabelle 1980

Statistische Jahrbücher für die Bundesrepublik Deutschland, 1979-1983

Wirtschaft und Statistik

Zensus im Produzierenden Gewerbe, Heft 2, Bergbau und Verarbeitendes Gewerbe

[8] ten Raa, Th.: Applied Dynamic Input-Output with Distributed Activities. European Economic Review 30 (1986a), 805-831

[9] ten Raa, Th.: Dynamic Input-Output Analysis with Distributed Activities. Review of Economics and Statistics 68 (1986b), 300-310

[10] Volkonsky, V.: The Role of Intersectoral Models in Determining the Discount Rate. In: Polenske, K.R. and I.V. Skolka (ed.): Advances in Input-Output-Analysis. Proceedings of the Sixth International Conference on Input-Output-Techniques, Vienna, April 22-26, 1974. Cambridge/Mass. 1976

DYNAMIC SYSTEMS AND INPUT–OUTPUT THEORY

Albrecht Ritschl, University of Munich[1]

1. Introduction

Over the last decades increasing attention has been paid to the fact that in describing dynamic phenomena, scientists of various different fields developed formal models which revealed to be more or less identical. This discovery (see e.g. v.Bertalanffy [2]) led to the introduction of General System Theory as a pure theory of dynamic systems. From the environmental discussion regarding modern large–scale production it is obvious that the process of production of commodities can be regarded as an open system [4]. It should thus be expected that the economic theory of interindustry relations, that is, input–output analysis, provides models which are formally equivalent to those of system theory. However, Livesey [13] [14] has shown that Leontief's [11] dynamic input–output system appears as inconsistent when analyzed from a system–theoretic point of view. These disappointing results have led to the suspicion of Kalman [6] that mathematical economics as a whole provides models which are inherently inconsistent.

In the following it shall be tried to give an explanation of why Leontief's dynamic system (like almost all economic systems regarding dynamics with embedded statics) shows such unsatisfactory results. In Sec. 2 we introduce a dynamic production system on an intuitive level. It is shown that the introduction of growth imposes some fundamental restrictons on the process of production. This reasoning is formalized in Sec. 3, in which we introduce the dynamic input–output system of Goodwin [5] and Metzler [17]. Employing a result of Nikaidô [20], we show that this system is consistent in the sense of system theory if all produced goods are basic commodities in the sense of Sraffa [23]. Sec. 4 applies this result to Leontief's dynamic system. It is shown that the inconsistency of this model is caused by its failure to take into account the dynamic structure of commodity production. If such dynamics are reintroduced into the system, its inconsistency is partly removed. Sec. 5 examines the problem somewhat further. Some problems involved with durable capital goods are sketched, and it is shown that, as a rule, systems like that of Leontief are necessarily confined to steady states with no fluctuations.

[1] I am grateful to Dr. Reinhard Neck, University of Economics, Vienna, for his helpful criticism of my conference paper, which lead to an almost complete revision of the text, though not of its conclusions.

2. Outline of the Problem

Consider a simple case: A consumer good whose quantity is denoted as C is produced in three successive steps which themselves require the products of their foregoing stages as inputs to production. We denote the quantities of these intermediate goods as X_1 and X_2, respectively. Assuming a very simple structure of production, we suppose that X_1 is produced solely from natural resources with the aid of labor, whereas X_2 requires inputs of X_1 The consumer good is then manufactured from intermediate goods X_2. We thus assume a unidirectional production chain which leads from the extraction of raw materials from nature to the final production of consumer goods. This is represented graphically in Figure 1:

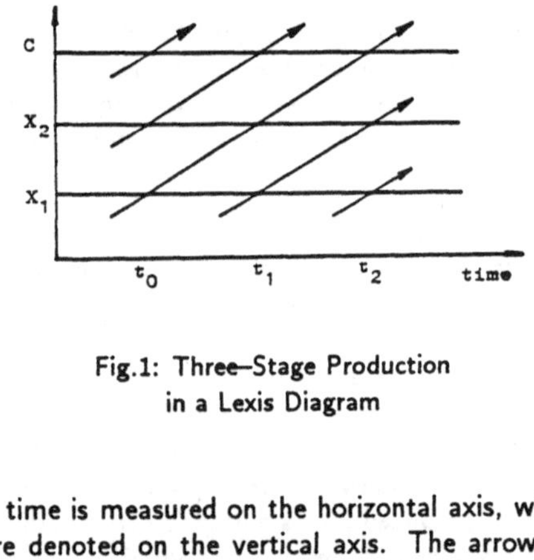

Fig.1: Three–Stage Production
in a Lexis Diagram

In this diagram, time is measured on the horizontal axis, whereas the stages of production are denoted on the vertical axis. The arrows going from the lower left to the upper right represent the flow of intermediate commodities as conforming to our assumptions. The products which leavie stage 1 at time t_0 undergo further processing in stage 2, from which they leave as semi–finished goods at time t_1. These goods are then used in stage 3 to produce consumer goods, which are themselves finished at time t_2. Apparently, this flow of goods through the production chain can be understoof as a flow through time. Since no given particle can be found at two different points in space simultaneously, a certain time span is required to turn raw materials into finished goods, and it is clear that this intertemporal character of production is independent of the time scale applied. As long as production velocities are not infinitely high, there will always be a time scale which is sensitive enough to depict the time structure of the production process correctly.

Thus far, nothing has been said about quantities. Suppose that t_0, t_1 etc. are equidistant points in time and that it takes, say, 10 units of intermediate good 2 to produce 1 unit of the finished good, while 5 units of raw products 1 are used per unit of the intermediate product 2 (i.e. we assume technical coefficients $a_{12} = 5$ and $a_{23} = 10$). Equilibrium then requires that the outputs of intermediate goods are $X_1 = 500$ at time t_0 and $X_2 = 100$ at time t_1, respectively; these support a production of 10 units of the consumer good at time t_2. As long as outputs are stationary over time, the technical relations a_{12} resp. a_{23} appear in both the cross- section data and the dynamic (cohort) analysis of the production chain:

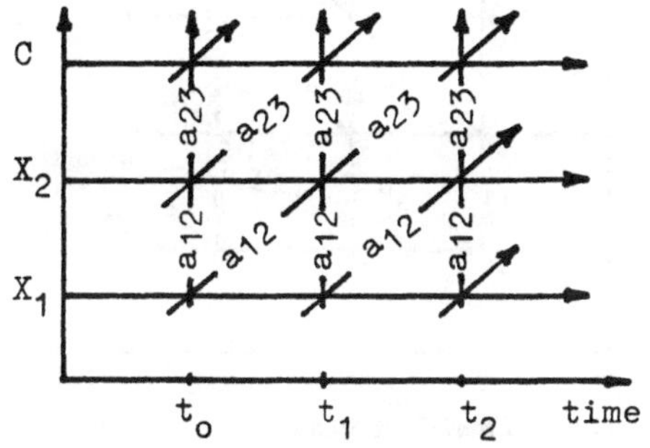

Fig. 2: Cross–Section Data and Cohort Analysis in the Stationary State

Figure 2 applies two concepts of data analysis to our three– stage production chain. First, a cross–section observation of outputs is taken at a given point in time, for example at time t_1. Second, a given production lot is followed through the production chain from primary production in stage 1 to the fabrication of finished consumer goods in stage 3. As is easily verified, with constant rates of output the technical input– output coefficients a_{12} and a_{23} are obtained by either method of data analysis.

This is no longer true if exponential growth is introduced. Assume that the output of consumer goods is to grow between t_2 and t_3 by, say, 10% (i.e. $C(t_2) = 10$, $C(t_3) = 11$). This requires that deliveries by the foregoing industries must have increased prior to t_3, for otherwise the increase in output could not have been supported. In equilibrium, the output of stage 2 at time t_2

must have been $X_2 = 110$ units, which in turn requires that stage 1 supplied $X_1 = 550$ units at time t_1. In the same way, the output of primary products X_1 in the following point of time, t_2, must equal 555.5 units if growth is to continue.

If we now observe cross-section data of production, for example at time t_2, the input–output coefficients which we obtain are biased estimators of the underlying technical ones. Taking a_{23} as an example, we note that its estimator $\tilde{a}_{23} \equiv X_{2,t_2}/C_{t_2} = 110/10 = 11$. Similarly, $\tilde{a}_{12} = 5.5$. It is apparent that this bias is equal to the proportionate factor of growth g which in our example is $g = 1.1$. This result is illustrated graphically in Figure 3:

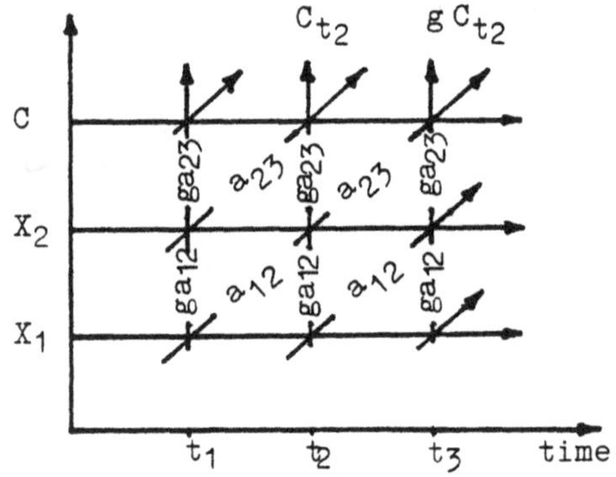

Fig. 3: Cross–Section and Cohort Analysis in the State of Exponential Growth

In this figure, outputs are assumed to grow by factor g. As before, the stages of production are connected by the technical coefficients a_{ij}. Once again, it is possible to observe cross–section coefficients, but these are now biased against the underlying technical coefficients. Given steady–state growth, these biases are systematic, since we always have $\tilde{a}_{ij} = g\, a_{ij}$. This result, of course, applies as well if the production chain comprises loops and feedbacks, that is, if products from stages near to the final production are recycled into productions on earlier stages of the system.

3. A Dynamic System of Production

In this section, the concepts given on an intuitive level in Section 2 will be introduced formally. This serves as a preparatory exercise for the examination of Leontief's dynamic system in Section 4. As well, it enables us to highlight some of the relationships between this problem and certain concepts of system theory.

In the Lexis diagrams employed so far, we have assumeed that each production step is of duration $\Delta t = t_i - t_{i-1}$. This may be rationalized in several ways, see my [22] for details. One method is to follow von Neumann [19] and to conceptually decompose longer production chains into as many elementary processes as there are time steps required. This implies the introduction of auxiliary intermediate "shadow" products. Though not unambiguous, this convention suffices for the considerations to follow.

As before, let a_{ij} be the technical input coefficient of commodity i in production j. i may or may not be used for private consumption, and γ_i is the ratio of consumption of i to total output of i. In equilibrium, the output of i at time t must suffice to both provide inputs to commodities finished at time $t + 1$ and meet consumer demands:

$$x_{i,t} \stackrel{!}{=} \sum_{i=1}^{n} a_{ij} x_{j,t+1} + \gamma_i x_{i,t} \quad i,j = 1,\ldots,n \qquad (3.1a.)$$

or, in matrix notation:

$$x_t = A x_{t+1} + C x_t \qquad (3.1b.)$$

where C is a diagonal matrix with elements γ_i and x_t is the n–dimensional vector of outputs. (3.1.) has been proposed by [5] and [17].

The output of consumer goods may be written separately:

$$c_t = C x_t \qquad (3.2.)$$

c_t is a semipositive n–dimensional vector. There are, of course, external inputs r_k, $k = 1,\ldots,m$, which can be related to commodity outputs by a matrix of resource inputs D:

$$r_t = D x_{t+1} \qquad (3.3.)$$

(3.3.) is usually omitted in economic input–output theory, for naturally the interest focuses on the properties of the flow of intermediates rather than the inputs of external materials, of which many need not themselves be traded commodities.

Not all commodities i will appear with strictly positive input coefficients a_{ij} everywhere, nor will they necessarily have positive consumption ratios. The following cases may be distinguished:

1. Pure consumption goods: Commodity l is said to be a pure consumption good if $a_{lj} = 0$, $j = 1, \ldots, n$. Then it follows from (3.1a.) that $\gamma_l = 1$.

2. Pure intermediate goods: If commodity h has a consumption rate of zero, i.e. $\gamma_h = 0$ it is a pure intermediate good. Then there must be some j such that $a_{hj} > 0$.

3. Basic commodities, Sraffa [23]: A commodity k is a basic commodity if $a_{kj} > 0$ for some j and if there is a chain such that $a_{km} \cdot a_{mp} \cdot a_{po} \cdot \ldots \cdot a_{tu} \cdot a_{uv} > 0$ for all $v = 1, ..., n$.

Apparently, definitions 1) and 2) exclude one another. Definition 3) requires that no basic commodity is a pure consumption good. This implies that a pure intermediate good may be a basic commodity, but not that all intermediate goods are necessarily basic commodities, for the latter property also requires that a basic commodity is a direct or indirect input to all other commodities. This definition plays a role when solutions to (3.1.) are examined.

Clearly, (3.1.) will not always have a solution such that the strict equality holds. But positive solutions do exist if vector x_{t+1} is a scalar multiple of vector x_t and if the matrix of technical coefficients A is both semipositive and indecomposable. Equation (3.1.) then takes the form:

$$x_t = wAx_t + Cx_t, \quad x_{t+1} = wx_t$$

or

$$(\frac{1}{w}I - (I - C)^{-1}A) = 0 \tag{3.4.}$$

and the Frobenius theorem can be applied (see in this context [20]). Then $g > 0$, the proportionate factor of growth (or decline), is the reciprocal of the dominant eigenvalue and x is the the respective eigenvector of matrix $(I - C)^{-1}A$.

Matrix A is both semipositive and indecomposable if all goods are basic commodities, for it is then impossible to transform A into a nonnegative triangular matrix. If e.g. A was upper– block triangular, there would be a commodity i of order $i > j$ which would in no way serve as an input to commodity i, since all coefficients a_{ig}, $g < i$ are zero. This contradicts the definition of a basic commodity. From this it follows also that matrix $(I - C)^{-1}$ is positive. No basic commodity can be a pure consumption good, hence $0 \leq \gamma_i < 1$ for all i and $C < I$. Trivially, the inverse of a positive diagonal matrix like (I- C) is positive as well.

Nikaidô [20] examined the conditions under which system (3.1.) can be transformed into a system of ordinary forward difference equations. Solve (3.1b.)

for x_{t+1} to find:

$$x_{t+1} = A^{-1}(I - C)x_t \tag{3.5.}$$

If a solution to (3.4.) exists, A is invertible (and we may infer from (3.4.) that if $(I - C)^{-1}A$ is semipositive, its inverse $A^{-1}(I - C)$ must be semipositive as well). Together with (3.3.), (3.5.) constitutes a forward–recursive system, whereas (3.1.) together with (3.2.) forms a backward–recursive system. This is graphically represented in Figure 4:

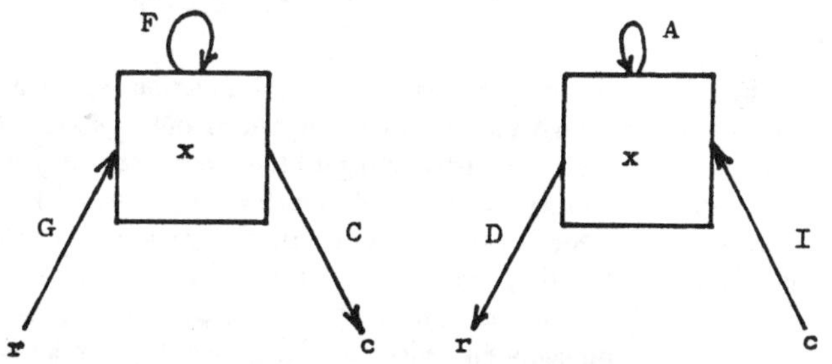

Fig. 4: Forward–Recursive and Backward–Recursive Systems of Production

Figure 4 illustrates two ways of system representation. In either case, the system is viewed as a black box whose state is denoted by vector x. An open system communicates with its environment by inputs and outputs. In formal representation, this means that there are input and output vectors which are connected with the state vector by vector–valued functions. System theory examines the conditions under which the behavior of the state variables can be inferred or controlled by its outputs and inputs. (The standard reference of the state–space approach is [7]. There are several textbooks of the topic with applications to economics, see [1], [3], and [15].)

The first diagram denotes inputs as r and outputs as c, which is intuitively appealing; resource inputs enter the system for further processing, in the course of which intermediate goods x are produced. Some of these leave the system (i.e. the sphere of production) as consumer goods c, others are recycled as productive inputs. Eqs. (3.5.) and (3.2.) can be viewed as a description of this system: (3.5.) relates the state vector at time $t + 1$ to the state vector at time t (which in the left diagram of Figure 4 is represented by F), (3.2.) describes mapping C of Figure 4, which makes system outputs dependent on the state vector. (3.5.) neglects the effects that inputs have on the state vector.

Note that the concepts of "output" in system theory and in input–output theory are different: In system theory, output is the signal to the environment emitted by the system (we denoted it as c). Outputs in that sense can be observed but not controlled. On the other hand, the notion of output in input–output theory refers to the magnitude of the state variables, that is, to the elements of x. System theory often assumes that x is not directly observable.

Now it is possible to logically reverse the causal direction within a system. Assume that the variables which we first denoted as outputs are subject to external control. Then it makes sense to regard these outputs as logical inputs. Similarly, the system inputs may be viewed as its logical outputs. This concept of reversed causation underlies the second diagram in Figure 4 as well as eqs. (3.1.) and (3.3.). In this diagram, system "outputs" are now represented by r, which is regarded as dependent on x: $D : x \mapsto r$. System "inputs", on the other hand, are only implicitly embedded in (3.1.), but they can be expressed as $I : c \mapsto x$. The state transformation is given by the mapping $A : x \mapsto x$, this is, the technical input–output matrix. Since matrix A determines x_t by recursion to the state of the system in the subsequent point of time, x_{t+1}, such a system is called backward recursive. Written out in full, the equations of motionn thus are:

$$x_t = Ax_{t+1} + c_t \qquad \text{where } c_t = Cx_t$$
$$r_t = Dx_{t+1}$$

(where of course the time structure of the second equation presents logical difficulties). This reversed order of causation has an economic meaning: Following the Keynesian debates of that time, input–output theory has been concerned with assigning commodity production to a given rate of consumer demand c_t, see Leontief [10], and the rate of demand itself was regarded as subject to macroeconomic control.

The forward– and backward–recursive forms of a given system are not necessarily equivalent. Where they are, the state–transition matrix must be invertible, this is, nonsingular. System theory requires as a test of consistency that a given system should be both controllable and observable, and a necessary condition for that is the nonsingularity of matrix A (or matrix F in the 'forward' case). We have used Nikaidô's [20] results to show that basic–commodity systems in the sense of Sraffa [23] meet this requirement. It is easily shown that they are also controllable and observable, [22]. But these concepts are of doubtful value for such steady–state systems. See below Sec. 4.

There is a class of economic input–output models which lack invertible state–transition matrices. To these we shall now turn.

4. Singular Input–Output Systems

Recall that in Section 2, two different ways of data analysis were examined. Where the input–output systems of Section 3 imply dynamic analysis, we now consider a system which is based on cross–section analysis. At any given point in time, the output of commodity i is divided between its further destinations. Denoting the delivery to production j as $x_{ij,t}$, we may express the technical coefficient a_{ij} as given by: $x_{ij,t} = a_{ij}x_{j,t+1}$: 500 tyres are delivered today, and 100 cars are finished tomorrow. Cross–section analysis relates the deliveries of today to the outputs of today. We denote the estimated input–output coefficient which we obtain by cross– section analysis as $\hat{a}_{ij}$. If output grows or decreases by factor g, the true technical coefficient is related to its estimator by:

$$a_{ij} = \frac{x_{ij,t}}{gx_{j,t}}$$

Since the cross–section estimator of a_{ij} is defined as:

$$\hat{a}_{ij} = \frac{x_{ij,t}}{x_{j,t}} \tag{4.1.}$$

we have:

$$\hat{a}_{ij} = ga_{ij} \tag{4.2.}$$

Consider the steady state where all productions grow or fall by g. Defining $\hat{A}$ as the matrix of estimated input–output coefficients, we find that the system:

$$x_t = wAx_t + c_t$$

or:

$$x_t = (I - wA)^{-1}c_t \tag{4.3.}$$

can be represented by its cross–section estimation:

$$x_t = (I - \hat{A})^{-1}c_t \tag{4.4.}$$

(4.4.) is Leontief's [10] static open system. Leontief himself is aware of the cross–section nature of his input–output coefficients. Turning to the problem of growth he states that if production is to expand from time t to time $t+1$, intermediate products must be supplied at time t in excess of reproduction demands at time t. Thus, he continues, the input–output coefficients obtained through cross–section at t must be divided by the growth factor to obtain the underlying technical coefficients. In our notation this is:

$$\frac{\hat{a}_{ij}}{w} = a_{ij}$$

which is just a reformulation of (4.2.). In a publication that appeared as early as 1928, Leontief [9] in fact employs a Lexis diagram to visualize the process of growth.

In economics, system (4.4.) is often misinterpreted as a system of technical coefficients. This is especially misleading when economic growth is considered, for it compells analysts to think of growth as a sequence of mutually unconnected processes which are inherently static. Leontief [11] [12] has proposed a dynamic system which indeed admits such misinterpretation. Systems of that kind have become known as "dynamics with embedded statics" (which in fact comes close to a selfcontradiction). In Leontief's dynamic system, stocks and flows are discriminated. At time t (which is introduced as a period of time, though Leontief employs continuous time) flows of commodities are produced as governed by (4.4.). In addition to (4.4.), stocks of commodities s are accumulated, where s is an n–dimensional vector. The amount accumulated is (in discrete time, which we shall adopt as to avoid contradictions):

$$I = s_{t+1} - s_t$$

Here I is to denote investment in stocks. The expenditure equation then becomes:

$$x_t = \bar{A}x_t + (s_{t+1} - s_t) + c_t \tag{4.5.}$$

with $\bar{A}$ as Leontief's input–output matrix. Stocks s_t are used as durable means of production such that:

$$s_t = Bx_t \tag{4.6.}$$

Wear and tear is neglected. Insert this into (4.5.) to find:

$$x_t = \bar{A}x_t + B(x_t + 1 - x_t) + c_t \tag{4.7.}$$

which is Leontief's dynamic model in reduced form.

It has been attempted to convert this model into the forward–recursive form. However, this requires that matrix B is invertible. If B is invertible the solution is:

$$x_{t+1} = B^{-1}(I - A - C + B)x_t \qquad \text{where } c_t = Cx_t \tag{4.8.}$$

This case, however, is regarded as unlikely, for it is not assured that all goods find their use as durable means of production (i.e. some rows of B are empty). As well some goods may be produced without durable goods (i.e. some columns of B are empty). This has become known as the singularity problem in Leontief's dynamic model, see [[8] [13] [14] [16] [18]].

Two ways out of this impasse have been proposed. The first is to just provide a backward recursion for x_t, Leontief [12]:

$$x_t = (I - A + B)^{-1}[Bx_{t+1} + c_t] \tag{4.9.}$$

where

$$c_t = Cx_t$$

is, once again, regarded as given exogenously (the assumption that $(I-A+B)$ is nonsingular is not harmful whereas the singularity assumption on B is). But this method yields no information regarding outputs at time $t+1$ if the current rate of output is known. The second approach is to consider lower–dimensional subsystems of (4.8.) such that there is a matrix $\tilde{B}$ given by:

$$B = \begin{bmatrix} \tilde{B} \\ 0 \end{bmatrix} \tag{4.10.}$$

Then a suitable subvector of stocks $\tilde{s}$ may be defined. If now the system is partitioned, one obtains after rearrangement:

$$\begin{bmatrix} \tilde{B} \\ 0 \end{bmatrix} x_{t+1} = \begin{bmatrix} G \\ H \end{bmatrix} x_t - \begin{bmatrix} D_1 \\ D_2 \end{bmatrix} c_t \tag{4.11.}$$

Only in well–behaved cases, $\tilde{B}$ is obtained from B by just rearranging rows. Then matrices G and H are mere submatrices of $I - \bar{A} + B = \begin{bmatrix} G \\ H \end{bmatrix}$, and matrices D_1 and D_2 together form the identity matrix. Otherwise, transformations of B must be made as to assure $\tilde{B}$ to be of full rank $r < n$. This is where matrices D_1 and D_2 are needed. Assuming that matrix:

$$Z \equiv \begin{bmatrix} \tilde{B} \\ H \end{bmatrix}$$

is nonsingular and denoting its inverse as:

$$Z^{-1} = [EF] \quad ,$$

(4.11.) can be transformed into:

$$\tilde{s}_{t+1} = GE\tilde{s}_t + (GFD - C)c_t \qquad (4.12.)$$

See [15]. Then a lower–dimensional subsystem is found which presents no singularity problems. But there still remains the basic logical slippery of the model, its failure to take respect of demands for flow input in its relation to extended reproduction.

Suppose it were possible to produce flow commodities x within a certain time span t, after which all production ceases and starts anew in $t+1$ (possibly on an expanded scale). Stock commodities s are the only means of transferring wealth through time. This is what Leontief's model says. In the context of the Lexis diagram, we obtain a series of mutually unconnected processes of production:

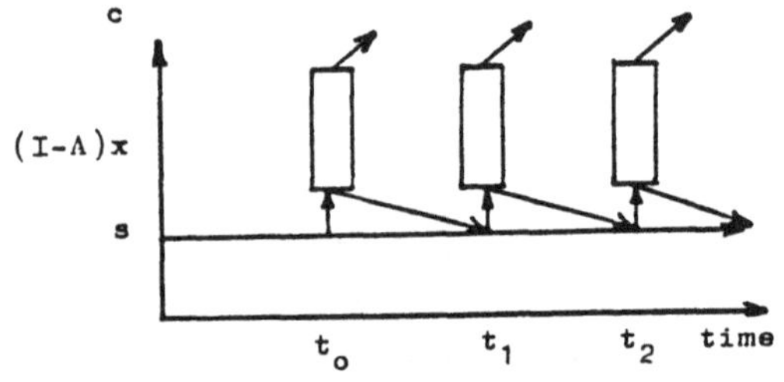

Fig. 5: Intertemporal Structure of Production
in the System of Leontief

In Figure 5, two production cycles t and $t+1$ are depicted. Let periods t and $t+1$ be called "summer" and the period in between, "winter", then Figure 5 is a seasonal model of, say, an agrarian economy. We have no useful information as to the time structure of production within "summer", since there are no intermediate products inherited from the past. For if inputs Ax_t were regarded as the outputs of the year before, we would enter into a contradiction (see eq. (4.13.) below): Inputs Ax_t just suffice to continue production on a constant level, while we seek to explain the phenomenon of growth. It is only known that the process yields net outputs $(I - \bar{A})x$. These are used up completely by consumption c_t and accumulation of stocks Δs_t. For this reason it is impossible to build up inventories of flows. If a given commodity i survives "summer", it does so by entering into stock s_i. Leontief describes a one–stage production where only stock inputs are used. Or, if it describes the intertemporal input–output structure of the economy, the production process must necessarily be stationary.

Recall that the basic equation of Leontief's dynamic system requires net output to be sufficient to meet the demands for stock accumulation and consumption. From (4.5.) we have:

$$(I - \bar{A})x_t = \Delta s_t + c_t = C + I \tag{4.13.}$$

Now if $\bar{A}$ was a matrix of technical coefficients then the LHS of (4.13.) would represent equilibrium net output in a stationary environment: x_t is what the economy produces, and $\bar{A}x_t$ is the vector of inputs required for just that. Since inputs precede outputs in time, in the absence of imports the output of "summer" t, $\bar{A}x_t$, imopses an upper boundary to next period's output:

$$\bar{A}x_t = \bar{A}x_{t+1} \tag{4.14.}$$

and trivially $x_t = x_{t+1}$. (In the context of Figure 5: The "summers" must be linked by input–output arrows such as in Figure 1 above etc.) For this reason Δs_t on the RHS of (4.13.) must be zero, and the whole problem collapses into a stationary state with no accumulation. $\bar{A}$ can thus be no matrix of technical coefficients (though it is commonly held to be), or s can be no vector of durable stocks. What is missing in (4.5.) and (4.13.) is the increment in demand for flow inputs which is necessary to support reproduction on an expanded scale:

$$A(x_{t+1} - x_t) \tag{4.15.}$$

where now A represents the true technical matrix. Assume now that $\bar{A}$ is a matrix of observed cross–section data. This would conform to Leontief's early [10] view of input–output coefficients. Then we are entitled to subtract (4.15.) from the LHS of (4.14.) such that:

$$(I - A)x_t - A(x_{t+1} - x_t) = \Delta s_t + c_t \tag{4.16.}$$

where the LHS is gross production as diminished by the flow–input demands for extended reproduction. In the case of steady–state growth this procedure is even more obvious, for then:

$$\underbrace{x_t - Ax_{t+1}}_{\text{LHS (4.16.)}} = (I - wA)x_t = \underbrace{(I - \hat{A})x_t}_{\text{LHS (4.13.)}}$$

from (4.3.) and (4.4.). (4.16.) means that the static inverse $(I - \hat{A})$ must be replaced by the steady–state dynamic inverse $(I - wA)$ if the model is to be capable of growth (for this interpretation see also Pasinetti [21]).

Now the system has the desired features. Replace Δs in (4.16.) by (4.6.) to obtain:

$$x_t - Ax_{t+1} = Bx_{t+1} - Bx_t + c_t$$

or:

$$(A + B)x_{t+1} = (I - B)x_t + c_t$$

If all goods are basic commodities, then A will be invertible and so will $(A + B)^1$. Therefore:

$$x_{t+1} = \underbrace{(A + B)^{-1}(I - B)}_{F} \; x_t + \underbrace{(A + B)^{-1}}_{G} \; c_t \qquad (4.17.)$$

where all matrices have rank n. (4.17.) means that outputs at time $t + 1$ are determined by both intermediate inputs and consumption at time t. Since matrices F and G are of full rank n, each of the higher–order matrices:

$$F^G; F^3G; \ldots; F^{n-1}G$$

of the multiplier chain which connects outputs with past consumption are of rank n, so (4.17.) is completely controllable. Note that this result remains unaffected if c_t is only semipositive, this is, if some commodities are pure intermediate goods.

There are, however, additional obstacles to a general applicability of Leontief's dynamic system. First, it is doubtful from an economic point of view whether systems like (4.17.) or the systems of Sec. 3 can be employed to describe the evolution of output in time, for they implicitly assume that markets always clear. Though not impossible, it is hard to imagine that economic agents react on disequilibria without changing the technique of production, accumulating inventories, and the like. Therefore, matrices like F and G of (4.17.) may

---

[1] If there are pure consumption goods, we might alternatively consider a lower–dimensional subsystem of basic commodities.

still be derived, but they will now be time dependent, and it will no longer be possible to regard the coefficients of A as purely technical input coefficients (for changes in A may merely reflect fluctuations of inventories). A second impasse is its way of treating durable capital. It would go beyond the scope of this paper to show this here, see however my [22] for a discussion.

REFERENCES:

[1] AOKI, M.: *Optimal Control and System Theory in Dynamic Economic Analysis*. New York: North Holland, 1976.

[2] BERTALANFFY, L.v.: *General System Theory*, New York: Braziller, 2 1969.

[3] CASTI, J.L.: *Dynamical Systems and Their Applications: Linear Theory*. New York: Academic Press, 1977.

[4] GEORGESCU-ROEGEN, N.: *The Entropy Law and the Economic Process*, Cambridge, Mass.: Harvard University Press, 1971.

[5] GOODWIN, R.M.: "Dynamical Coupling with Especial Reference to Markets Having Production Lags", *Econometrica*, 18 (1947), 355–374.

[6] KALMAN, R.E.: "Dynamic Econometric Models: A System-Theoretic Critique" in *New Quantitative Techniques for Economic Analysis*, ed. by G.P. Szegö. New York: Academic Press, 1982.

[7] — ; P.L. FALB; M.A. ARBIB: *Topics in Mathematical System Theory*, New York: McGraw-Hill, 1968.

[8] KENDRICK, D.: "On the Leontief Dynamic Inverse", *Quarterly Journal of Economics*, 86 (1972), 693–696.

[9] LEONTIEF, W.W.: "Die Wirtschaft als Kreislauf", *Archiv für Sozialwissenschaft und Sozialpolitik*, 60 (1928), 577-628.

[10] — : *The Structure of the American Economy 1919–1939*, 2nd ed. New York: Oxford University Press, 1951.

[11] — et al.: *Studies in the Structure of the American Economy*. New York: Oxford University Press, 1953.

[12] — : "The Dynamic Inverse", in: *Contributions to Input-Output Analysis*, Vol. I, ed. by A.P. Carter; A. Bródy. Amsterdam: North Holland, 1970.

[13] LIVESEY, D.A.: "Control Theory and Input-Output Analysis", *International Journal of Systems Science* 2 (1971), 307–318.

[14] — : "The Singularity Problem in the Dynamic Input-Output Model", *International Journal of Systems Science*, 4 (1973), 437–440.

[15] LUENBERGER, D.G.: *Introduction to Dynamic Systems*. New York: Wiley, 1979.

[16] LUENBERGER,D.G., A.ARBEL: "Singular Dynamic Leontief Systems", *Econometrica*, 45 (1977), 991–995.

[17] METZLER, L.A.: "A Multiple–Region Theory of Income and Trade", *Econometrica*, 18 (1950), 329–354.

[18] MEYER, U.: "Singuläre dynamische Leontiefsche Systeme", *Zeitschrift für die gesamte Staatswissenschaft*, 135 (1979), 81–92.

[19] NEUMANN, J.v.: "Über ein ökonomisches Gleichungssystem und eine Verallgemeinerung des Brouwerschen Fixpunktsatzes", *Ergebnisse eines mathematischen Kolloquiums*, 8 (1937), 73–83.

[20] NIKAIDÔ, H.: "Some Dynamic Phenomena in the Dynamic Leontief Model of the Reversely Lagged Type", *Review of Economic Studies*, 29 (1962), 313–323.

[21] PASINETTI, L.: *Lectures on the Theory of Production*. New York: Columbia University Press, 1977.

[22] RITSCHL, A.: *Prices and Production. Elements of a System–Theoretic Perspective*, forthcoming.

[23] SRAFFA, P.: *Production of Commodities by Means of Commodities*, Cambridge: Cambridge University Press, 1960.

Technological Change and Employment

An Outline of the Theoretical Approach Used by the Bremen Research Unit "Technologischer Wandel und Beschäftigung"

by

Dieter Kattermann and Heinz D. Kurz

University of Bremen

I. Introduction

The purpose of the present paper is to give a brief outline of the analytical approach adopted by the Research Unit "Technologischer Wandel und Beschäftigung" at the University of Bremen (FRG), in an attempt to investigate the employment effects of the introduction of new information technologies in the West German economy. [1] The paper thus summarizes some of the theoretical work carried out by the members of the Research Unit, thereby drawing heavily on the series of *Working Papers* published by the Unit, in particular [4, 5, 7, 8, and 9]. Since the basic model is still subject to change, a complete description of the model cannot yet be provided. With regard to some areas of investigation it must suffice to indicate the direction of our efforts.

The structure of the paper is as follows. Section II describes the problem under consideration and hints at the method underlying our approach. Section III briefly motivates the adoption of a dynamic input-output framework for the analysis. What is called the 'core' of the model, that is, the formulation of (private) investment demand and the treatment of the process of the diffusion of a new technique, is expounded in some detail in Section IV. Section V deals with private consumption demand and Section VI with foreign trade and government expenditures. The volume and composition of employment is dealt with in Section VII. The concluding Section VIII addresses the problem of whether the dynamic system under discussion is well defined.

II. The Problem

As is well known, according to Schumpeter [16, p. 684) "the controversy that went on throughout the nineteenth century and beyond, mainly in the form of argument pro and con 'compensation' [of the labour displacing effects of technological change], is dead and buried." Schumpeter was mistaken. Apart from heated public debates between compensation pessimists, mainly from the trade unions, and compensation optimists, mainly from the employers' associations and the government, there is a growing concern among social scientists about the consequences of technological change. Particularly, it is the so-called 'microelectronics revolution' and the associated specter of technological unemployment that is responsible for the current revival of interest in this problem.

There are three aspects of the problem under discussion that have been distinguished since the inception of systematic economic analysis of technological change in classical political economy:

(1) What are the employment consequences of different forms of technological change?

(2) What is the impact of different forms of technological change on the distribution of income and wealth?

(3) How do different forms of technological change affect the characteristics or quality of work performed?

Our main concern will be with questions (1) and (3). In particular, it is our aim to provide an analytical framework within which labour displacement and compensation can be consistently discussed. Our aim is not to forecast the time profile of employment in the West-German economy until the year 2000. We are rather concerned with simulating exclusively those employment effects that accompany the diffusion of a new, well-defined microelectronic-based 'best-practice technique' (*bpt*) in the West-German economy. Factors other than those related to the diffusion process will not be taken into account; they will be considered 'frozen in'.

We shall make use of the scenario technique. In particular, we shall distinguish between transitions of the economy from the 'old' to the 'new' technique characterised by different speeds of the diffusion process. The reference scenario, which serves as a baseline, will be the one in which after the initial period no additional change in the methods of production (caused by the ongoing diffusion process) will take place. This generates a time path of employment with which we compare those paths that involve a more or less rapid substitution of the new *bpt* for the old technique. The new *bpt* itself is taken to be immutable, i.e., we do not enter into speculations about further developments of technological knowledge. The identification of the going *bpt* is a major task of our project and involves the evaluation of the literature on modern technology, branch studies and expert interviews.

Our study exhibits several limitations. The most important one (which it shares with other input-output studies referred to in the next section) is to be seen in the fact that the model is essentially formulated in terms of quantities and does not adequately deal with prices. In particular, it will be assumed that the 'new' technique is superior to the 'old' one in the sense that unit costs per unit of output are smaller. Since under competitive conditions entrepreneurs will adopt whichever method of production is cheapest, it can in fact be expected, then, that the economic system will 'gravitate' towards a situation, in which the old technique is completely superseded by the new one. [2]

III. The Dynamic Input-Output Framework

It has been argued that in order to take into account not only the primary effects of technological change on employment but also the secondary effects, an input-output framework of the analysis appears to be appropriate. Two studies have recently been carried out which make use of input-output analysis. There is first the study investigating the impact of microelectronics in Austria commissioned by the Austrian Ministry of Science and Research [1], and second the study by the Institute of Economic Analysis, New York University, on the impact of automation on workers in the US economy (Leontief and Duchin [11]). While the former contribution uses a static input-output framework, the latter uses a dynamic one. [3]

As is well known, in the conventional dynamic input-output model net investment demand is treated as a purely passive factor which has to adjust in such a way that a vector of final demand is generated, which is compatible with the continuous full utilization of productive capacity. Unless the system happens to be on the path of balanced growth from the beginning onwards, the model with a fixed capital matrix B tends to generate trajectories, which sooner or later involve negative outputs of some products (problem of 'causal indeterminacy'). Therefore, the conventional dynamic input-output model is of limited use only.

If investment demand is to be assigned a more active role in the economic process and if, at the same time, the problem of causal indeterminacy is to be avoided, a different type of model needs to be developed. In particular, it is to be allowed that production satisfying total effective demand will generally not be associated with the 'normal' or 'planned' utilization of the capacities in existence. Such 'disequilibria' are in fact to be expected to occur in the context of our investigation, since there is no presumption that the substitution of one technique for another is effectuated smoothly. More precisely, we shall allow that, in a specific sense and within certain limits, productive capacity may be *under-* or *overutilized*. [4] For the sake of simplicity (and with a view to the empirical data available) we shall not attempt to model rationing schemes or the accumulation and decumulation of inventories. Hence it will be assumed that in each period the various industries of the economy will produce exactly the quantities demanded of the different products, irrespective of whether this entails the under- or overutilization of capacity. (On the problem of the demand for a commodity exceeding the maximum capacity in existence, see Section VII below .)

Our approach thus bears some resemblance to the models by Meyer and Schumann [13], Meyer ([12], Ch.7), Duchin and Szyld [2] and Leontief and Duchin [11]. However, while these contributions are restricted to the case of non-negative *net* investment, we presuppose non-negative *gross* investment only. This implies that productive capacity installed in a particular sector can decrease over time, provided there is not sufficient replacement investment to make good the wear and tear of plant and equipment. By this token we attempt to capture the phenomenon of declining industries, i.e., industries whose products are rendered obsolete by product and process innovations.

As regards the modelling of investment demand, we advocate a two-steps decision hypothesis. *First*, entrepreneurs are assumed to form expectations about future demand. *Second*, as to the capacities required to match these levels of demand, entrepreneurs are assumed to decide in each period how much of the expected output is to be provided by means of the old and the new technique, respectively. By these two steps (and by aggregation over sectors) we arrive at a vector of investment demand ('by origin').

IV. The 'Core' of the Model: Private Investment Demand

Each sector j of the economy (j=1,2, ...,n) is assumed to have at its disposal one or two single product methods of production each of which is characterised by fixed coefficients of production. The coefficients of what will be called the 'old' and the 'new' technique will be indicated by the superscripts I and II, respectively. Thus, in an otherwise conventional notation, a_{ij}^{m} and l_{kj}^{m} (i,j=1,2,...,n; k=1,2,...,s; m=I,II) represent the amounts of circulating means of production and different kinds of labour required per unit of output of the different commodities if technique m is adopted. Similarly, the b_{ij}^{m} designate requirements of durable capital goods per unit of capacity in sector j under technique m.

In the course of time technique II will gradually replace technique I. For a while the two methods of production will co-exist in many sectors, with ever larger weights to be assigned to the new method. However, it is equally possible that a sector uses a single method only throughout the period under consideration. This is the case (i) if a new sector emerges producing an entirely new commodity (e.g., word processors) and (ii) if there do exist sectors, in which no technological alternatives are available. [5], [6]

As was explained above, b_{ij}^m indicates the quantity of capital good i required by technique m at the beginning of the production period in order to provide the capacity to produce one unit of commodity j; the coefficients refer to 'normal' capacity utilization. The model is formulated in terms of stocks and not in terms of marginal capital coefficients. Let $S_{ij}^m(t)$ be the stock of capital good i installed in sector j producing with technique m at the beginning of period t. We may now define the productive capacity associated with each of the two techniques. Let $K_j^m(t)$ represent sector j's level of production obtainable with technique m at 'normal' capital utilization, i.e.

$$K_j^m(t) := \min \{S_{ij}^m(t)/b_{ij}^m : b_{ij}^m > 0\} \qquad (1)$$

(The minimum condition refers to all i which imply a positive stock coefficient.) The sector's total capacity is of course given by

$$K_j(t) := \sum_{m=I}^{II} K_j^m(t) \quad . \qquad (2)$$

The implementation lag (or gestation period) of new investment is uniform throughout the economy and equal to one period (year). Investment goods which are produced in period t serve to create new capacity in period t+1. The capacity desired in sector j in period t+1, $K_j^*(t+1)$ [7], thus has to be decided on (planned) at the end of period t-1. These decisions are assumed to depend on the demand for the sectoral output expected for t+1, which in turn is assumed to depend on the information available in period t-1. [8] This information concerns the already observed values of the state variables of the system in t-1 and in preceding periods. In short, we assume that there exist (continuous and real) functions F, such that

$$K_j^*(t+1) = F_j[\ \ldots\ (t-1)] \qquad (3)$$

(The notation is meant to express that the arguments in F_j refer to magnitudes in t-1 and/or still earlier periods.) As to the specification of F_j, one possibility, due to Duchin and Szyld [2] and properly modified to mirror our assumption of a one-period gestation lag, is

$$F_j := \left[\min \left\{1+g_j, \frac{x_j(t-1)+x_j(t-2)}{x_j(t-2)+x_j(t-3)}\right\}\right]^2 x_j(t-1) \quad .$$

x_j is gross output realised in sector j in the respective period and g_j indicates the (exogenously given) maximum (expected) rate of growth of the sector. Our own dynamic explorations take this function and similar formulations as a starting point only. Further developments are discussed in the contribution of Franke and Weghorst [6] to this volume.

Concerning the technological make-up of the desired total capacity $K_j{}^*(t)$ of sector j, parameter $\alpha_j(t)$, $0 \le \alpha_j(t) \le 1$, informs about that part of $K_j{}^*(t)$ which is planned to be effectuated in terms of the new technique, i.e.

$$K_j{}^{II*}(t) = \alpha_j(t) K_j{}^*(t) \qquad (4)$$

$$K_j{}^{I*}(t) = [1-\alpha_j(t)]\, K_j{}^*(t) \quad .$$

The $\alpha_j(t)$'s are exogenous scenario parameters. They describe the speed and pattern of the introduction and diffusion of the new technique. In general we will have $\alpha_j(t+1) \ge \alpha_j(t)$. $\alpha_j(t) = 1$ implies that in period t sector j's entire capacity will embody the new method of production. A particular scenario would be $\alpha_j(t) = \alpha_j = 0$, the "no diffusion" scenario. It would show us the economy's behaviour in the absence of innovative efforts. For some purposes it will serve as reference scenario.

Associated with the capacities $K_j{}^{m*}(t)$ are the capital stocks $S_{ij}{}^{m*}(t)$ which sector j has to have at its disposal at the beginning of period t, where

$$S_{ij}{}^{m*}(t) = b_{ij}{}^m K_j{}^{m*}(t) \quad . \qquad (5)$$

Before proceeding with the main argument a few additional assumptions underlying our model should be mentioned.

First, we shall assume that the installation of the desired capacities is always possible, i.e.

$$K_j{}^m(t) \ge K_j{}^{m*}(t), \quad S_{ij}{}^m(t) \ge S_{ij}{}^{m*}(t)$$

(for all t and all i,j=1,2,...,n; m=I,II). It is not implied that the relevant capital goods will always be available in the right proportions. In particular, there is the possibility of excess capital goods in the case where the (desired) decrease of a sector's capacity is more rapid than the wear and tear of one or several of its capital goods.

Second, it will be assumed that within a given sector partly used capital goods can be transferred costlessly between techniques, whereas transfer between sectors and foreign trade in used capital goods is not possible. Moreover, it will be assumed that there is no foreign trade in used capital goods. By this token some of the potentially superfluous capital goods may be re-employed, provided the particular goods under consideration are also required by the new method of production (e.g., buildings and general implements).

Third, the wear and tear of fixed capital items (in efficiency units) is described by means of the parameters $\delta_{ij}{}^{m}$, where $0 < \delta_{ij}{}^{m} < 1$; $\delta_{ij}{}^{m}$ designates the proportion of the stock of capital good i installed in sector j using technique m which 'evaporates' during the unit period.

We are now in a position to determine investment demand in an unambiguous way:

$$z_{ij}{}^{m}(t) := S_{ij}{}^{m*}(t+1) - (1-\delta_{ij}{}^{m})S_{ij}{}^{m}(t) \qquad (6)$$

$$= b_{ij}{}^{m} K_j{}^{m*}(t+1) - (1-\delta_{ij}{}^{m})S_{ij}{}^{m}(t)$$

i,j=1,2,...,n; m=I,II. These magnitudes can be positive, negative or nil. In the second case, part of the stock of good i will become idle or the items that are in excess with respect to one technique will be transferred partly or totally to the other technique, provided in the latter there is need for this particular capital good.

It is of course equally possible that the sum of $z_{ij}{}^{I}(t)$ and $z_{ij}{}^{II}(t)$ is negative. This would imply that sector j sells good i in the market. However, since by assumption a capital good once installed never can leave the respective sector, sector j's total gross investment demand for product i is given by

$$z_{ij}(t) := \max\{0,\ z_{ij}{}^{I}(t) + z_{ij}{}^{II}(t)\} \quad .$$

(Clearly, a positive $z_{ij}(t)$ is compatible with a negative net investment demand.)

The sum of all sectoral gross investment demands for product i gives total gross investment demand for that product, i.e.,

$$z_i(t) = \sum_{j=1}^{n} z_{ij}(t) \geq 0 \quad . \qquad (7)$$

We could now distinguish between different constellations concerning the respective values of $z_{ij}{}^{I}(t)$ and $z_{ij}{}^{II}(t)$. Taking into account the assumptions expounded above, it can, however, be shown that all possible constellations may be summarized in the following expression

$$S_{ij}^{m}(t+1) = (1-\delta_{ij}^{m})S_{ij}^{m}(t)$$

$$- \min \{-z_{ij}^{m}(t), \max \{0, z_{ij}^{m'}(t)\}\} , \qquad (8)$$

$i,j=1,2,\ldots,n;\ m=I,II;\ m' \neq m.$

(The case, in which a sector has only a single technique m at its disposal is also covered by expressions (6)-(8). In this case $z_{ij}^{m'}(t) = 0$, where $m' \neq m$.)

The model of investment behaviour explained so far provides a framework within which one of the major compensation effects referred to in the literature can be dealt with: compensation due to new investment demand as it is contemplated by the so-called 'argument concerning the production of machinery'.

V. Private Consumption Demand

As the very name of a second argument pro compensation, 'purchasing power compensation theorem', already indicates, compensation is seen to become effective in a more roundabout way, that is, via changes in incomes and prices. Thus, it is widely held that technological progress is generally associated with rising nominal incomes and/or falling prices, i.e., rising real incomes. Rising real incomes however will spur consumption demand and thereby exert a positive influence on the overall level of employment.

Technological change tends to bring about remarkable structural changes in consumption demand. For the sake of simplicity we shall adopt an extremely simple formulation of consumption demand, which allows however, albeit in an elementary way, to take into account structural shifts.

Consumption demand will be assumed to be a function of current value added. Let c^{o} be the n-vector of autonomous consumption and c^{a} the n-vector defining the structure of additional consumption. Both vectors may but need not be taken to be constant. Let $Y(t)$ designate nominal national income in period t and $\tau(t)$ the proportion of income spent on additional consumption, where $0 \leq \tau(t) \leq 1$. With $p(t)$ as the n-vector of prices ruling in the given period, the vector of consumption demand $c(t)$ amounts to

$$c(t) = \frac{\tau(t)Y(t)}{p(t)c^{a}}\, c^{a} + c^{o} \quad . \qquad (9)$$

It can be shown (Franke and Kattermann [4], part 8) that (9) may be written in the form

$$c(t) = C(t)x(t) + c^{o} \qquad (10)$$

where C(t) is a non-negative nxn square matrix. The elements of C(t) are defined as follows

$$c_{ij}(t) := \frac{\tau(t) c_i^{a} v_j(t)}{p(t) c^{a}} \qquad (11)$$

where v_j (t) are the sectoral value-added coefficients which depend, among other things, on the technology mix in the different sectors of the economy. Clearly, in an obvious notation,

$$Y(t) = v(t)x(t)$$

Preliminary explorations in a simplified framework (Franke and Kattermann [5]) have shown that this formulation of consumption demand does not imply partial or complete compensation of technology-related job losses without further qualifications concerning the prevailing kind of technical change. The need for a closer inspection of the techniques involved arises from the fact that technological change affects Y in a direct and indirect way (via v and x, respectively).

VI. Foreign Trade and Government Demand

Before turning to the output solution of the model as well as the subsequent formulation of the demand for labour, a few remarks concerning the remaining two components of final demand, i.e. exports and government expenditures, are in place. In addition to the two arguments pro compensation already mentioned, a third line of reasoning states that (rapid) technological change at home improves competitiveness of domestic firms and industries and thus leads to positive employment effects via increasing domestic exports which outweigh possible increases in imports (cf., for example, Whitley and Wilson [17]).

Without the possibility of resorting to a world model, our treatment of sectoral exports and imports has to rely heavily on parametric terms: On the one hand, exports, being independent of endogenous variables, have to be assigned some "autonomous" growth factor. On the other hand, changes in export demand due to technical change cannot even in principle be based on domestic innovative activities alone - in contradistinction to other components of final demand. Only a direct comparison with innovative activities in "the rest of the world" leads to a conclusive formulation for export demand. The comparative term takes the form of a correction of the above mentioned autonomous growth factor. A special feature of this formulation is that the reference path referred to in section V implies a gradual decrease of export demand. This simply means that domestic technological inactivity will eventually depress domestic production. With $y_i^{ex}(t)$ as export demand for commodity i, G_i as growth factor, t indicating the time period, $q_i>0$ as a parameter as well as with $\alpha_i^d(t)$ and $\alpha_i^f(t)$ as domestic vs. foreign diffusion parameters, we express export demand as

$$y_i^{ex}(t) = y_i^{ex}(0)G_i^t \; [1 + q_i(\alpha_i^d(t)-\alpha_i^f(t))] \; . \quad (12)$$

Comparative innovative achievements also affect the extent to which domestic intermediate and final demand other than export demand (resulting from the application of new as well as old techniques) is to be satisfied out of domestic or foreign production. These magnitudes have to be split up into domestically produced vis-à-vis foreign sources for their satisfaction. Changes of the respective domestic shares, which are multiplicative factors of coefficients a_{ij}^m, b_{ij}^m, c_i^a, and c_i^o, are directly dependent on comparative innovative efforts, much in the same way as the changes in sectoral exports discussed above: With $d_{ij}(t)$ and $m_{ij}(t)$ as domestic and imported quantities of commodity i per unit of commodity j, respectively, and with $w_{ij}(t)$ as the share of domestic production, all in period t, we have, taking intermediate coefficients as an example and disregarding the m-superscript,

$$a_{ij}(t) = w_{ij}(t)d_{ij}(t) + [1-w_{ij}(t)]m_{ij}(t) \quad (13)$$

and

$$w_{ij}(t) = w_{ij}(0) + q_{ij}[\alpha_i^d(t)-\alpha_i^f(t)] \quad , \quad (14)$$

where $w_{ij}(0)$ refers to the domestic share of the start period and $q_{ij}>0$ is a parameter. Again we can see the implications of the "no diffusion" reference scenario: $w_{ij}(t)$ decreases steadily, such that virtually <u>all</u> demand components in addition to the above mentioned export demands are negatively affected. Taken to the extreme, therefore, in an open economy without sufficient innovative activity production would cease altogether.

We now turn to the last component of final demand, government expenditures. In contrast to the components discussed in the above sections and paragraphs, no particular argument concerning displacement and reemployment of labour in the course of technical change is known to focus on government expenditures. This is probably due to the numerous noneconomic constraints which feed into expenditure decisions. We therefore content ourselves with extrapolations of government expenditures' share in gross domestic product, of investment and consumption shares in government expenditures and of individual capital goods' shares in government investment. Information on public investment in microelectronics equipment will be used to produce scenarios affecting the shares of the latter as well as those of government investment and consumption demand.

VII. The Determination of Employment

The solution of the dynamic input-output model for a given time period consists in a vector of gross outputs, which exactly satisfies the final and intermediate demands for the various products. Post-multiplying the going matrix of employment coefficients L(t) by x(t) yields the amounts of employment of the different kinds of labour. Clearly, the coefficients of the matrix L(t), l (t), have to be defined analogously to the other technical coefficients of the system of production, thereby taking account of the prevalent technology mix.

Let us first determine the vector of sectoral demands. For the sake of brevity we define the nxn matrix

$$Q(t) := A(t) + C(t) \quad . \tag{15}$$

Furthermore, we define $h^{o}(t)$ as the n-vector which comprises (net) exports and government demand. Hence with a gross output vector x(t) total demand is given by $Q(t)x(t) + c^{o} + h^{o}(t) + z(t)$. Following the premise, expounded above, that the vector of actual outputs will generally tend to match exactly the vector of commodities in demand, the output to be produced in period t, $x^{*}(t)$, is given by

$$\begin{aligned} x^{*}(t) &= Q(t)x^{*}(t) + c^{o} + h^{o}(t) + z(t) \\ &= [I - Q(t)]^{-1} \; [c^{o} + h^{o}(t) + z(t)] \quad . \end{aligned} \tag{16}$$

(It can be shown that under economically sensible assumptions, $\tau(t) < 1$ - as defined in section V - is sufficient to guarantee that matrix $[I - Q(t)]^{-1}$ exists and is non-negative; cf. Franke and Kattermann [4].)

In general there are good reasons to suppose that $x^{*}(t)$, given the non-negativity of $[I - Q(t)]^{-1}$, is non-negative even if $h^{o}(t)$ should happen to exhibit some negative components. As far as demand is not met by domestic supplies, (intermediate and final) imports have to be subtracted from (final) demand so that the sign of, say, $h^{o}(t)$ cannot be determined a priori.

Finally, the problem of the technical feasibility of solution (16) needs to be considered. In fact, it cannot be excluded that in some sectors the capacities available, $K_{j}(t)$, do not suffice to produce $x^{*}(t)$. We shall deal with this problem in the following way. First, we introduce some elasticity into the system of production by rendering flexible the degree of capacity utilization. Let $\Gamma_j \geq 0$ designate the margin of additional output available per unit of productive capacity in sector j due to the maximum possible utilization of capacity beyond the normal level. Hence actual gross output x(t) is determined by

$$x_j(t) = \min\{x_j^*(t),\ (1+\Gamma_j)K_j(t)\}\ , \qquad j=1,2,\ldots n \tag{17}$$

Of course, the upper limits Γ_j in (17) should become effective in exceptional situations only. (The main candidate among the equations to be subjected to modifications appears to be the behavioral function (3).)

Second, since we shall assume that the target output vector $x^*(t)$ will not be revised because of domestic capacity constraints, any remaining bottlenecks will be overcome by additional imports.

The vector of the levels of employment of the different kinds of labour in period t, e(t), is now given by

$$e(t) = L(t)x(t) \tag{18}$$ [9]

If we set aside the heterogeneity of labour and measure the total labour input indiscriminately in labour time, we get the total volume of labour performed, V(t), i.e.

$$V(t) = i^s L(t)x(t) = \mu(t)x(t), \tag{19}$$

where i^s is the s-dimensional unit vector. The vector of sectoral volumes of labour performed, v(t), is

$$v(t) = \mu(t)\langle x(t)\rangle \tag{20}$$

with $\langle x(t)\rangle$ as the diagonal matrix constructed of x(t). The number of people employed in the different sectors performing the different kinds of labour can be expressed as the sxn-matrix N(t),

$$N(t) = L(t)x(t)\ \langle h(t)\rangle^{-1}\ , \tag{21}$$

with $\langle h(t)\rangle$ as the diagonal matrix of the average working hours per employed person specific to the different sectors. The sectoral allocation of people employed is given by the vector n(t),

$$n(t) = i^s N(t), \tag{22}$$

and the total number of people employed by

$$E(t) = n(t)i^n = \mu(t)\ \langle h(t)\rangle^{-1}\ x(t)\ . \tag{23}$$

Clearly, other things being equal, the time path of E(t) depends on the time paths of average sectoral working hours.

VIII. The Dynamic Process

It remains to be shown that for given initial conditions all endogenous variables can indeed be determined period after period and that the respective time series are unique. The central variables are (i) the technique-specific sectoral capacities $K_j{}^{I}$ and $K_j{}^{II}$, (ii) the stocks of capital goods determining these capacities, $S_{ij}{}^{I}$ and $S_{ij}{}^{II}$, and (iii) the realised levels of production, x_j.

The equations relevant for the dynamics of the system are equations (1)-(8) and (16)-(17). When the process is started in period t=0, the initial stocks of capital goods, $S_{ij}{}^{I}(0)$ and $S_{ij}{}^{II}(0)$, and the magnitudes entering functions F_j have to be given from outside the system. This allows the determination of $K_j{}^{*}(1)$ and then via equations (3)-(8) the stocks $S_{ij}{}^{I}(1)$ and $S_{ij}{}^{II}(1)$. Thereafter $K_j{}^{I}(1)$, $K_j{}^{II}(1)$, $K_j(1)$ and eventually $x_j(1)$ in equation (17) can be determined. Next, equation (3) can be fed with the relevant data, which yields $K_j{}^{*}(2)$, and so on and so forth. The exploration of the dynamic behaviour properly speaking is the subject of the contribution by Franke and Weghorst [6] to this volume.

Notes

1 Financial support by the Stiftung Volkswagenwerk is gratefully acknowledged.

2 However, as is well known from the discussion of the problem of the choice of technique in linear models of production (cf., for example, Pasinetti [14]), the question which technique is cost-minimizing cannot generally be decided independently of relative prices and income distribution.

5 For a critical examination of the Leontief/Duchin model see, for example, Hagemann, Kattermann and Kurz [7].

4 For a discussion of the elasticity of the industrial systems due to the variability of the rates of capital utilization, see, for example, Kurz [10].

5 In the first case we may introduce the convention that the associated coefficients of the old technique are nil, while in the second case the coefficients of the new technique are nil.

6 In general, our sectoral breakdown will not be detailed enough to embrace case (i). In contrast to the assumption in theoretical analysis, empirical input-output sectors are made up of a multitude of goods.

7 In what follows, variables with an asterisk generally refer to desired magnitudes.
8 This one-period implementation lag is a working hypothesis which may, in the course of subsequent dynamical investigations, be submitted to modifications.
9 For the following see Duffner [3].

REFERENCES

[1] BUNDESMINISTERIUM FÜR WISSENSCHAFT UND FORSCHUNG (ed.): *Mikroelektronik, Anwendungen, Verbreitung und Auswirkungen am Beispiel Österreichs*. Wien and New York: Springer Verlag, 1981.

[2] DUCHIN, F., AND D. SZYLD: "A dynamic input-output modell with assured positive outputs," *Metroeconomica*, 37 (1985), 269 - 282.

[3] DUFFNER,.H.: "Die Einbeziehung des Arbeitsmarktes in ein dynamisches Input-Output-Modell der Bundesrepublik Deutschland," *Working Papers*, 8 (1987), Research Unit "Technologischer Wandel und Beschäftigung", University of Bremen.

[4] FRANKE, R., AND D. KATTERMANN: "Investitionen und Technikdiffusion in einem dynamischen Input-Output-Modell," *Working Papers*, 7 (1987), Research Unit "Technologischer Wandel und Beschäftigung", University of Bremen.

[5] -------: "Konsumnachfrage und technologische Arbeitslosigkeit in einem Input-Output-Modellrahmen," *Working Papers*, 11 (1987), Research Unit "Technologischer Wandel und Beschäftigung", University of Bremen.

[6] FRANKE, R., AND W. WEGHORST: "On a Stabilizing but Chaos Generating Mechanism in Behavioural Functions: Obervations in a Multiplier-Accelerator Model," this volume.

[7] HAGEMANN, H., D. KATTERMANN, AND H.D. KURZ: "Technological Change and Employment in Advanced Industrial Countries. The Case of the Federal Republic of Germany", *Working Papers*, 1 (1985), Research Unit "Technologischer Wandel und Beschäftigung", University of Bremen.

[8] Kalmbach, P.: "Beschäftigungseffekte des technologischen Wandels - Ansatzpunkte ihrer Quantifizierung im Rahmen der Input-Output-Analyse," *Working Papers*, 2 (1986), Research Unit "Technologischer Wandel und Beschäftigung", University of Bremen.

[9] KATTERMANN, D.: "Investitions- und Konsumnachfrage in einem dynamischen Input-Output-Modell zur Bestimmung technologischer Arbeitslosigkeit," *Working Papers*, 4 (1986), Research Unit "Technologischer Wandel und Beschäftigung", University of Bremen.

[10] KURZ, H.D.: "Normal Positions and Capital Utilization," *Political Economy*, 2 (1986), 37 - 54.

[11] LEONTIEF, W., AND F. DUCHIN: *The Future Impact of Automation on Workers*. New York: Oxford University Press, 1986.

[12] MEYER, U.: *Dynamische Input-Output-Modelle*. Königstein (Taunus): Athenäum, 1980.

[13] MEYER, U. AND J. SCHUMANN: "Ansätze zur Weiterentwicklung des dynamischen Input-Output-Modells," in *Neuere Entwicklungen in den Wirtschaftswissenschaften*, ed. by E. Helmstädter. Berlin: Duncker und Humblot, 1978.

[14] PASINETTI, L.L.: *Lectures on the Theory of Production*. London and Basingstoke: Macmillan, 1977.

[15] SCHUMANN, J.: "Möglichkeiten und Bedeutung einer teilweise endogenen Erklärung des privaten Konsums und der privaten Investition im statischen offenen Input-Output-System," *Jahrbücher für Nationalökonomie und Statistik*, 189 (1975), 387 - 409.

[16] SCHUMPETER, J.A.: *History of Economic Analysis*. London: Oxford University Press, 1954.

[17] WHITLEY, J. AND R. WILSON: "Quantifying the Employment Effects of Micro-Electronics," *Futures*, (1982).

DYNAMISCHE WIRTSCHAFTSTHEORIE

herausgegeben von Peter Flaschel und Michael Krüger

Band 1 Reiner Franke: Production prices and dynamical processes of the gravitation of market prices. 1987.

Band 2 Peter Flaschel/Michael Krüger (eds.): Recent Approaches to Economic Dynamics. Conference Volume (Bielefeld: 7-9th October, 1987). 1988.

Printed by
CPI books GmbH, Leck